Mathematical Aspects of Signal Processing

Signal processing requires application of techniques and tools to extract information from a given set of data, or conversely, to generate data with some desirable property. For any comprehensive understanding of signal processing, a study of mathematical methods is beneficial because signal processing is associated in significant ways with the interpretation, interrelation and applicability of essential mathematical concepts. The intricacies of signal processing get appreciated better if one understands the relationship between mathematical concepts and signal processing applications.

This text offers detailed discussion on mathematical concepts and their interpretations in the field of signal processing. The purpose of the book is to demonstrate how advanced signal processing applications are structured on basic mathematical concepts. It identifies four areas of mathematics including function representation, generalized inverse, modal decomposition and optimization for detailed treatment. It illustrates how basic concepts transcend into mathematical methods, and mathematical methods perform signal processing tasks. While presenting theorems of mathematics, the emphasis is on applications and interpretations. In order to introduce the computational side of signal processing algorithms, MATLAB programs are referred for simulation, and graphical representations of the simulation results are provided with the concepts wherever required. A number of signal processing applications and practice problems are also embedded in the text, for better understanding of the concepts.

The book targets senior undergraduate and graduate students from multiple disciplines of science and engineering, interested in computations and data processing. Researchers and practicing engineers will be equally benefited by its use as reference.

Pradip Sircar is a Professor of Electrical Engineering at the Indian Institute of Technology, Kanpur. During his teaching career, spanning over three decades, he has guided more than 50 master's and doctoral theses. He has served on the editorial board of several journals, and has contributed to the quality assessment of higher education in India.

Mathematical Aspects of Signal Processing

Mathematical Aspects of Signal Processing

Pradip Sircar

CAMBRIDGE
UNIVERSITY PRESS

University Printing House, Cambridge CB2 8BS, United Kingdom
One Liberty Plaza, 20th Floor, New York, NY 10006, USA
477 Williamstown Road, Port Melbourne, vic 3207, Australia
4843/24, 2nd Floor, Ansari Road, Daryaganj, Delhi – 110002, India
79 Anson Road, #06–04/06, Singapore 079906

Cambridge University Press is part of the University of Cambridge.

It furthers the University's mission by disseminating knowledge in the pursuit of education, learning and research at the highest international levels of excellence.

www.cambridge.org
Information on this title: www.cambridge.org/9781107175174

First published 2016

Printed in India by Thomson Press India Ltd., New Delhi 110001

A catalogue record for this publication is available from the British Library

Library of Congress Cataloguing in Publication data

Names: Sircar, Pradip, author.
Title: Mathematical aspects of signal processing / Pradip Sircar.
Description: Daryaganj, Delhi, India ; New York : Cambridge University Press, 2016. | Includes bibliographical references and index.
Identifiers: LCCN 2016019491 | ISBN 9781107175174 (hardback)
Subjects: LCSH: Signal processing--Mathematics.
Classification: LCC TK5102.9 .S5626 2016 | DDC 621.382/20151--dc23 LC record available at https://lccn.loc.gov/2016019491

ISBN 978-1-107-17517-4 Hardback

Additional resources for this publication at www.cambridge.org/9781107175174

Dedicated to the memory of my parents
Amiya (Putul) and Nalini Ranjan Sircar

Contents

Figures

Tables

Foreword

Signal processing is a fundamental discipline and plays a key role in many fields including astronomy, biology, chemistry, economics, engineering, and physics, etc. Signal processing is based on fundamental mathematical tools and methods. As a result, a deeper understanding of these mathematical aspects is crucial towards successful application of signal processing in practice and to carry out advanced research. The aim of the book *Mathematical Aspects of Signal Processing* by Professor Pradip Sircar is to provide the reader with a comprehensive review, along with typical applications, of four specific mathematical topics: Function representation, generalized inverse, modal decomposition, and optimization in signal processing. This book is based on the extensive experience of the author as a researcher and educator in signal processing, and the presented materials have been tested with many students over a number of years at the Indian Institute of Technology, Kanpur, India. In my opinion, this book will be a very good text for an upper division and/or graduate level course on the subject. Researchers and professional engineers will also find this book very useful as a reference.

Sanjit K. Mitra
Santa Barbara, California
January 2016

Preface

I have been using the material of this book for more than two decades while teaching a foundation level course on mathematical methods in signal processing being offered to the final-year undergraduate and first-year postgraduate students of Electrical Engineering at Indian Institute of Technology (IIT) Kanpur. Signal processing being an application area of mathematics, in order to motivate students to get involved in research of signal processing, it was felt that mathematics has to be presented in an informal and thought-provoking manner. In fact, the application area of signal processing is primarily concerned with the interpretation, interrelation, and applicability of mathematical concepts, whereas a formal course of mathematics is structured on treatise of existence, uniqueness, and convergence of mathematical results. Therefore, one has to relook at mathematics and find ways to make mathematical concepts go beyond abstractions and formal proofs.

My experimentation in teaching the above mentioned course has been quite rewarding. I have identified four areas of mathematics, namely, function representation, generalized inverse, modal decomposition, and optimization for our discussions. Mathematical concepts are frequently borrowed from these four areas for signal processing applications. However, the list of areas is not complete. Fundamental results of each area are presented in the class, followed by an interactive discussion of what can be their plausible implications. Finally, signal processing applications are showcased as illustrations of basic principles. Thus, the course serves the purpose of demonstrating how advanced signal processing applications are structured on basic mathematical concepts. I know a good number of students who credited the course, and subsequently decided to work on a research problem related to signal processing applications. The book will motivate students to develop a career in the broad area of signal processing as teachers, researchers or professionals.

During the year 1998–99, while I was visiting École Nationale Supérieure des Télécommunications (ENST) Paris, I started framing the idea of writing a book on the course material. I realized soon that it will not be easy to write a book on this subject. My lectures are more like discussion sessions, and the comprehensions are informal. In a few months time, I jotted down an outline of the proposed course, and drafted some sections of a chapter. However, I could not continue my endeavour after coming back to Kanpur because of other priorities. Nevertheless, the course material was constantly updated over the years.

While selecting the chapters of the book, it was felt that a course on statistical signal processing which requires a background in linear algebra, probability and random processes is offered in the majority of postgraduate programs around the world. Similarly, a foundation course on Fourier, Gabor and wavelet transforms and their applications in signal processing is available in the curriculum at most departments in various universities. However, the scopes of function representation and generalized inverse in signal processing applications are yet to be thoroughly revealed. In the same way, the uses of eigenfunction expansion and mathematical programming in signal processing applications are yet to be fully exploited. These factors were considered while formalizing the contents of this book.

In 2010, I visited the Center for Remote Sensing of Ice Sheets (CReSIS), a research center funded by National Science Foundation (NSF). The center is attached to the University of Kansas at Lawrence, Kansas. The researchers of the center are engaged in some challenging signal processing tasks with real data collected from the planet's polar regions. While participating in various discussion sessions here, I felt the need of having a reference which will show the genesis of many sophisticated signal processing algorithms. It was during this time that I realized again that I should complete writing this book as soon as possible. The bridge between mathematics and signal processing should always be in sight when we are dealing with sophisticated signal processing techniques.

Finally, it should be emphasized that the proposed book does not intend to replace standard texts on mathematics. The purpose of the book is to demonstrate how advanced signal processing applications are structured on basic mathematical concepts. Once students or researchers see the connection, they really get motivated for more exploration. The experiment was very successful at IIT Kanpur. It should be pointed out that the approach taken in this book is the forward approach where we first learn concepts from mathematics, and then look for applications in signal processing. It is felt that the reverse approach of finding mathematical tools for solving signal processing problems does not exploit mathematics to its true potential. It is likely that the book will provide a very distinct flavor.

Acknowledgments

Writing a book is a great learning experience. While learning and framing various underlying concepts leading to different fields of knowledge, one realizes how much one must have acquired from one's teachers. It is natural to have a deep sense of gratitude to all our teachers whose contributions are invaluable. I express my gratitude to Professor Tapan K. Sarkar for all the knowledge that I have acquired from him.

I am grateful to Professor Sanjit K. Mitra for all the encouragements that I received from him every now and then. His enthusiasm and appreciation never failed to boost my spirit. I am indebted to Professor Yves Greiner for all his support in my professional career. He always stood by my side as a true friend. I am very thankful to Professor Sivaprasad Gogineni who entrusted me with a big project which provided ample opportunity for learning.

While teaching the course with the same material at IIT Kanpur, I often got various requests and suggestions from my students. Their critical assessment of the course was extremely helpful in improving the quality of the course material. I sincerely acknowledge this contribution from my students in designing the content of the book. I am very thankful to all my colleagues of our institute for providing a congenial environment where ideas can grow with cherished academic freedom.

I greatly appreciate the kind cooperation that I received from Manish Choudhary of Cambridge University Press from inception to completion of this project. Likewise I thank the editorial team at CUP for all the good work that they have done.

1 Paradigm of Signal Processing

1.1 Introduction

Signal processing is a branch of applied mathematics. The primary aim of signal processing is to extract information from a given-set of data, a procedure known as *signal analysis*, or conversely, to use information in generating data with some desirable property, a procedure known as *signal synthesis*. Mathematical methods are developed based on concepts from mathematics with the purpose of signal processing as an area of application. Various mathematical concepts find use in signal processing, and a systematic introduction of those mathematical concepts to the students who will become researchers and professionals in this area, has become an essential part of the curricula worldwide.

When the mathematical ideas are introduced in a course of mathematics, they are presented as abstractions. Theories are built in structured manner, based on the present state of knowledge and the aspiration of the mathematician to extend knowledge in some desired direction. A mathematician is often not concerned of the areas of application of the developed theories. In contrast, a researcher engaged in signal processing often considers an established theory and identifies ways how it can be adopted and applied in extracting information inherent in data or for generating data with some desirable property.

It is the purpose of the present book to introduce the mathematical concepts and their interpretations related to signal processing. It is illustrated how basic concepts mould into mathematical methods, and mathematical methods perform signal processing tasks. We take up the case study of spectral estimation of a signal in the next section. The spectral estimation, by definition, refers to computing the energy spectral density of a signal with finite energy or the power spectral density of a signal with finite power. In the case study, however, we generalize our tasks by stating that primarily we are interested in finding the frequency-content of a signal, and computation of the energy-level or power-level at each frequency is our secondary purpose. This case study will illustrate various connections of signal processing with mathematics.

1.2 A Case Study: Spectral Estimation

Suppose that we want to know what frequencies are present in a signal $g(t)$ which is observed over some duration of time. If the signal is periodic with period T, then it satisfies the relation $g(t) = g(t+T)$, for the minimum of such non-zero positive values of T.

Using the Fourier series expansion of a periodic signal, we can express the real-valued signal $g(t)$ as

$$g(t) = a_0 + \sum_{k=1}^{\infty}\left[a_k \cos(2\pi k f_0 t) + b_k \sin(2\pi k f_0 t)\right] \tag{1.1}$$

where $f_0 = \frac{1}{T}$ is the fundamental frequency, and $k f_0$ is the kth harmonic frequency. The Fourier series coefficients a_0, a_k and b_k are given by

$$a_0 = \frac{1}{T}\int_{-T/2}^{T/2} g(t)dt$$

$$a_k = \frac{2}{T}\int_{-T/2}^{T/2} g(t)\cos(2\pi k f_0 t)dt, \quad k = 1, 2, \ldots, \infty \tag{1.2}$$

$$b_k = \frac{2}{T}\int_{-T/2}^{T/2} g(t)\sin(2\pi k f_0 t)dt, \quad k = 1, 2, \ldots, \infty$$

The upper and lower bounds of the integrals are not important for a periodic signal so long the time duration of integration remains as one period T.

The Fourier series expansion of a periodic complex-valued signal $g(t)$ with period T is expressed as

$$g(t) = \sum_{k=-\infty}^{\infty} c_k \exp(j2\pi k f_0 t) \tag{1.3}$$

where $f_0 = \frac{1}{T}$ is the fundamental frequency, and $k f_0$ is the kth harmonic frequency. The complex exponential Fourier series coefficients c_k are given by

$$c_k = \frac{1}{T}\int_T g(t)\exp(-j2\pi k f_0 t) \tag{1.4}$$

To ensure that the coefficients c_k are finite, and the infinite series of (1.3) converges to $g(t)$ at every point t (in the average sense at the point of discontinuity), the signal $g(t)$ must satisfy the three Dirichlet conditions and the main among those conditions is that the signal must be absolutely integrable over one period of the signal, i.e.,

$$\int_T |g(t)|\, dt < \infty \tag{1.5}$$

For a real-valued signal $g(t)$, (1.3) gets a special form with $c_k^* = c_{-k}, k = 1, 2, \ldots, \infty$, and c_0 being real. Writing $c_k = |c_k|\exp(j\theta_k)$, the signal of (1.3) gets the form

$$g(t) = d_0 + \sum_{k=1}^{\infty} d_k \cos(2\pi k f_0 t + \theta_k) \quad (1.6)$$

where $d_0 = c_0 = |c_0|$ and $d_k = 2|c_k|$, $k = 1, 2, \ldots, \infty$. Substituting $d_0 = a_0$, $d_k \cos\theta_k = a_k$ and $-d_k \sin\theta_k = b_k$ in (1.6), we get the same signal of (1.1).

For aperiodic signals of finite duration, the signal representation consists of frequency components over a continuous domain, and the Fourier sum becomes the Fourier integral on the frequency-axis. The Fourier analysis is extended from periodic to aperiodic signals with the argument that an aperiodic signal can be thought of as a periodic signal with infinite period. As the period becomes larger and larger, the fundamental frequency becomes smaller and smaller, and in the limit, the harmonic frequencies form a continuum.

For the frequency domain representation of signals, we define the Fourier transform pair as

$$G(f) = \int_{-\infty}^{\infty} g(t) \exp(-j2\pi ft)dt \quad (1.7)$$

$$g(t) = \int_{-\infty}^{\infty} G(f) \exp(j2\pi ft)df \quad (1.8)$$

While the Fourier transform of $g(t)$, $F[g(t)] = G(f)$ shows the complex spectrum of the signal $g(t)$ at any frequency f, the inverse Fourier transform $F^{-1}[G(f)] = g(t)$ provides a means for reconstructing the signal. Once again, to guarantee that $G(f)$ is finite at every value of f and the integral of (1.8), which is over infinite duration, converges to $g(t)$ at every point of t (in the average sense at the point of discontinuity), the signal $g(t)$ must satisfy the three Dirichlet conditions. The main among those conditions is that the signal must be absolutely integrable, i.e.,

$$\int_{-\infty}^{\infty} |g(t)| dt < \infty \quad (1.9)$$

Although the periodic signals are neither absolutely integrable nor square integrable, i.e.,

$$\int_{-\infty}^{\infty} |g(t)|^2 dt < \infty \quad (1.10)$$

which is the condition for a signal being of finite energy, we can still compute the Fourier transform for periodic signals provided we permit inclusion of the impulse function(s) in the Fourier transform. The area under an impulse function $\delta(t)$ is unity, and the inverse Fourier transform of a train of impulses converges to a periodic signal:

$$\int_{-\infty}^{\infty} \left[\sum_{k=-\infty}^{\infty} c_k \delta(f - kf_0) \right] \exp(j2\pi ft)df = \sum_{k=-\infty}^{\infty} c_k \exp(j2\pi k f_0 t) \quad (1.11)$$

A periodic signal considered in (1.11) has infinite energy, but it has finite power.

One important property of the Fourier transform is that the transform preserves energy of a signal in time and frequency domains. The energy of the signal $g(t)$ is given by

$$\int_{-\infty}^{\infty} |g(t)|^2 \, dt = \int_{-\infty}^{\infty} |G(f)|^2 \, df \tag{1.12}$$

The energy spectral density $E_g(f) = |G(f)|^2$ is the measure of the energy of the signal contained within a narrow band of frequency around f divided by the width of the band. The autocorrelation function $r_g(\tau)$ of the signal defined by

$$r_g(\tau) = \int_{-\infty}^{\infty} g^*(t) g(t+\tau) dt \tag{1.13}$$

and the energy spectral density $E_g(f)$ of the signal form the Fourier transform pair, i.e.,

$$E_g(f) = \int_{-\infty}^{\infty} r_g(\tau) \exp(-j2\pi f \tau) d\tau \tag{1.14}$$

For a signal $g(t)$ with finite power, we define the autocorrelation function $r_g(\tau)$ as

$$r_g(\tau) = \lim_{T \to \infty} \frac{1}{2T} \int_{-T}^{T} g^*(t) g(t+\tau) dt \tag{1.15}$$

and the power spectral density $P_g(f)$ is given by the Fourier transform of $r_g(\tau)$, i.e.,

$$P_g(f) = \int_{-\infty}^{\infty} r_g(\tau) \exp(-j2\pi f \tau) d\tau \tag{1.16}$$

which is the measure of the power of the signal contained in the frequency range between f and $f + \Delta f$, divided by Δf, and setting $\Delta f \to 0$.

Let us now investigate the effect of finite length of observation time of the signal on the estimation of frequency. When the signal $g(t)$ is available for the time duration $t \le \left|\frac{T}{2}\right|$ for processing, we compute the Fourier transform of the windowed signal as follows,

$$G_W(f) = F[g(t)w(t)] = G(f) * W(f) \tag{1.17}$$

where $*$ stands for the convolution operation, and the window function is

$$w(t) = \begin{cases} 1, & t \le \left|\frac{T}{2}\right| \\ 0, & \text{elsewhere} \end{cases} \tag{1.18}$$

and its Fourier transform is given by

$$W(f) = \frac{\sin(\pi fT)}{\pi f} = T\,\text{sinc}(fT) \tag{1.19}$$

The maximum of the sinc function occurs at $f = 0$, and its first zero-crossings appear at $f = \pm\frac{1}{T}$. When the signal $g(t)$ comprises of two complex exponentials at frequencies f_1 and f_2, and its Fourier transform is

$$G(f) = \delta(f - f_1) + \delta(f - f_2) \tag{1.20}$$

we compute the Fourier transform of the windowed signal as

$$G_W(f) = W(f - f_1) + W(f - f_2) \tag{1.21}$$

Thus, the two peaks at $f = f_1$ and $f = f_2$ in the plot of the square-magnitude of the Fourier transform can be resolved provided the difference of frequencies, $f_1 \sim f_2 \geq \frac{1}{T}$. In other words, the resolution of estimation of frequency by the method based on the Fourier transform is inversely proportional to the duration of observation time of the signal. This condition on the resolution of frequency can be of concern when the signal is observable only for short duration of time.

We shall now consider discrete-time signal for processing. Even for those cases where the signal $g(t)$ is observable over continuous time, we need to sample the signal over some discrete points of the time axis before a digital computer can be employed for signal processing. When a band-limited signal is sampled at a uniform time-interval T_s, the spectrum of the discrete-time signal becomes periodic with period T_s. The discrete-time Fourier transform $G_d(f)$ of the sequence $\{g_n = g(nT_s)\}$ is related to the continuous-time Fourier transform $G(f)$ as follows,

$$\begin{aligned} G_d(f) &\triangleq \sum_{n=-\infty}^{\infty} g_n \exp(-j2\pi fnT_s) \\ &= \frac{1}{T_s}\sum_{k=-\infty}^{\infty} G(f + kf_s) \end{aligned} \tag{1.22}$$

where $f_s = \frac{1}{T_s}$ is the sampling rate.

Assuming that $\frac{f_s}{2} \geq B$ where B is the bandwidth of the signal, i.e., $|G(f)|$ is non-zero only for $|f| \leq B$, we can low-pass filter $G_d(f)$ with gain T_s to get $G(f)$, and then compute the inverse Fourier transform to obtain the signal $g(t)$. In the process, we get the formula for reconstructing the original signal from its sampled sequence as given below,

$$g(t) = \sum_{n=-\infty}^{\infty} g_n\,\text{sinc}(f_s t - n), \quad -\infty \leq t \leq \infty \tag{1.23}$$

When B is comfortably less than $\frac{f_s}{2}$, some other formula for reconstructing the signal may be more desirable because in that case, a generating function can be involved which is localized on the time-axis, unlike the sinc function which stretches for all time, $-\infty < t < \infty$. Note that the sinc function is the inverse Fourier transform of the rectangular window. By choosing a smooth transition-band between pass-band and stop-band of low-pass filtering, we get an appropriate kernel function in (1.23), which is time-limited.

Using the notation $g[n] = g(nT_s)$ for the discrete-time signal, we compute the autocorrelation function $r_g[l]$ of the signal with finite power as

$$r_g[l] = \lim_{M\to\infty} \frac{1}{2M+1} \sum_{n=-M}^{M} g^*[n]g[n+l] \tag{1.24}$$

The power spectral density $P_g(f)$ is now defined as

$$P_g(f) = \sum_{l=-\infty}^{\infty} r_g[l]\exp(-j2\pi fl), \quad -\frac{1}{2} \le f \le \frac{1}{2} \tag{1.25}$$

where the frequency variable f is the discrete frequency defined as the relative value in comparison with the sampling rate f_s. Thus, if f and f_s are known, then the frequency in Hertz can be determined.

When a finite number of signal samples $\{g[n]; n = 0,1,\ldots,N-1\}$ are available for processing, we compute the discrete Fourier transform (DFT) given by

$$G[k] = \sum_{n=0}^{N-1} g[n]\exp(-j2\pi kn/N), \quad k = 0,1,\ldots,N-1 \tag{1.26}$$

and the inverse DFT given by

$$g[n] = \frac{1}{N}\sum_{k=0}^{N-1} G[k]\exp(j2\pi kn/N), \quad n = 0,1,\ldots,N-1 \tag{1.27}$$

Since the total range of discrete frequency is sampled by the integer k at N points, the resolution of discrete frequency is $\frac{1}{N}$. When multiplied by $f_s = \frac{1}{T_s}$, the resolution of frequency in Hertz becomes $\frac{1}{T}$, where T is the duration of sampling in seconds. Once again, two close frequencies can be resolved by the DFT only if they are more than $\frac{1}{T}$ apart, which gives a condition similar to what we have obtained for the case of continuous-time signal.

When some finite number of signal samples $g[n]$ are available, we can compute the autocorrelation function $r_g[l]$ accurately only for some limited number of lags. In this case, we do not use (1.25) to compute the power spectral density $P_g(f)$ of the signal. For accurate estimation of $P_g(f)$ by (1.25), we need the sequence $r_g[l]$ for a large number of lags. Therefore, for accurate estimation of $P_g(f)$ with a short length of the sequence $r_g[l]$, we introduce the approach of model-based signal processing.

In the model-based signal processing approach, we first assume a model for the signal, and then we use the model for extracting information about the signal. Basically, a model identifies a class of signals, and signal processing becomes the way of extracting features relevant to the class. Often a signal is identified as the output of a system with known input, and the model-based signal processing is done via system identification. A parametric model is the one which represents the signal in terms of a set of unknown parameters, which may be the coefficients of a set of known simple functions. Alternatively, a parametric model can be formulated by using a set of parameters related by some simple mathematical expression. For non-parametric modelling, we assume the form of the signal before extracting the signal properties.

A model-mismatch problem occurs where the signal does not fit the model used for it. For instance, if an algorithm is developed to estimate the pitch-frequency or the pitch-period of a signal assuming that the underlying process is periodic, then the algorithm cannot provide any sensible results when applied for an aperiodic signal. For parametric modelling, parsimony of parameters is an important point to ensure that parameter estimation can be done accurately. An error analysis can reveal how closely the signal is fitting the parametric model. We shall address many issues relevant to signal modelling in a later chapter when we consider various concepts of function representation.

In our approach of signal modelling, let the signal $g[n]$ be represented by an autoregressive (AR) process of order J defined by the linear difference equation

$$g[n] = -\sum_{k=1}^{J} \alpha_k g[n-k] + e[n] \tag{1.28}$$

where $\{-\alpha_k; k = 1,2,\ldots,J\}$ are the AR coefficients and $e[n]$ is a zero-mean uncorrelated random sequence. Taking the z-transform of the sides, (1.28) can be rewritten in terms of the system function of an all-pole filter as follows,

$$\frac{G(z)}{E(z)} = \frac{1}{A(z)} = \frac{1}{\sum_{k=0}^{J} \alpha_k z^{-k}} \tag{1.29}$$

where $G(z)$ and $E(z)$ are the z-transforms of the response $g[n]$ and the excitation $e[n]$ respectively, and $\alpha_0 = 1$.

By definition of the AR process, the autocorrelation function of $e[n]$ is given by

$$r_e[l] = \sigma^2 \delta[l] \tag{1.30}$$

with σ^2 being an unknown constant, and the cross-correlation of $g[n]$ and $e[n]$ is given by

$$r_{ge}[l] = \lim_{M \to \infty} \frac{1}{2M+1} \sum_{n=-M}^{M} g^*[n-l] e[n] = \sigma^2 \delta[l] \tag{1.31}$$

Multiplying both sides of (1.28) by $g^*[n-l]$ and performing the time averaging as in (1.24) or (1.31), we get the following relation

$$r_g[l] = -\sum_{k=1}^{J} \alpha_k \, r_g[l-k] + \sigma^2 \delta[l] \tag{1.32}$$

Assuming that the sequence of autocorrelation functions $\{r_g[l]; l=-J,\ldots,0,\ldots,J\}$ is available for processing, from (1.32) we write the matrix equation

$$\begin{bmatrix} r_g[0] & r_g[-1] & \cdots & r_g[-(J-1)] \\ r_g[1] & r_g[0] & \cdots & r_g[-(J-2)] \\ \vdots & \vdots & \ddots & \vdots \\ r_g[J-1] & r_g[J-2] & \cdots & r_g[0] \end{bmatrix} \begin{bmatrix} \alpha_1 \\ \alpha_2 \\ \vdots \\ \alpha_J \end{bmatrix} = -\begin{bmatrix} r_g[1] \\ r_g[2] \\ \vdots \\ r_g[J] \end{bmatrix} \tag{1.33}$$

which is solved for the vector containing the AR coefficients α_k. Next we estimate the variance of excitation σ^2 as

$$\sigma^2 = r_g[0] + \sum_{k=1}^{J} \alpha_k r_g[-k] \tag{1.34}$$

The power spectral density $P_g(f)$ of the output of the AR process is related to the power spectral density $P_e(f)$ of the input as follows,

$$P_g(f) = \frac{1}{|A(f)|^2} P_e(f) \tag{1.35}$$

where $A(f) = A(z)\big|_{z=e^{j2\pi f}} = \sum_{k=0}^{J} \alpha_k e^{-j2\pi kf}$ and $P_e(f) = \sigma^2$ from (1.30). Rewriting (1.35), we get

$$P_g(f) = \frac{\sigma^2}{\left|1 + \sum_{k=1}^{J} \alpha_k e^{-j2\pi kf}\right|^2} \tag{1.36}$$

Note that although the autocorrelation sequence $r_g[l]$ is available for the lags $-J \le l \le J$, we can extrapolate the sequence $r_g[l]$ for lags $J+1, J+2, \ldots$ by using the relation (1.32) with the estimated values of α_k. This implied extension of the autocorrelation sequence beyond the window of available lags is the reason why the computation of $P_g(f)$ by (1.36) does not put any restriction on resolution of frequency. If we compare the present situation with the other option of computing $P_g(f)$ by (1.25) using the autocorrelation sequence of available lags, then we see that the advantage of the model-based approach is to remove the restriction on resolution of frequency.

In an approach of fitting an equivalent model, we propose that the signal samples $\{g[n]; n=0,1,\ldots,N-1\}$ be fitted into a linear predictor of order J. Then, the predicted signal samples $\hat{g}[n]$ are given by

$$\hat{g}[n] = -\sum_{k=1}^{J} \alpha_k g[n-k], \quad n \ge J \tag{1.37}$$

where now $\{-\alpha_k; k=1,2,\ldots,J\}$ are the prediction coefficients. The prediction error sequence $e[n]$ is expressed as

$$
\begin{aligned}
e[n] &= g[n] - \hat{g}[n] \\
&= g[n] + \sum_{k=1}^{J} \alpha_k g[n-k], \; n \geq J
\end{aligned}
\tag{1.38}
$$

and the power of prediction error ρ is given by

$$
\begin{aligned}
\rho &= \frac{1}{N-J} \sum_{n=J}^{N-1} |e[n]|^2 \\
&= \frac{1}{N-J} \sum_{n=J}^{N-1} \left(g^*[n] + \sum_{l=1}^{J} \alpha_l^* g^*[n-l] \right) \left(g[n] + \sum_{k=1}^{J} \alpha_k g[n-k] \right)
\end{aligned}
\tag{1.39}
$$

Setting $\frac{\partial \rho}{\partial \alpha_l^*} = 0$ for minimization of ρ, we get

$$
\frac{1}{N-J} \sum_{n=J}^{N-1} g^*[n-l] \left(g[n] + \sum_{k=1}^{J} \alpha_k g[n-k] \right) = 0 \quad \text{for} \quad l = 1, 2, \ldots, J \tag{1.40}
$$

Defining the autocorrelation functions for finite-length signal sequence appropriately, (1.40) is rewritten as

$$
r_g[l] + \sum_{k=1}^{J} \alpha_k \, r_g[l-k] = 0 \quad \text{for} \quad l = 1, 2, \ldots, J \tag{1.41}
$$

Substituting (1.41) in (1.39), we get the minimum prediction-error power $\rho_{\min}$ as

$$
\rho_{\min} = r_g[0] + \sum_{k=1}^{J} \alpha_k \, r_g[-k] \tag{1.42}
$$

Comparing (1.41) and (1.42) with (1.32) and (1.34), we find that an AR process and a linear predictor, both having the same order, essentially model a signal in the same way so far the inter-relationship of the autocorrelation functions is expressed, and hence, the power spectral density of the signal will be represented in the same form whether the signal is modelled by an AR process or a linear predictor. Now, taking the z-transform of both the sides, (1.38) can be rewritten in the form of the system function of the prediction-error filter (PEF) as follows,

$$
\frac{E(z)}{G(z)} = A(z) = \sum_{k=0}^{J} \alpha_k z^{-k} \quad \text{with} \quad \alpha_0 = 1 \tag{1.43}
$$

which is the inverse-system of the AR process of same order. The zeros of the PEF evaluated by extracting the roots of the polynomial equation, $A(z) = 0$, are related to the frequencies f_i present in the signal as $z_i = \exp(j2\pi f_i); i = 1, 2, \ldots, J$.

Although the estimation of frequencies of a signal by plotting the AR-PSD or by computing the zeros of the PEF polynomial equation can provide optimal resolution with finite data-length, the resolution can be adversely affected with presence of noise. It has been shown empirically that the resolution of the AR spectral estimator decreases with the decrease of the signal-to-noise ratio (SNR). The reason for this degradation can be given as a model-mismatch problem. In other words, the noisy signal does not fit the all-pole model, unless we use a large model order. An AR process with a large model-order models the signal as well as the additive noise. However, the use of a large model-order will require the autocorrelation sequence to be computed for a large number of lags, which in turn, will require that a larger number of signal samples be available for processing. Therefore with finite data-length, the resolution of the AR spectral estimator may not be acceptable at low SNR level.

We now describe the approach of parametric modelling for spectral estimation. Let the sequence of signal y_n be represented by a weighted sum of complex exponentials with or without additive noise,

$$y_n = \sum_{i=1}^{M} b_i z_i^n + w_n \,;\; n = 0, 1, \ldots, N-1 \tag{1.44}$$

where $b_i = |b_i| \angle \theta_i$ is a complex number denoting in polar coordinate the amplitude and phase of the ith complex exponential, $z_i = \exp(j2\pi f_i)$, f_i being the frequency, and w_n is the noise which may be present in the signal samples. While estimating the frequency-parameters of the model, we set initially the noise sequence $w_n = 0$, and later, we modify the estimation procedure to take account of the presence of noise.

Letting $w_n = 0$ in (1.44), we form the polynomial equation

$$a_M \prod_{i=1}^{M} (z - z_i) = 0 \;\Rightarrow\; a_M z^M + a_{M-1} z^{M-1} + \cdots + a_1 z + a_0 = 0 \tag{1.45}$$

whose roots $z_i, i = 1, 2, \ldots, M$ can be extracted for frequency estimation provided the unknown coefficients $a_0, a_1, \ldots, a_M$ are determined first. Utilizing the N samples of the sequence y_n, we can write the following matrix equation according to (1.45),

$$\mathbf{Ya} = \mathbf{0} \;\Rightarrow\; \begin{bmatrix} y_0 & y_1 & \cdots & y_M \\ y_1 & y_2 & \cdots & y_{M+1} \\ \vdots & \vdots & & \vdots \\ y_{N-M-1} & y_{N-M} & \cdots & y_{N-1} \end{bmatrix} \begin{bmatrix} a_0 \\ a_1 \\ \vdots \\ a_M \end{bmatrix} = \mathbf{0} \tag{1.46}$$

where $N \geq 2M$ to ensure that the matrix $\mathbf{Y}$ has at least M rows. In order to solve (1.46), we assume, without loss of generality, that one of the coefficients $a_0, \ldots, a_M$ is unity. Setting $a_M = 1$, we rewrite (1.46) as follows,

$$\begin{bmatrix} y_0 & y_1 & \cdots & y_{M-1} \\ y_1 & y_2 & \cdots & y_M \\ \vdots & \vdots & & \vdots \\ y_{N-M-1} & y_{N-M} & \cdots & y_{N-2} \end{bmatrix} \begin{bmatrix} a_0 \\ a_1 \\ \vdots \\ a_{M-1} \end{bmatrix} = -\begin{bmatrix} y_M \\ y_{M+1} \\ \vdots \\ y_{N-1} \end{bmatrix} \;\Rightarrow\; \mathbf{Y}_0 \bar{\mathbf{a}} = \mathbf{y} \tag{1.47}$$

Note that with the sequence y_n having M distinct frequencies, the matrix $\mathbf{Y}_0$ will be full-rank. Hence, the solution of (1.47) will be given by

$$\bar{\mathbf{a}} = \begin{cases} \mathbf{Y}_0^{-1}\mathbf{y} & \text{when } N = 2M \\ \left(\mathbf{Y}_0^H\mathbf{Y}_0\right)^{-1}\mathbf{Y}_0^H\mathbf{y} & \text{when } N \geq 2M+1 \end{cases} \tag{1.48}$$

where $\mathbf{Y}_0^{-1}$ is the inverse and $\left(\mathbf{Y}_0^H\mathbf{Y}_0\right)^{-1}\mathbf{Y}_0^H$ is the pseudo-inverse of $\mathbf{Y}_0$.

Once the coefficients are known, the polynomial equation (1.45) can be formed and its roots can be extracted for estimation of frequencies. The procedure described above is the Prony's method which belongs to the class of super-resolution techniques. In spectral estimation, a super-resolution technique does not impose any theoretical restriction on the resolution of frequencies, which implies that two frequencies to be resolved can be arbitrarily close. However, we may need sophisticated signal processing means to resolve such close frequencies.

We now describe another method to compute the coefficient vector $\mathbf{a}$ of (1.46). The matrix $\mathbf{Y}$ of (1.46) is formed with $N \geq 2M$, such that there are at least M rows. The rank of matrix $\mathbf{Y}$ is M when the underlying signal contains M distinct frequencies. Hence, the estimate of the autocorrelation matrix $\mathbf{R} = \mathbf{Y}^H\mathbf{Y}$, which is of dimension $(M+1)\times(M+1)$, will have rank equal to M. An eigenvalue decomposition of the matrix $\mathbf{R}$ provides the eigenvalues λ_i and the corresponding eigenvectors $\mathbf{v}_i$ as follows,

$$\mathbf{R} = \sum_{i=1}^{M+1} \lambda_i \mathbf{v}_i \mathbf{v}_i^H \tag{1.49}$$

where the ordered eigenvalues are $\lambda_1 \geq \lambda_2 \geq \cdots \geq \lambda_M > \lambda_{M+1} = 0$. It is apparent that the coefficient vector $\mathbf{a}$ of (1.46) is given by

$$\mathbf{a} = \gamma \mathbf{v}_{M+1} \tag{1.50}$$

where γ is a scale factor, and $\mathbf{v}_{M+1}$ is the eigenvector corresponding to the eigenvalue λ_{M+1} of $\mathbf{R}$. Post-multiplying (1.46) by $\mathbf{Y}^H$, we get

$$\mathbf{Ra} = \mathbf{0} \tag{1.51}$$

which shows that $\mathbf{a}$ is the eigenvector corresponding to the eigenvalue $\lambda_{\min} = 0$.

Once the coefficient vector $\mathbf{a}$ is determined by the eigenvalue decomposition of the matrix $\mathbf{R}$, the polynomial equation (1.45) can be formed and its roots $z_i = \exp(j2\pi f_i)$ can be extracted for estimation of frequencies. The above procedure where the minimum eigenvalue and the corresponding eigenvector of a correlation matrix is used to find the frequencies of sinusoids is known as the Pisarenko's method, which also belongs to the class of super-resolution techniques.

We now describe a method to estimate the z_i-parameters of (1.44) directly. Initially, we consider the no-noise case and set $w_n = 0$ in the parametric model of the signal in (1.44). We form two matrices

$$\mathbf{Y}_0 = \begin{bmatrix} y_0 & y_1 & \cdots & y_{M-1} \\ y_1 & y_2 & \cdots & y_M \\ \vdots & \vdots & & \vdots \\ y_{N-M-1} & y_{N-M} & \cdots & y_{N-2} \end{bmatrix} \tag{1.52}$$

and

$$\mathbf{Y}_1 = \begin{bmatrix} y_1 & y_2 & \cdots & y_M \\ y_2 & y_3 & \cdots & y_{M+1} \\ \vdots & \vdots & & \vdots \\ y_{N-M} & y_{N-M+1} & \cdots & y_{N-1} \end{bmatrix} \tag{1.53}$$

where $N \geq 2M$ as required. We then study the rank-properties of $\mathbf{Y}_0$, $\mathbf{Y}_1$, and the matrix pencil $(\mathbf{Y}_0 - \lambda\mathbf{Y}_1)$ where λ is the pencil parameter. When the signal is consisting of M distinct frequencies, each of the data matrices $\mathbf{Y}_0$ and $\mathbf{Y}_1$ will be of rank M. Therefore, for an arbitrary value of the λ-parameter, the matrix pencil $(\mathbf{Y}_0 - \lambda\mathbf{Y}_1)$ will also be of rank M. However, when the pencil parameter is set at $\lambda = z_i^{-1}$ for $i = 1, 2, \ldots, M$, the rank of matrix pencil $(\mathbf{Y}_0 - \lambda\mathbf{Y}_1)$ will become $M-1$. Under this condition, the ith component will be missing from the data, and the resulting sequence will have $M-1$ frequencies.

Therefore, in order to estimate the z_i-parameters of (1.44), we have to determine the rank-reducing pencil value $\lambda = z_i^{-1}$ of the matrix pencil $(\mathbf{Y}_0 - \lambda\mathbf{Y}_1)$. There are various approaches of finding the rank-reducing pencil values. The simplest among those approaches is to solve the determinant equation as follows,

$$\left|\mathbf{Y}_1^+\mathbf{Y}_0 - \lambda\mathbf{I}\right| = 0 \tag{1.54}$$

where $\mathbf{Y}_1^+$ is the inverse of $\mathbf{Y}_1$ computed when $N = 2M$ and it is the pseudo-inverse of $\mathbf{Y}_1$ when $N \geq 2M+1$. Note that we are using a standard technique of extracting the roots of a polynomial equation for computing the eigenvalues of a matrix in (1.54). The procedure described above is known as the Matrix Pencil method, and the method together with the parametric modelling of (1.44), can provide super-resolution performance of spectral estimation.

When the signal samples are corrupted with noise, the three super-resolution techniques described above cannot provide accurate estimation of the frequencies present in the signal. The parametric model that is used for spectral estimation does not fit the noisy samples. In this situation, we use extended-order modelling to avoid the model-mismatch problem. The idea is to model the noisy signal samples with L complex exponentials for L sufficiently larger than M, where M is the number of distinct frequencies present in the signal. Thus with extended-order modelling, L–M components are used to model the additive noise. Note that the estimated noise-frequencies can be easily separated from the estimated signal-frequencies by checking the corresponding amplitudes. The noise-amplitudes will be negligible compared to the signal-amplitudes assuming the noise-level to be low.

With extended-order modelling, we extract the roots $z_i = \exp(j2\pi f_i)$ of the polynomial equation given by

$$\prod_{i=M+1}^{L}(z-z_i)\left[a_M z^M + a_{M-1}z^{M-1} + \cdots + a_1 z + a_0\right] = 0$$
$$\Rightarrow c_L z^L + c_{L-1}z^{L-1} + \cdots + c_1 z + c_0 = 0 \tag{1.55}$$

for the Prony's and the Pisarenko's methods. Setting $c_L = 1$, we write the matrix equation

$$\begin{bmatrix} y_0 & y_1 & \cdots & y_{L-1} \\ y_1 & y_2 & \cdots & y_L \\ \vdots & \vdots & & \vdots \\ y_{N-L-1} & y_{N-L} & \cdots & y_{N-2} \end{bmatrix} \begin{bmatrix} c_0 \\ c_1 \\ \vdots \\ c_{L-1} \end{bmatrix} = -\begin{bmatrix} y_L \\ y_{L+1} \\ \vdots \\ y_{N-1} \end{bmatrix} \Rightarrow \mathbf{Y}_{0e}\overline{\mathbf{c}} = \mathbf{y}_e \tag{1.56}$$

similar to (1.47) in this case, where $M \le L \le N - M$ to ensure that the matrix $\mathbf{Y}_{0e}$ has at least M rows and at least M columns. In order to find the solution of (1.56), we need to compute a matrix $\mathbf{Y}_{0e}^{+}$ such that the desired solution

$$\overline{\mathbf{c}} = \mathbf{Y}_{0e}^{+}\mathbf{y}_e \tag{1.57}$$

satisfies the following two conditions:

1. The sum of square errors $\sum_{i=0}^{N-L-1}\left(\sum_{l=0}^{L-1} c_l\, y_{i+l} + y_{i+L}\right)^2$ is least, and
2. The norm of the solution vector, $\|\overline{\mathbf{c}}\|$ is minimum.

We call the desired solution as the minimum-norm least-squares solution. Note that the effective rank of the matrix $\mathbf{Y}_{0e}$ is M, which depends on the number of signal components. Hence, the matrix $\mathbf{Y}_{0e}$ will be rank-deficient when typically it will have more than M rows and more than M columns. The effective rank of $\mathbf{Y}_{0e}$ is decided by the significant eigenvalues of the matrix $\mathbf{Y}_{0e}^{H}\mathbf{Y}_{0e}$ as explained below.

Therefore, the conventional techniques for computation of the inverse or the pseudo-inverse of a matrix cannot be applied to obtain $\mathbf{Y}_{0e}^{+}$ which will be known as the generalized inverse of the matrix $\mathbf{Y}_{0e}$. In the sequel, we discuss how to compute the generalized inverse of a matrix which may be square, over-determined or under-determined, and full-rank or rank-deficient.

We now investigate the second method based on the eigenvalue decomposition to compute the coefficient vector $\mathbf{c} = [c_0, c_1, \ldots, c_L]^T$. To solve the homogeneous matrix equation

$$\mathbf{Y}_e^H\mathbf{Y}_e\mathbf{c} = \mathbf{0} \;\Rightarrow\; \mathbf{R}_e\mathbf{c} = \mathbf{0} \tag{1.58}$$

where $\mathbf{Y}_e = \begin{bmatrix} y_0 & y_1 & \cdots & y_L \\ y_1 & y_2 & \cdots & y_{L+1} \\ \vdots & \vdots & & \vdots \\ y_{N-L-1} & y_{N-L} & \cdots & y_{N-1} \end{bmatrix}$, we find the ordered eigenvalues $\lambda_1 \geq \lambda_2 \geq \cdots \geq \lambda_M >> \lambda_{M+1} \geq \cdots \geq \lambda_{L+1} \approx 0$ and the corresponding eigenvectors $\mathbf{v}_1, \mathbf{v}_2, \ldots, \mathbf{v}_{L+1}$ of the autocorrelation matrix $\mathbf{R}_e$ whose effective rank is M. The eigenvalues $\lambda_1, \ldots, \lambda_M$ represent the power of the M sinusoidal signal components, and they are called the principal eigenvalues. Assuming low noise level the eigenvalues $\lambda_{M+1}, \ldots, \lambda_{L+1}$ will be negligible compared to the first M eigenvalues of the matrix.

The solution of (1.58) will be given by

$$\mathbf{c} = \sum_{i=1}^{L-M+1} \gamma_i \mathbf{v}_{M+i} \tag{1.59}$$

where $\mathbf{v}_{M+1}, \ldots, \mathbf{v}_{L+1}$ are the eigenvectors corresponding to the non-principal eigenvalues. The scale factors $\gamma_1, \ldots, \gamma_{L-M+1}$ can be determined by setting typically $c_0 = 1$. In a later chapter, we discuss various aspects of singular-value and eigenvalue decompositions of matrices and its applications in signal processing.

When the eigenvalue decomposition of the autocorrelation matrix $\mathbf{R}_e$ is available, a popular method of estimation of frequency is to plot the *spectral estimator* $P(f)$ given by

$$P(f) = \frac{1}{\sum_{i=M+1}^{L+1} \left| \mathbf{z}^H \mathbf{v}_i \right|^2} \tag{1.60}$$

where $\mathbf{z} = \left[1, z, \ldots, z^{N-1}\right]^T$, $z = \exp(j2\pi f)$, and $\mathbf{v}_{M+1}, \ldots, \mathbf{v}_{L+1}$ are the eigenvectors of the noise subspace. It can be shown that with $f = f_i$, the ith sinusoidal frequency present in the signal, and $z = z_i$, the term $P(f)$ theoretically tends to infinity. Hence, in practice, there will be M peaks in the plot of $P(f)$ at the M frequencies of the underlying signal. By using the relation

$$\mathbf{I} = \sum_{i=1}^{L+1} \mathbf{v}_i \mathbf{v}_i^H \tag{1.61}$$

where $\mathbf{I}$ is the identity matrix, an alternative *spectral estimator* $Q(f)$ can be derived as

$$Q(f) = \sum_{i=1}^{M} \left| \mathbf{z}^H \mathbf{v}_i \right|^2 \tag{1.62}$$

where $\mathbf{v}_1, \ldots, \mathbf{v}_M$ are the eigenvectors of the signal subspace. The plot of $Q(f)$ will show M peaks at the M frequencies of the signal. The estimation of frequency using the spectral estimators of (1.60) or (1.62) is known as the multiple signal classification (MUSIC) algorithm.

We now modify the matrix pencil method when the signal samples are corrupted with noise. Using a model order L sufficiently larger than M for the signal consisting of M sinusoids, we form the data matrices

$$\mathbf{Y}_{0e} = \begin{bmatrix} y_0 & y_1 & \cdots & y_{L-1} \\ y_1 & y_2 & \cdots & y_L \\ \vdots & \vdots & & \vdots \\ y_{N-L-1} & y_{N-L} & \cdots & y_{N-2} \end{bmatrix} \tag{1.63}$$

and

$$\mathbf{Y}_{1e} = \begin{bmatrix} y_1 & y_2 & \cdots & y_L \\ y_2 & y_3 & \cdots & y_{L+1} \\ \vdots & \vdots & & \vdots \\ y_{N-L} & y_{N-L+1} & \cdots & y_{N-1} \end{bmatrix} \tag{1.64}$$

where $M \leq L \leq N-M$, but typically L is much larger than M. The effective rank of the data matrices $\mathbf{Y}_{0e}$ and $\mathbf{Y}_{1e}$ will be M for the underlying signal having M distinct frequencies. Therefore, the matrix pencil $\left(\mathbf{Y}_{0e} - \lambda \mathbf{Y}_{1e}\right)$ will also be of rank M for an arbitrary value of the λ-parameter. However, the rank of the matrix pencil $\left(\mathbf{Y}_{0e} - \lambda \mathbf{Y}_{1e}\right)$ will become $M-1$ when the pencil parameter is set at $\lambda = z_i^{-1}$ for $i = 1, 2, \ldots, M$.

We determine the rank-reducing pencil values by solving the determinant equation

$$\left|\mathbf{Y}_{1e}^{+}\mathbf{Y}_{0e} - \lambda \mathbf{I}\right| = 0 \tag{1.65}$$

where $\mathbf{Y}_{1e}^{+}$ is the generalized inverse of $\mathbf{Y}_{1e}$. The computation of the eigenvalues of the matrix $\mathbf{Y}_{1e}^{+}\mathbf{Y}_{0e}$ in (1.65) will provide M principal eigenvalues $\left\{z_i^{-1}; i = 1, \ldots, M\right\}$ related to the signal frequencies $f_i = \frac{1}{2\pi}\arg\left(z_i\right)$ and $L-M$ eigenvalues close to zero. In the sequel, we shall discuss how to use the singular value decomposition of the matrix $\mathbf{Y}_{1e}$ to compute the generalized inverse $\mathbf{Y}_{1e}^{+}$ taking contributions from the signal components and rejecting that from the noise components.

The three super-resolution techniques described above are capable of determining the frequencies present in a signal accurately even with a short data-length. When the signal is corrupted with noise, we need to use extended-order modelling, which in turn requires that the data-length be moderate. It is to be noted that the Prony's, the Pisarenko's and the Matrix Pencil methods basically exploit the geometrical properties of the modelled complex-exponential signal for estimating the frequency-parameters without taking any consideration of the noise statistics. In order to design an optimal estimator for the parameters of the modelled signal we make use of the known or partially known probability distribution of the random noise. The maximum likelihood estimator finds the unknown parameters by maximizing the likelihood function which is the probability of observed values of the signal samples as derived from the known or partially known probability density function (PDF) of the noise.

We write the parametric model of the signal sequence as given by (1.44) in vector-matrix form,

$$\begin{bmatrix} y_0 \\ y_1 \\ \vdots \\ y_{N-1} \end{bmatrix} = \begin{bmatrix} 1 & 1 & \cdots & 1 \\ z_1 & z_2 & \cdots & z_M \\ \vdots & \vdots & & \vdots \\ z_1^{N-1} & z_2^{N-1} & \cdots & z_M^{N-1} \end{bmatrix} \begin{bmatrix} b_1 \\ b_2 \\ \vdots \\ b_M \end{bmatrix} + \begin{bmatrix} w_0 \\ w_1 \\ \vdots \\ w_{N-1} \end{bmatrix} \Rightarrow \mathbf{y} = \mathbf{Zb} + \mathbf{w} \tag{1.66}$$

Assuming complex zero-mean multivariate Gaussian distribution for the noise vector $\mathbf{w}$, the likelihood function is given by

$$p(\mathbf{y}-\mathbf{Zb}) = \frac{1}{\pi^N \det(\mathbf{R}_{ww})} \exp\left[-(\mathbf{y}-\mathbf{Zb})^H \mathbf{R}_{ww}^{-1} (\mathbf{y}-\mathbf{Zb})\right] \tag{1.67}$$

where $\mathbf{R}_{ww}$ is the autocorrelation matrix of the noise sequence. When the noise is assumed to be white, $\mathbf{R}_{ww} = \sigma_w^2 \mathbf{I}$, where σ_w^2 is the unknown power spectral density of the noise process. In this case, the right hand side of (1.67) reduces to

$$p(\mathbf{y}-\mathbf{Zb}) = \frac{1}{\pi^N \sigma_w^{2N}} \exp\left[-\frac{1}{\sigma_w^2} (\mathbf{y}-\mathbf{Zb})^H (\mathbf{y}-\mathbf{Zb})\right] \tag{1.68}$$

In order to find the maximum likelihood estimate (MLE) for the parameters σ_w^2, $\mathbf{b}$, and $\{z_i; i=1,\ldots,M\}$, we need to maximize the PDF of $\mathbf{y}$ as given by (1.68). Note that the problem of maximization of $p(\mathbf{y})$ with respect to $\mathbf{b}$ or $\{z_i; i=1,\ldots,M\}$ is equivalent to the dual problem of minimization of the term

$$Q = (\mathbf{y}-\mathbf{Zb})^H (\mathbf{y}-\mathbf{Zb}) \tag{1.69}$$

This term is quadratic in $\mathbf{b}$, and setting $\frac{\partial Q}{\partial \mathbf{b}} = \mathbf{0}$, we find the MLE of $\mathbf{b}$ as

$$\hat{\mathbf{b}} = \left(\mathbf{Z}^H \mathbf{Z}\right)^{-1} \mathbf{Z}^H \mathbf{y} \tag{1.70}$$

which can be computed provided the MLEs of $\{z_i; i=1,\ldots,M\}$ are available. Substituting (1.70) in (1.69) and simplifying, we get

$$Q = \mathbf{y}^H \left[\mathbf{I} - \mathbf{Z}\left(\mathbf{Z}^H \mathbf{Z}\right)^{-1} \mathbf{Z}^H\right] \mathbf{y} \tag{1.71}$$

which is to be minimized now with respect to $\{z_i; i=1,\ldots,M\}$. Alternatively, we need to maximize the term

$$P = \mathbf{y}^H \mathbf{Z}\left(\mathbf{Z}^H \mathbf{Z}\right)^{-1} \mathbf{Z}^H \mathbf{y} \tag{1.72}$$

to find the MLEs of $\{z_i; i=1,\ldots,M\}$. The M-dimensional optimization problem of P or Q is not tractable except for the case of $M=1$. The terms P and Q are highly non-linear functions of z_i, and hence, finding the MLEs of z_i has remained a major challenge.

We shall now describe an iterative method for obtaining the MLEs of z_i. This approach is based on converting the non-linear optimization problem into a standard quadratic minimization problem. Note that the signal sequence of (1.44) obeys the 'special' autoregressive moving-average (ARMA) process given by

$$\begin{aligned} &a_M y_{n+M} + a_{M-1} y_{n+M-1} + \cdots + a_1 y_{n+1} + a_0 y_n \\ &\quad = a_M w_{n+M} + a_{M-1} w_{n+M-1} + \cdots + a_1 w_{n+1} + a_0 w_n \end{aligned} \tag{1.73}$$

where the coefficients $a_0, a_1, \ldots, a_M$ form the polynomial equation of (1.45) whose roots are $\{z_i, i=1,2,\ldots,M\}$. We form the $N \times (N-M)$ Toeplitz matrix $\mathbf{A}$ as follows:

$$\mathbf{A} = \begin{bmatrix} a_0^* & 0 & \cdots & 0 \\ \vdots & \ddots & \ddots & \vdots \\ a_M^* & & \ddots & 0 \\ 0 & \ddots & & a_0^* \\ \vdots & \ddots & \ddots & \vdots \\ 0 & \cdots & 0 & a_M^* \end{bmatrix} \tag{1.74}$$

At first, note that the rank properties of the matrices $\mathbf{A}$ and $\mathbf{Z}$ are as given below:

$\text{rank}(\mathbf{A}) = N-M$, $\text{rank}(\mathbf{Z}) = M$, and the $N \times N$ matrix $[\mathbf{Z}, \mathbf{A}]$ is non-singular. Moreover, we have $\mathbf{A}^H \mathbf{Z} = \mathbf{0}$ using (1.45) for each element of the matrix. It follows that the columns of $\mathbf{A}$ are orthogonal to those of $\mathbf{Z}$. In other words, the columns of $\mathbf{A}$ span the orthogonal complement to the range space of $\mathbf{Z}$. Under the above condition, a standard result from matrix algebra related to the projection matrices is the following relation,

$$\mathbf{P}_Z + \mathbf{P}_A = \mathbf{I} \;\Rightarrow\; \mathbf{Z}\left(\mathbf{Z}^H\mathbf{Z}\right)^{-1}\mathbf{Z}^H + \mathbf{A}\left(\mathbf{A}^H\mathbf{A}\right)^{-1}\mathbf{A} = \mathbf{I} \tag{1.75}$$

where $\mathbf{P}_Z$ and $\mathbf{P}_A$ are the projections onto the subspaces of $\mathbf{Z}$ and $\mathbf{A}$, respectively. Using (1.75) in (1.71) we get

$$Q = \mathbf{y}^H \mathbf{A}\left(\mathbf{A}^H\mathbf{A}\right)^{-1}\mathbf{A}^H\mathbf{y} \tag{1.76}$$

which can be rewritten as

$$Q = \mathbf{a}^H \mathbf{Y}\left(\mathbf{A}^H\mathbf{A}\right)^{-1}\mathbf{Y}^H\mathbf{a} \tag{1.77}$$

where the coefficient vector $\mathbf{a}$ and the data matrix $\mathbf{Y}$ are as defined in (1.46). The term Q is to be minimized now with respect to the coefficients $\{a_i; i=0,1,\ldots,M\}$. Once the MLEs of the coefficients are known, the polynomial equation of (1.45) can be formed and its roots are extracted to find the MLEs of z_i.

We present an iterative method to obtain the solution of the minimization problem of (1.77). We start with enlisting the constraints that the acceptable solution of the coefficient vector $\mathbf{a} = [a_0, a_1, \ldots, a_M]^T$ must satisfy. Note that the constraints can be implemented in the iterative technique either implicitly or explicitly and an appropriate formulation can be developed for the respective purpose. We enlist the constraints and describe the basic algorithm as follows:

Constraints

1. $\mathbf{a} \neq \mathbf{0}$ is the 'non-triviality' constraint, which ensures that the matrix $\mathbf{A}$ is full-rank and $(\mathbf{A}^H\mathbf{A})$ is invertible.
2. $\mathbf{a}$ must provide a stable polynomial; i.e., all the roots of the polynomial equation:

$$a_M z^M + a_{M-1} z^{M-1} + \cdots + a_1 z + a_0 = 0,$$

   must be inside or on the unit circle.
3. $\mathbf{a}$ has the 'mirror symmetry' property:

$$a_i = a^*_{M-i} \text{ for } i = 0, 1, \ldots, M,$$

   when the underlying signal consists of pure sinusoids.
4. $\mathbf{a}$ is real,

$$\operatorname{Im}(a_i) = 0 \text{ for } i = 0, 1, \ldots, M,$$

   when the underlying signal consists of real sinusoids.

Basic Algorithm

Step-1 Initialization: $k = 0$ and $\mathbf{a}_{(0)} = \mathbf{a}_0 \neq \mathbf{0}$,
where $\mathbf{a}_0$ is an acceptable initial solution.

Step-2 Compute the matrix: $\mathbf{C}_X^{(k)} = \mathbf{Y}^H \left(\mathbf{A}_{(k)}^H \mathbf{A}_{(k)}\right)^{-1} \mathbf{Y}$,
where $\mathbf{A}_{(k)}$ is formed with $\mathbf{a}_{(k)}$.

Step-3 Solve the quadratic minimization problem:

$$\arg\min \ \mathbf{a}_{(k+1)}^H \mathbf{C}_X^{(k)} \mathbf{a}_{(k+1)},$$

subject to the given constraints on $\mathbf{a}_{(k+1)}$.

Step-4 Set iteration: $k = k + 1$.

Step-5 Check convergence: For small ε,

$$\left\|\mathbf{a}_{(k-1)} - \mathbf{a}_{(k)}\right\| < \varepsilon ?$$

Yes – done, go to *Step-6*; No – go to *Step-2*.

Step-6 Find the roots of the polynomial equation:

$$a_M z^M + a_{M-1} z^{M-1} + \cdots + a_1 z + a_0 = 0.$$

In a later chapter, we discuss how to solve optimization problems with constrains on the variables of optimization. We use mathematical programming approach to obtain solutions of such problems employing iterative techniques. We consider the issue of convergence in an iterative method. Note that the constant ε used for checking convergence in Step-5 above is preset according to the accuracy limit of the computing machine.

1.3 Outline Plan

Signal processing broadly refers to the subject which deals with various techniques for extraction of information from a given-set of data, or conversely, for utilization of information to generate data with some desirable property. These two procedures of signal processing are known as signal analysis and signal synthesis, respectively. Wide varieties of mathematical ideas are adopted and applied for the purpose of signal processing. For effective learning and to get motivated in research in the area of signal processing, students need to go through a treatise on mathematical methods, which explain fundamental concepts and illustrate how engineering problems are solved in practice employing techniques derived from fundamental concepts. The end-users may be conversant with computer-codes or algorithms, but the researchers need to know the theoretical background as well as the practical applications.

The book identifies four areas of mathematics, viz., function representation, generalized inverse, modal decomposition, and optimization, and covers these topics for basic concepts and their interpretations, and consequently, their applications for signal processing purpose have been demonstrated. Coverage of each topic is not exhaustive, rather we mainly focus on concepts which are central or for which some signal processing applications are known.

In the chapter of function representation, we explore the ways how digital data can be represented by polynomials or sinusoidal functions. Two ways of getting function representation, viz., interpolation and approximation have distinctly different characteristics. In digital signal processing, the function that is implicitly generated due to the use of interpolation or approximation indeed represents the underlying signal. We often get substantially different results in signal processing depending on whether interpolation or approximation technique is used for digital data.

The chapter on generalized inverse concentrates on solution of matrix equation or linear operator equation, where the matrix is singular or the operator is not invertible. The other cases of concern are where a matrix, not necessarily a square matrix, has null space or a linear operator is ill-conditioned, meaning that the input–output relationship is extremely sensitive to inaccuracies in data. We compute generalized inverse in such cases to get a meaningful solution.

The chapter on modal decomposition introduces to the concept of principal component analysis. The singular value decomposition (SVD) of the data matrix decomposes into 1-dimensional signal subspaces along which the data show maximum variation or maximum information. This allows us to separate signal subspace from noise subspace, one signal subspace from another signal subspace, and several schemes of signal processing become easily implementable. The SVD technique can provide high accuracy of solution even when the data given are incomplete and noise-corrupted. The eigenfunction expansion of a real function decomposes the function into its constituents. It is demonstrated how this decomposition can reveal the intrinsic properties of a function.

In the chapter on optimization, we get accustomed with the principles and procedures of optimization. Variety of signal processing methods is developed based on maximization or minimization of some objective function chosen during problem formulation. Maximum likelihood estimation, maximum entropy spectral analysis, minimum error-variance approximation are such methods where the concept of optimization plays the central role. Some optimization problems are not easily solvable. Depending on the difficulty of such problems, we often settle for non-optimal solution whose accuracy is close to the optimal solution. Diversified problems can be formulated as problems of mathematical programming, and their solutions can be obtained by iterative methods. A detailed coverage of mathematical programming is included in the book with some examples of real-life problems.

While presenting theorems of mathematics in the following chapters, the priority is given on applications and interpretations. For the proof of a mathematical theorem, often standard textbooks of mathematics are referred, unless the proof encompasses some idea which may help in its interpretation and provides some insight of its trend. It is expected that our approach will motivate students to develop mathematical methods stemming from mathematical abstractions and to utilize mathematical methods for signal processing applications.

Bibliography

The list is indicative, not exhaustive. Similar references can be searched in the library.

Arfken, G. B., and H. J. Weber. 2001, *Mathematical Methods for Physicists*. 5th ed., San Diego: Harcourt Academic Press.

Bresler, Y., and A. Macovski. 1986. 'Exact Maximum Likelihood Parameter Estimation of Superimposed Exponential Signals in Noise.' *IEEE Trans. Acoustics, Speech, Signal Processing*. 34 (5): 1081–89.

Candy, J. V. 2006. *Model-Based Signal Processing*. Hoboken, NJ: John Wiley.

Hess, W. 1983. *Pitch Determination of Speech Signals: Algorithms and Devices*. Springer-Verlag.

Hildebrand, F. B. 1956. *Introduction to Numerical Analysis*. New York: McGraw-Hill.

Horn, R. A., and C. R. Johnson. 1985. *Matrix Analysis*. Cambridge: Cambridge University Press.

Hua, Y., and T. K. Sarkar. 1990. 'Matrix Pencil Method for Estimating Parameters of Exponentially Damped/Undamped Sinusoids in Noise.' *IEEE Trans. Acoustics, Speech, Signal Processing*. 38 (5): 814–24.

Kay, S. M. 1988. *Modern Spectral Estimation: Theory and Application*. Englewood Cliffs, NJ: Prentice Hall.

Kumaresan, R., L. L. Scharf, and A. K. Shaw. 1986. 'An Algorithm for Pole-Zero Modeling and Spectral Analysis.' *IEEE Trans. Acoustics, Speech, Signal Processing*. ASSP-34 (3) 637–40.

Madisetti, V. K., ed. 2010. *The Digital Signal Processing Handbook*. 2nd ed., Boca Raton, FL: CRC Press.

Marple, L. 1977. 'Resolution of Conventional Fourier, Autoregressive, and Special ARMA Methods of Spectrum Analysis.' *Proc. IEEE Int. Conf. Acoustics, Speech, Signal Process*. ICASSP '77, Connecticut, 2: 74–77.

McWhirter, J. G., ed. 1990. *Mathematics in Signal Processing II*. Oxford: Clarendon Press.

Oppenheim, A. V., A. S. Willsky, and S. H. Nawab. 1997. *Signals and Systems*. 2nd ed., New Delhi: Prentice Hall of India.

Pisarenko, V. F. 1973. 'The Retrieval of Harmonics from a Covariance Function.' *Geophysics J. Roy. Astron. Soc.* 33 (3): 347–66.

Prony, R. 1795. 'Essai éxperimental et analytique ....' *J. l'École Polytechnique* 1 (22): 24–76.

Schmidt, R. O. 1979. 'Multiple Emitter Location and Signal Parameter Estimation.' *Proc. RADC Spectrum Estimation Workshop*, New York, 243–58. (Reprint: *IEEE Trans. Antennas Propagation*, 34 (3): 276–80, 1986).

Weaver, H. J. 1989. *Theory of Discrete and Continuous Fourier Analysis*. New York: John Wiley.

2 Function Representation

2.1 Polynomial Interpolation and Lagrange Formula

Consider a continuous real-valued function $f(x)$ defined over an interval $[a, b]$ of the real axis x. We want to find an n-degree polynomial $p(x)$ given by

$$p(x)=\sum_{j=0}^{n} c_j x^j \tag{2.1}$$

to represent $f(x)$ for $a \le x \le b$. To determine the $(n + 1)$ coefficients $\{c_j\}$, we evaluate the function $f(x)$ on $(n + 1)$ distinct points $\{x_i; i = 0, 1, \ldots, n\}$, $x_i \in [a, b]$, and equate $p(x_i) = f(x_i)$ resulting the following matrix equation

$$\begin{bmatrix} 1 & x_0 & \cdots & x_0^n \\ 1 & x_1 & \cdots & x_1^n \\ \vdots & \vdots & \ddots & \vdots \\ 1 & x_n & \cdots & x_n^n \end{bmatrix} \begin{bmatrix} c_0 \\ c_1 \\ \vdots \\ c_n \end{bmatrix} = \begin{bmatrix} f(x_0) \\ f(x_1) \\ \vdots \\ f(x_n) \end{bmatrix} \tag{2.2}$$

which is to be solved for the coefficients $\{c_j\}$. The $(n + 1) \times (n + 1)$ matrix of (2.2) is the Vandermonde matrix whose determinant has the value $\Delta = \prod_{0 \le j < i \le n} (x_i - x_j)$, and $\Delta \ne 0$ for distinct points $\{x_i\}$. Thus, the matrix is invertible and the coefficients $\{c_j\}$ can be determined uniquely.

The Lagrange interpolation formula provides an n-degree polynomial representation for a function $f \in \mathscr{C}[a,b]$, $\mathscr{C}[a,b]$ being the space of real-valued functions that are continuous on $[a, b]$, without requiring the inversion of a matrix. The representative polynomial $p(x)$ is given by

$$p(x)=\sum_{k=0}^{n} f(x_k) g_k(x) \tag{2.3}$$

where $\{f(x_k)\}$ are the discrete values of the function evaluated on the distinct points $\{x_k\}$ and the generating polynomials $g_k(x) \in \mathscr{P}_n$, $\mathscr{P}_n$ being the space of algebraic polynomials of degree at most n. The polynomials $g_k(x)$ are chosen such that

$$g_k(x_i) = \delta_{ki} = \begin{cases} 1, & k = i \\ 0, & k \neq i \end{cases} \tag{2.4}$$

and this property ensures that the conditions of interpolation given by

$$p(x_i) = f(x_i) \quad \text{for } i = 0, 1, \dots, n \tag{2.5}$$

are satisfied. The generating polynomials $g_k(x)$ are expressed as

$$g_k(x) = \prod_{\substack{i=0 \\ i \neq k}}^{n} \frac{(x - x_i)}{(x_k - x_i)}, \quad a \leq x \leq b \tag{2.6}$$

and the polynomials have the property (2.4) as required. The identity given by

$$\sum_{k=0}^{n} g_k(x) = 1, \quad a \leq x \leq b \tag{2.7}$$

is useful for checking the values $\{g_k(x); k = 0, 1, \dots, n\}$ while employing the Lagrange interpolation formula. More generally, we have the identity

$$\sum_{j=0}^{n} x_k^j \, g_k(x) = x^j \quad \text{for } j = 0, 1, \dots, n \tag{2.8}$$

which is useful in deriving some important properties of polynomial interpolation.

Now, suppose that there are two polynomials $p \in \mathscr{P}_n$ and $q \in \mathscr{P}_n$, both satisfy the interpolation conditions (2.5). Then, the difference polynomial $(p - q) \in \mathscr{P}_n$, and it attains zero at $(n + 1)$ distinct points $\{x_i\}$. However, a polynomial of degree at most n that attains zero at more than n distinct points is identically zero. Therefore, we get $p = q$, and there is only one polynomial that satisfies the interpolation condition.

Theorem 2.1

Let the function $f \in \mathscr{C}[a, b]$ be evaluated at $(n + 1)$ distinct points $\{x_i; i = 0, 1, \dots, n\}$ on $[a, b]$. Then, there exists only one polynomial $p \in \mathscr{P}_n$ such that $p(x_i) = f(x_i)$ for $i = 0, 1, \dots, n$.

For any function $f \in \mathscr{C}[a, b]$ and fixed points $\{x_0, \dots, x_n\}$ in $[a, b]$, denote $L_n f = p \in \mathscr{P}_n$ such that $p = f$ at $\{x_0, \dots, x_n\}$, then it is easy to show that L_n is a linear operator, that is, $L_n(\lambda f) = \lambda L_n(f)$ and $L_n(f + g) = L_n f + L_n g$. Moreover, $L_n f = f$ if and only if $f \in \mathscr{P}_n$, which indicates that L_n is a projection.

Now, the error of interpolation $e(x)$ is given by

$$e(x) = f(x) - p(x), \quad a \leq x \leq b \tag{2.9}$$

where the polynomial $p \in \mathscr{P}_n$ satisfies the interpolation conditions (2.5). Since by definition, $e(x)$ is zero at $(n + 1)$ distinct points $\{x_i\}$, it should be clear that $e(x)$ is identically zero unless it has element of $\mathscr{P}_{n+1}$. We assume that the function $f \in \mathscr{C}^{(n+1)}[a, b]$, $\mathscr{C}^{(n+1)}[a, b]$ being the space of real-valued functions that have continuous $(n + 1)^{\text{th}}$ order derivatives on $[a, b]$.

Theorem 2.2

Let the function $f \in \mathscr{C}^{(n+1)}[a,b]$ be represented by the polynomial $p \in \mathscr{P}_n$ that satisfies $p(x_i) = f(x_i)$ for $(n+1)$ distinct points $\{x_i\}$, $x_i \in [a,b]$. Then, for any $x \in [a,b]$, the error $e(x) = f(x) - p(x)$ is given by

$$e(x) = \frac{1}{(n+1)!}\left\{\prod_{i=0}^{n}(x - x_i)\right\} f^{(n+1)}(\xi) \tag{2.10}$$

where ξ is a point in $[a, b]$ whose exact location depends on x.

For a quick verification of the above theorem, consider $f(x) = x^{n+1}$. Then, $e(x) = x^{n+1} - p(x) = \prod_{i=0}^{n}(x - x_i)$, which is the same expression that is obtained from (2.10). The use of the above theorem is two-fold. Firstly, the theorem provides a clue of how a recursion formula for polynomial interpolation can be derived. Given the polynomial interpolation of f in $\mathscr{P}_n$ and the function value $f(x_{n+1})$ at an additional point $x_{n+1} \in [a,b]$, we want to find the polynomial interpolation of f in $\mathscr{P}_{n+1}$. This problem is considered in a later section. Secondly, since the error of interpolation depends on the interpolating points $\{x_i\}$, we need to investigate whether by distributing these points properly on $[a, b]$, we can have some control over interpolation error. This issue is discussed in detail in the next section.

2.2 Error in Polynomial Interpolation and Chebyshev Interpolating Points

In this section, we focus on how to reduce error in polynomial interpolation. We may think that as the number of interpolating points is increased and the degree of interpolating polynomial is enhanced, we shall get the representation of a function with higher accuracy. However, the application of polynomial interpolation to a particular function $f(x) = 1/(1+x^2)$, $-5 \le x \le 5$, as presented by Runge, shows strikingly different results.

The interpolating points are spaced uniformly at $x_i = -5 + 10i/n$ for $i = 0, 1, \ldots, r$, and the interpolating polynomials $p_n(x)$ are obtained by using the Lagrange formula. Figure 2.1 (a) & (b) show the plots of $p_4(x)$, $p_8(x)$ and $p_{16}(x)$, together with the plot of $f(x)$. It is seen that the absolute errors in the middle of the last intervals between two interpolating points near the two ends can multiply by a large factor as n is increased by a factor of two. Therefore, any attempt to improve the accuracy of polynomial interpolation by increasing the number of interpolating points may be in vain.

We now consider Theorem 2.2 to find a solution of the above problem. It is found that large error of interpolation at points near the two ends of the function domain is occurring due to the product function $q(x) = \prod_{i=0}^{n}(x - x_i)$, whose value may be large at some x for large n when the interpolating points $\{x_i\}$ are spaced uniformly. Since the function $q(x)$ depends on the points $\{x_i\}$, it will be desirable to choose the points $\{x_i\}$ such that the ∞-norm $\|q(x)\|_\infty = \max_{a \le x \le b}|q(x)|$ is minimized. The ∞-norm is also called the Chebyshev norm.

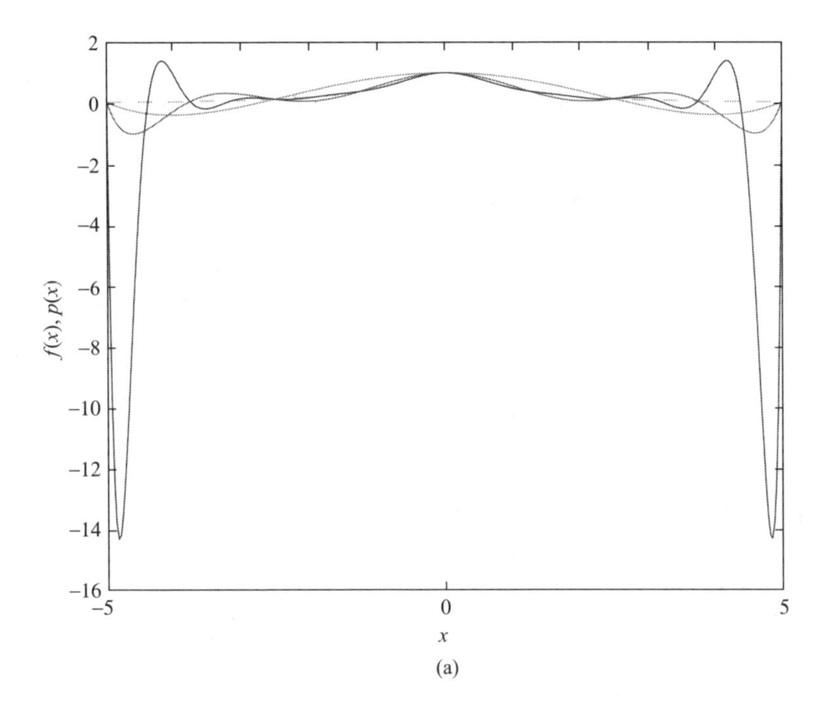

(a)

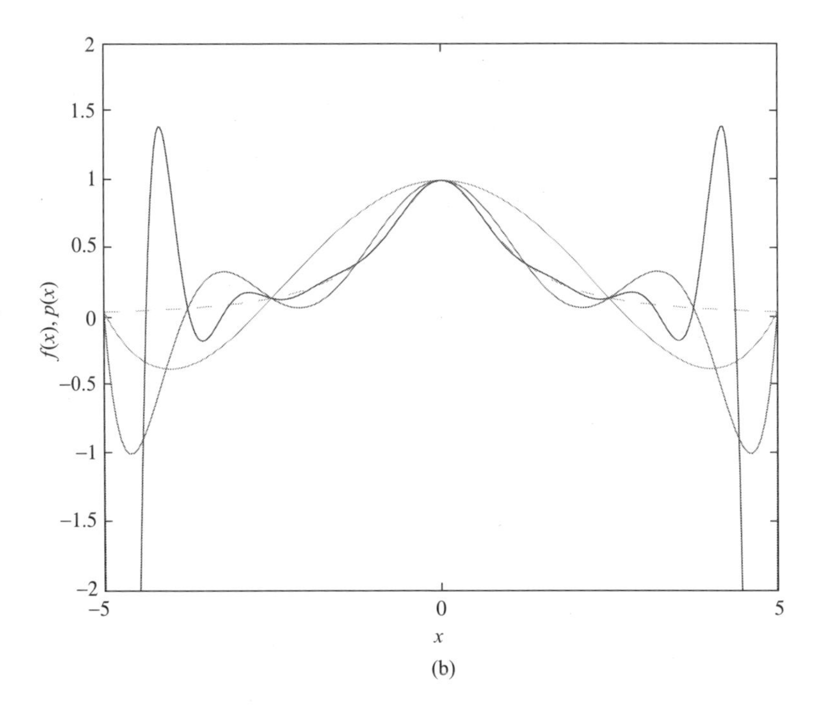

(b)

FIGURE 2.1 Polynomial interpolation: $f(x)=\frac{1}{1+x^2}$ (dashed line); $p_4(x), p_8(x), p_{16}(x)$ (solid lines) (a) Plots of polynomials with uniformly spaced interpolating points (b) Zoomed plot

Theorem 2.3

The Chebyshev norm of the function $q(x)=\prod_{i=0}^{n}(x-x_i)$ is minimized on $[-1,1]$ when the points $\{x_i; i=0,1,\ldots,n\}$ are chosen as $x_i=\cos\left[\frac{(2i+1)\pi}{2(n+1)}\right]$.

The interpolating points $\{x_i\}$ defined in Theorem 2.3 are known as the Chebyshev interpolating points. For a function $f\in\mathscr{C}[a,b]$, the Chebyshev interpolating points are

$$x_i=\frac{a+b}{2}+\frac{b-a}{2}\cos\left[\frac{(2i+1)\pi}{2(n+1)}\right],\quad i=0,1,\ldots,n \tag{2.11}$$

which minimizes the maximum absolute error of polynomial interpolation. We consider the application of polynomial interpolation to the function $f(x)=1/(1+x^2)$ over $-5\le x\le 5$ again, by using the interpolating points given by (2.11). Figure 2.2 shows the plots of $p_4(x)$, $p_8(x)$ and $p_{16}(x)$, together with the plot of $f(x)$. Now, we see that the maximum absolute error is decreasing with an increase of the number of interpolating points.

It will be interesting to look at the genesis of the Chebyshev interpolating points. The nth degree Chebyshev polynomial $T_n(x)$ defined over $[-1, 1]$ is given by

$$T_n(x)=\cos(n\theta),\ x=\cos\theta \tag{2.12}$$

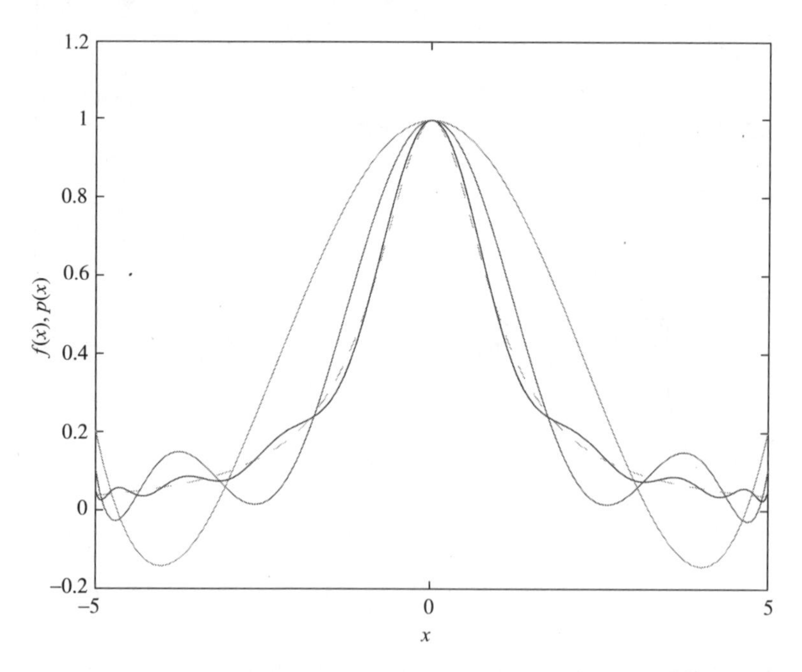

FIGURE 2.2 Polynomial interpolati on: $f(x)=\frac{1}{1+x^2}$ (dashed line); $p_4(x), p_8(x), p_{16}(x)$ (solid lines); Plots of polynomials with Chebyshev interpolating points

By expanding cos $(n\theta)$ in powers of cos θ and replacing x for cos θ, we get

$$T_0(x)=1,\ T_1(x)=x,\ T_2(x)=2x^2-1,\ T_3(x)=4x^3-3x,$$
$$T_4(x)=8x^4-8x^2+1,\ T_5(x)=16x^5-20x^3+5x,\ldots \tag{2.13}$$

Moreover, by using the trigonometric identity

$$\cos\left[(n+1)\theta\right]+\cos\left[(n-1)\theta\right]=2\cos\theta\cos(n\theta) \tag{2.14}$$

we write the recurrence relation

$$T_{n+1}(x)=2xT_n(x)-T_{n-1}(x),\quad -1\le x\le 1 \tag{2.15}$$

Now, note that the points $\{x_i\}$ defined in Theorem 2.3 are the roots of the polynomial T_{n+1} (x). Therefore, by using these interpolating points, we set the product function $q(x)=T_{n+1}(x)/2^n$, which ensures that the error of interpolation $e(x)$ is a multiple of $T_{n+1}(x)$. Since the absolute value of $T_{n+1}(x)=\cos\left[(n+1)\cos^{-1}x\right]$ is bounded by unity, the absolute error will be bounded by the Chebyshev norm of the $(n+1)^{\text{th}}$ order derivative $f^{(n+1)}(x)$ over the interval $[a, b]$, scaled by a constant factor.

2.3 Gregory-Newton Divided-Difference Formula and Hermite Interpolation

Consider a function $f\in\mathscr{C}[a,b]$ and a set of $(n+1)$ distinct points $\{x_i; i=0,1,\ldots,n\}$, $x_i\in[a,b]$, where the function values are evaluated. Let $p\in\mathscr{P}_n$ be the polynomial which satisfies the interpolation conditions: $p(x_i)=f(x_i),\ i=0,1,\ldots,n$.

Then, the coefficient of x^n in the polynomial p is defined to be a divided-difference of order n, and it is denoted by $f[x_0,x_1,\ldots,x_n]$. By definition, $f[x_0]$ is a divided-difference of order zero, and it is equal to the function value $f(x_0)$. Similarly, we have $f[x_i]=f(x_i)$.

There are several uses of divided-differences of a function. In this section, we shall explore how divided-differences can be used to develop a convenient method of polynomial interpolation. First, we interpret the divided-difference $f[x_0,x_1,\ldots,x_n]$ as the nth order derivative of f scaled by the factor $n!$ and evaluated at a point ξ as stated in the following theorem.

Theorem 2.4

Let the function $f\in\mathscr{C}^{(n)}[a,b]$ be evaluated at $(n+1)$ distinct points $\{x_i\}$, $x_i\in[a,b]$.

Then, the divided-difference of order n, $f[x_0,x_1,\ldots,x_n]$, is related to the nth order derivative of f as follows:

$$f[x_0,x_1,\ldots,x_n]=f^{(n)}(\xi)/n! \tag{2.16}$$

where ξ is a point in the smallest interval that contains $\{x_i\}$.

The above theorem, together with Theorem 2.2, suggest that if the polynomial $p_k \in \mathscr{P}_k$ interpolates the function values $\{f(x_i);\ i=0,1,\ldots,k\}$ and $k<n$, then the partial-error estimate is

$$f(x)-p_k(x) \approx \left\{\prod_{i=0}^{k}(x-x_i)\right\} f[x_0,x_1,\ldots,x_{k+1}] \tag{2.17}$$

Moreover, it is known that the divided-difference of order $(k+1)$, $f[x_0,x_1,\ldots,x_{k+1}]$, is defined to be the coefficient of x^{k+1} in the polynomial $p_{k+1} \in \mathscr{P}_{k+1}$, where p_{k+1} interpolates the function values $\{f(x_i);\ i=0,1,\ldots,k+1\}$. Therefore, we can derive a recursion formula for interpolation utilizing (2.17), and calculate the polynomials $\{p_k;\ k=0,1,\ldots,n\}$ in sequence, where p_{k+1} is obtained from p_k by addition of the correction term (negative error) comprising of the product term and the divided-difference of order $(k+1)$.

Theorem 2.5

Let the polynomial $p_k \in \mathscr{P}_k$ be satisfying the interpolation conditions

$$p_k(x_i)=f(x_i),\ i=0,1,\ldots,k \tag{2.18}$$

Then, the polynomial $p_{k+1} \in \mathscr{P}_{k+1}$ given by

$$p_{k+1}(x)=p_k(x)+\left\{\prod_{i=0}^{k}(x-x_i)\right\} f[x_0,x_1,\ \ldots,x_{k+1}],\quad a\le x\le b \tag{2.19}$$

satisfies the interpolation conditions

$$p_{k+1}(x_i)=f(x_i),\ i=0,1,\ldots,k+1. \tag{2.20}$$

An application of the above theorem inductively provides the interpolation formula as follows:

$$\begin{aligned} p_n(x)=\ & f[x_0]+(x-x_0)f[x_0,x_1]+(x-x_0)(x-x_1)f[x_0,x_1,x_2]+\cdots \\ & +\left\{\prod_{i=0}^{n-1}(x-x_i)\right\} f[x_0,x_1,\ldots,x_n], \qquad a\le x\le b \end{aligned} \tag{2.21}$$

where $f[x_0]=f(x_0)$. The above interpolation formula is known as the Gregory-Newton divided-difference formula. The main advantage of using the Gregory-Newton formula is that the order of interpolation, that is, the degree of the interpolating polynomial can be selected as required by truncating the polynomial series (2.21) at some desired order. This is particularly important when we cannot choose the set of points $\{x_i\}$ on which the function f is evaluated, and taking contribution of a far-away point in the interpolation process can adversely affect the accuracy of interpolation. The basic concept of piecewise polynomial interpolation or spline curve fitting is that the function f is represented locally by different polynomials of same lower degree in successive subintervals of $[a, b]$, where each subinterval comprises of a subset of interpolating points $\{x_i\}$.

We now present the recurrence relation for calculating the divided-differences of successive orders.

Theorem 2.6

The divided-difference of order $(k + 1)$, $f[x_i, x_{i+1}, \ldots, x_{i+k+1}]$, can be calculated using the relation

$$f[x_i, x_{i+1}, \ldots, x_{i+k+1}] = \frac{f[x_{i+1}, \ldots, x_{i+k+1}] - f[x_i, \ldots, x_{i+k}]}{x_{i+k+1} - x_i} \tag{2.22}$$

where $f[x_{i+1}, \ldots, x_{i+k+1}]$ and $f[x_i, \ldots, x_{i+k}]$ are the divided-differences of order k.

By applying the above theorem, we compute all the terms in the divided-difference table shown below:

$$\begin{array}{lllll}
x_0 : f[x_0] & & & & \\
 & f[x_0, x_1] & & & \\
x_1 : f[x_1] & & f[x_0, x_1, x_2] & & \\
 & f[x_1, x_2] & & \ddots & \\
x_2 : f[x_2] & & \vdots & & f[x_0, x_1, \ldots, x_n] \\
 & \vdots & & \cdot^{\cdot^{\cdot}} & \\
\vdots & & f[x_{n-2}, x_{n-1}, x_n] & & \\
 & f[x_{n-1}, x_n] & & & \\
x_n : f[x_n] & & & &
\end{array}$$

where the entries of the first column along with the row-tag, are taken from the function values $\{f(x_i);\ i = 0, 1, \ldots, n\}$ provided. Using the computed values of the divided-differences of various orders, we can express $p_n(x)$ of (2.21) as given by the Gregory-Newton interpolation formula.

To compute the successive order derivatives or integrals of the polynomial $p_n(x)$, it may be useful to convert $p_n(x)$ of (2.21) into $p_n(x)$ of (2.1), that is,

$$\begin{aligned}
p_n(x) = {} & D_0 f_0 + (x - x_0) D_1 f_0 + (x - x_0)(x - x_1) D_2 f_0 + \cdots \\
& + (x - x_0)(x - x_1) \cdots (x - x_{n-1}) D_n f_0 \\
\equiv {} & c_0 + c_1 x + c_2 x^2 + \cdots + c_{n-1} x^{n-1} + c_n x^n
\end{aligned} \tag{2.23}$$

where $D_0 f_0 = f[x_0]$, $D_1 f_0 = f[x_0, x_1]$, $D_2 f_0 = f[x_0, x_1, x_2]$, ..., $D_n f_0 = f[x_0, x_1, \ldots, x_n]$.

The coefficients c_j of the polynomial $p_n(x)$ can be calculated as follows:

We form the matrix $[b_{ij}]_{n \times (n+1)}$ by the recurrence relation shown below,

$$\begin{aligned} &b_{11} = -x_0\,,\; b_{12} = 1\,,\; b_{13} = \cdots = b_{1,n+1} = 0, \\ &b_{i1} = (-x_{i-1})\,b_{i-1,1} \qquad \text{for } i = 2, 3, \ldots, n \\ &b_{ij} = (-x_{i-1})\,b_{i-1,j} + b_{i-1,j-1} \quad \text{for } i = 2, 3, \ldots, n\,;\; j = 2, 3, \ldots, n+1 \end{aligned} \tag{2.24}$$

Then, the coefficients c_j are given by

$$\begin{aligned} c_0 &= D_0 f_0 + \sum_{i=1}^{n} (D_i f_0)\, b_{i1}\,, \\ c_j &= \sum_{i=1}^{n} (D_i f_0)\, b_{i,j+1} \quad \textit{for } j = 1, 2, \ldots, n \end{aligned} \tag{2.25}$$

2.3.1 Hermite Interpolation Method

In some cases, the derivative values may be known together with the function values at a set of distinct points. Accordingly, the interpolation conditions of (2.5) can be generalized as follows:

$$\begin{aligned} p^{(j)}(x_i) &= f^{(j)}(x_i), \quad i = 0, 1, \ldots, m;\; j = 0, 1, \ldots, l_i \\ \text{with} \quad n+1 &= \sum_{i=0}^{m} (l_i + 1) \end{aligned} \tag{2.26}$$

where $\{p^{(j)}(x_i)\}$ and $\{f^{(j)}(x_i)\}$ are the jth order derivatives evaluated at the distinct points $\{x_i\}$. By definition, the 0th order derivative is the function or polynomial itself. The Hermite interpolation problem is to calculate the polynomial $p \in \mathscr{P}_n$ that satisfies the interpolation conditions (2.26). Note that the number of unknown coefficients of p is equal to the number of known distinct data values, which ensures that the interpolating polynomial p is determined uniquely.

Theorem 2.7

Let the function f and its derivatives up to order l_i be evaluated at $(m + 1)$ distinct points $\{x_i\,;\; i = 0, 1, \ldots, m\}$ on $[a, b]$. Then, there exists only one polynomial $p \in \mathscr{P}_n$ such that

$$\begin{aligned} p^{(j)}(x_i) &= f^{(j)}(x_i), \quad i = 0, 1, \ldots, m;\; j = 0, 1, \ldots, l_i \\ \text{with} \quad n+1 &= \sum_{i=0}^{m} (l_i + 1). \end{aligned}$$

The interpolating polynomial p which satisfies the conditions (2.26) can be obtained by an interesting application of the Gregory-Newton divided-difference formula as described in the following algorithm.

Algorithm

Step 1 - Set up the divided-difference table with repeated entries of number l_i for the row $x_i : f[x_i]$;

Step 2 - If the lower-order divided-differences are not repeated, then, compute the higher-order divided-difference by the recurrence relation (2.22), else, compute the higher-order divided-difference by the definition

$$f[x_i, x_{i+1}, \ldots, x_{i+k+1}] = \frac{f^{(k+1)}(x_i)}{(k+1)!} \tag{2.27}$$

where $x_i = x_{i+1} = \cdots = x_{i+k+1}$.

Step 3 - Complete the divided-difference table by recursion of Step 2.

Step 4 - Apply the Gregory-Newton interpolation formula to obtain the polynomial $p(x)$ on the interval $[a, b]$.

2.4 Weierstrass Theorem and Uniform Convergence of Polynomial Approximation

By applying the method of interpolation, we can construct a polynomial p_n of degree at most n which agrees with a given continuous function f at $(n+1)$ distinct points on the interval $[a, b]$. However, if you look for the property that $p_n(x) \to f(x)$ as $n \to \infty$ for all points $x \in [a,b]$, then the interpolating polynomial p_n may not have this property. We have seen in the example of Runge that, for equally spaced interpolating points, increasing the degree of polynomial may actually worsen the accuracy of interpolation. Although by using the Chebyshev interpolating points, we can calculate the interpolating polynomials which seem to converge uniformly to the Runge's function, but this may not be a practical solution in many situations where the signal values evaluated on a collection of distinct but otherwise arbitrarily selected points are provided beforehand. Nevertheless, the search for polynomial with the property of uniform convergence continued due to the following theorem of Weierstrass.

Theorem 2.8 (Weierstrass)

For a function $f \in \mathscr{C}[a, b]$ and for any $\varepsilon > 0$, there exists a polynomial $p(x) = c_0 + c_1 x + \cdots + c_n x^n$, $a \le x \le b$, such that the bound on the error-norm

$$\|f - p\|_\infty \le \varepsilon \tag{2.28}$$

is satisfied.

A direct consequence of the above theorem is that $|f(x) - p(x)| \le \varepsilon$ for all points $x \in [a,b]$. Therefore, by choosing an appropriate value of ε and sufficiently large n, one can find a polynomial representation which is arbitrarily close to the function over the specified interval.

The proof of Theorem 2.8 depends on an interesting property of monotone operators as stated in the next theorem. First, the monotone operators are defined below:

The operator L is monotone, if for two functions $f, g \in \mathscr{C}[a,b]$, the inequality

$$f(x) \geq g(x), \quad a \leq x \leq b \tag{2.29}$$

implies that the polynomials $Lf, Lg \in \mathscr{C}[a,b]$ satisfy the condition

$$(Lf)(x) \geq (Lg)(x), \quad a \leq x \leq b \tag{2.30}$$

Moreover, if L is a linear operator, then it is monotone with the condition that for all non-negative functions $f \in \mathscr{C}[a,b]$, the polynomials $Lf \in \mathscr{C}[a, b]$ are also non-negative.

Monotone operators are useful for our discussion because, given an infinite sequence of linear monotone operators $\{L_i; i = 0, 1, 2, \ldots\}$, there is a simple test to verify whether the sequence of polynomials $\{L_i f\}$ converges uniformly to f for all $f \in \mathscr{C}[a, b]$. This simple test is given in the following theorem.

Theorem 2.9

For an infinite sequence of linear monotone operators $\{L_i; i = 0, 1, 2, \ldots\}$ on $\mathscr{C}[a, b]$, if the sequence of functions $\{L_i f\}$ converges uniformly to f for three functions $f(x) = 1, x, x^2$, then the sequence $\{L_i f\}$ converges uniformly to f for all $f \in \mathscr{C}[a,b]$.

The Bernstein operator defined on the interval [0,1] and expressed as shown below

$$(B_n f)(x) = \sum_{k=0}^{n} \frac{n!}{k!(n-k)!} x^k (1-x)^{n-k} f(k/n), \quad 0 \leq x \leq 1 \tag{2.31}$$

passes the test of uniform convergence as prescribed by the above theorem. The Bernstein operator is similar to the Lagrange operator, as described earlier, in two ways. Both operators are linear, and in both cases, the polynomial representations $p \in \mathscr{P}_n$ depend on the discrete values of the function f at $(n + 1)$ distinct points of $[a, b]$, where the intervals $[a, b]$ and $[0, 1]$ are interchangeable by scaling and shifting. However, unlike the Lagrange operator, the Bernstein operator is not a projection, that is, $B_n f$ may not equal f when $f \in \mathscr{P}_n$. Suppose that the function value $f(x_i) = 1$ at $x_i = k/n$, and $f(x_i) = 0$ at the points $\{x_i = i/n;\ i = 0, 1, \ldots, n;\ i \neq k\}$. Then, $B_n f(x)$ is a multiple of $x^k (1-x)^{n-k}$, which is positive non-zero at the points $\{x_i = i/n;\ i = 1, 2, \ldots, n-1\}$. Since the interpolation conditions are not necessarily satisfied in this case, we refer to $B_n f$ as the Bernstein approximating polynomial. The main advantage of Bernstein approximation over Lagrange interpolation is stated in the following theorem.

Theorem 2.10

For all functions $f \in \mathscr{C}[0, 1]$, the sequence $\{B_n f; n = 1, 2, 3, \ldots\}$ converges uniformly to f, where the operator B_n is defined by

$$(B_n f)(x) = \sum_{k=0}^{n} \frac{n!}{k!(n-k)!} x^k (1-x)^{n-k} f(k/n), \quad 0 \leq x \leq 1.$$

It follows from the above theorem that, for any function $f \in \mathscr{C}[0, 1]$ and for any $\varepsilon > 0$, there exists an integer n such that the inequality

$$\|f - B_n f\|_\infty \leq \varepsilon \tag{2.32}$$

holds. Hence, by letting $p = B_n f$, we can have a constructive proof of the Weierstrass theorem. One remarkable property of the Bernstein operator is that not only $B_n f$ converges uniformly to f, also the derivatives of $B_n f$ converge uniformly to the derivatives of $f \in \mathscr{C}^{(k)}[0, 1]$, for all orders of derivatives up to k inclusive.

Corollary 2.11

Consider a function $f \in \mathscr{C}^{(k)}[0, 1]$. Then, the following limits hold:

$$\lim_{n\to\infty} \left\| f^{(j)} - (B_n f)^{(j)} \right\|_\infty = 0 \quad \text{for } j = 0, 1, \ldots, k \tag{2.33}$$

where B_n is the Bernstein operator.

Despite of having very good features, the Bernstein operator is rarely used in practice to find an approximating polynomial, because the rate of convergence of $B_n f$ to f is often too slow. For example, when $f(x) = x^2, 0 \leq x \leq 1$, we get $\|B_n f - f\|_\infty = \frac{1}{4n}$. Thus, in order to achieve an accuracy of 10^{-4}, it is necessary to use $n = 2500$.

One important point to be mentioned is that, due to error in given data, the interpolation procedure may lead to an inconsistent set of equations. The approximation methods, as discussed in the next section, can provide an acceptable solution in this case.

2.5 Best Approximation in Normed Linear Space

Given a function $f \in \mathscr{C}[a, b]$, we want to find the approximating function $Lf \in \mathscr{A}$, where $\mathscr{A}$ is a finite-dimensional subspace in the function space $\mathscr{C}[a, b]$. In a linear space, the approximating function Lf is given in the form

$$Lf(A, x) = \sum_{j=0}^{n} a_j \varphi_j(x) \text{ with } \varphi_0(x) = 1, a \leq x \leq b \tag{2.34}$$

where the basis functions φ_j are linearly independent, and $A = \{a_j\}$ is the set of $(n+1)$ coefficients.

The choice of basis functions is primarily guided by the nature of the function to be approximated or by the characteristics of the linear space to be spanned. For example, when f is a real-valued continuous function, the basis functions $\{x^j;\ j = 0, 1, 2, \ldots\}$ can be chosen. Whereas, if f is real-valued continuous 2π-periodic function, then the basis functions $\{\cos(jx), \sin(jx); j = 0, 1, 2, \ldots\}$ will be the choice. Other than the primary considerations as stated above, there may be some analytical reason to choose a particular type of basis functions. For example, when $f \in \mathscr{C}[a, b]$, we may choose orthogonal polynomials, logarithmic functions, exponential functions, Bessel functions, etc. as basis, instead of monomials of form x^j, to have some advantage in analysis.

The function space $\mathscr{C}[a, b]$ is a normed linear space, and the p-norm of the function $f \in \mathscr{C}[a, b]$ is denoted by $\|f\|_p$ and defined by

$$\|f\|_p = \begin{cases} \left[\int_a^b |f(x)|^p \, dx \right]^{1/p}, & 1 \le p < \infty \\ \max_{a \le x \le b} |f(x)|, & p = \infty \end{cases} \tag{2.35}$$

The p-norm can be suitably defined in other normed linear spaces. For instance, when we consider the discrete function (vector) case, the corresponding linear (vector) space is $\mathscr{R}^m$, which is the space of real-valued m-component vectors. The p-norm of the vector $f = \{f_1, f_2, \ldots, f_m\}$ is defined by

$$\|f\|_p = \begin{cases} \left[\sum_{i=1}^{m} |f_i|^p \right]^{1/p}, & 1 \le p < \infty \\ \max_{i=1,2,\ldots,m} |f_i|, & p = \infty \end{cases} \tag{2.36}$$

A complete normed linear space is called a Banach space.

The best approximation $Lf(A^*) \in \mathscr{A}$ of the function f is defined by the relation

$$\left\| Lf(A^*) - f \right\| \le \left\| Lf(A) - f \right\| \quad \text{for all } Lf(A) \in \mathscr{A} \tag{2.37}$$

where a suitable norm is used as the distance measure for comparison. The following theorem ensures the existence of best approximation in a normed linear space.

Theorem 2.12

If $\mathscr{A}$ is a finite-dimensional subspace in a normed linear space $\mathscr{F}$, then, for every $f \in \mathscr{F}$, there exists an element $Lf(A^*) \in \mathscr{A}$ such that

$$\left\| Lf(A^*) - f \right\| \le \left\| Lf(A) - f \right\| \quad \text{for all } Lf(A) \in \mathscr{A}.$$

The choice of norm depends on the application of approximation, and usually, we consider the 1-norm, 2-norm, or the ∞-norm of the error function $e(x)$. The three norms are ordered as given in the theorem below when $e(x)$ is a real-valued continuous function.

Theorem 2.13

For all $e \in \mathscr{C}[a,b]$, the inequalities

$$\|e\|_1 \le (b-a)^{1/2} \|e\|_2 \le (b-a) \|e\|_\infty \tag{2.38}$$

hold.

As a consequence of the above theorem, when $\|e\|_\infty$ is restricted to a small value, the other two values, $\|e\|_1$ and $\|e\|_2$, are also small. However, the converse may not be true. Thus, any algorithm to obtain an approximation with small error in the ∞-norm also ensures that all other norms are small. The ∞-norm is often called the uniform or Chebyshev norm, and the corresponding error-criterion is known as the minimax criterion. The approximation using the 2-norm or least-squares error-criterion is perhaps easiest to obtain, because it gives a set of linear equations to be solved simultaneously. Besides, if the function values are inaccurate and the deviations are caused by additive Gaussian random error, then the least-squares error-criterion provides an optimal solution of the approximation problem. The 1-norm or least-absolute error-criterion has the property of rejecting outliers, and this property can be useful when some function values may be in gross errors. Moreover, the least-absolute error-criterion provides an optimal solution of approximation when the function values are deviated by additive Laplacian random error.

We now address the issue of uniqueness of best approximation. Consider a given function f as a point in a normed linear space $\mathscr{F}$, but f is not a point of the subspace $\mathscr{A}$ where the approximation is sought, and $\mathscr{A} \subset \mathscr{F}$. We define the ball of radius $r > 0$ centred at f as given below:

$$\mathscr{B}(f,r) \equiv \{g : \|g - f\| \le r,\ g \in \mathscr{F}\}, \tag{2.39}$$

Accordingly, if $r_1 > r_0$, then $\mathscr{B}(f,r_0) \subset \mathscr{B}(f,r_1)$. The unit balls in two-dimensional Euclidean space $\mathscr{R}^2$ for three different norms are shown in Figure 2.3. The shapes of these balls are interesting to note, because important clue related to the uniqueness of best approximation can be derived from them.

We now provide a geometric interpretation of best approximation as follows:

Since $f \notin \mathscr{A}$, we enlarge the ball $\mathscr{B}$ by increasing r from zero to some value r^* such that, the intersection of $\mathscr{B}$ and $\mathscr{A}$ is empty for $r < r^*$, and there are common points in $\mathscr{B}$ and $\mathscr{A}$ for $r \ge r^*$. Now, the question is whether for $r = r^*$, there will be one common

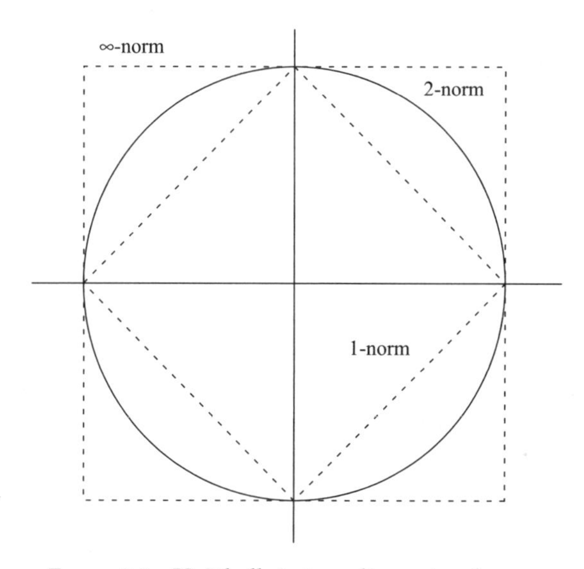

FIGURE 2.3 Unit balls in two-dimensional space

point in $\mathscr{B}$ and $\mathscr{A}$, or there will be several common points. It turns out that the answer to this question lies in the property of the ball that a particular norm induces.

Theorem 2.14

Let $\mathscr{F}$ be a normed linear space. Then, for any $f \in \mathscr{F}$ and for any $r > 0$, the ball

$$\mathscr{B}(f,r) \equiv \{g : \|g-f\| \le r,\ g \in \mathscr{F}\}$$

is convex.

A set $\mathscr{G}$ is said to be convex if, for all $g_0, g_1 \in \mathscr{G}$, the points $\{\theta g_0 + (1-\theta)g_1 ; 0 < \theta < 1\}$ are also in $\mathscr{G}$. Moreover, the set is strictly convex if for every distinct pair $g_0, g_1 \in \mathscr{G}$, the points $\{\theta g_0 + (1-\theta)g_1 ; 0 < \theta < 1\}$ are interior points of $\mathscr{G}$. Furthermore, if every ball $\mathscr{B}(f,r)$ is strictly convex, then the normed linear space is said to be strictly normed.

We now state the theorem related to the uniqueness of best approximation below.

Theorem 2.15

Let $\mathscr{F}$ be a strictly normed linear space, and $\mathscr{A}$ be a subspace in $\mathscr{F}$. Then, for any function $f \in \mathscr{F}$, there is at most one approximating function $Lf(A^*) \in \mathscr{A}$ such that

$$\|Lf(A^*) - f\| \le \|Lf(A) - f\| \quad \text{for all } Lf(A) \in \mathscr{A}.$$

The spaces of continuous functions with the p-norm, $1 < p < \infty$, are strictly normed. The 1-norm and ∞-norm do not give strictly normed spaces. This is another reason why the 2-norm distance measure is frequently used in approximation technique. If $\mathscr{A}$ is a subspace of a normed linear space whose norm is not strictly convex, then the uniqueness of best approximation depends on properties of $\mathscr{A}$ and f.

When for every $f \in \mathscr{F}$, $\mathscr{F}$ being a normed linear space, we have exactly one best approximation $Lf(A^*) \in \mathscr{A}$, defined by the best approximation condition (2.37), where $\mathscr{A}$ is a finite-dimensional subspace in $\mathscr{F}$, we have a linear operator $L : \mathscr{F} \to \mathscr{A}$. The operator L is a linear operator, because $L(\lambda f) = \lambda Lf$ for a real λ, and $L(f+g) = Lf + Lg$ for $f, g \in \mathscr{F}$. Moreover, the operator L is a projection operator, because $Lf = f$ for $f \in \mathscr{A} \subset \mathscr{F}$. Furthermore, the operator L has the property as stated below.

Theorem 2.16

Let $\mathscr{F}$ be a normed linear space and $\mathscr{A}$ be a finite-dimensional subspace in $\mathscr{F}$. If for every $f \in \mathscr{F}$, there is exactly one best approximation $Lf(A^*) \in \mathscr{A}$, then the operator L defined by the best approximation condition, is continuous.

A linear operator L is said to be bounded if there is a real number c such that for all $f \in \mathscr{F}$,

$$\|Lf\| \le c\|f\| \tag{2.40}$$

where $Lf \in \mathscr{A} \subset \mathscr{F}$, $\mathscr{F}$ being a normed linear space.

Lemma 2.17

A necessary and sufficient condition for a linear operator L to be bounded is that it be continuous.

Combining Theorem 2.16 and Lemma 2.17, we find that the linear operator defined by the best approximation condition, is bounded.

The norm of a bounded linear operator L is defined by

$$\|L\|_p = \sup_{\substack{f \in \mathscr{F} \\ f \neq 0}} \frac{\|Lf\|_p}{\|f\|_p} < \infty \tag{2.41}$$

or alternatively,

$$\|L\|_p = \sup_{\substack{f \in \mathscr{F} \\ \|f\|_p = 1}} \|Lf\|_p < \infty \tag{2.42}$$

where the operator norm is consistent with the function norm.

The uniform convergence of best approximation is guaranteed by the following theorem, because the approximation operators satisfy all the given conditions.

Theorem 2.18

Let $\mathscr{F}$ be a Banach space and $\mathscr{A}_n$ be a n-dimensional subspace in $\mathscr{F}$. Let $\{L_n; n = 0, 1, 2, \ldots\}$ be a sequence of bounded linear operators from $\mathscr{F}$ to $\mathscr{A}_n$ such that

$$L_n f = f \quad \text{for } f \in \mathscr{A}_n \tag{2.43}$$

Then,

$$\lim_{n \to \infty} \|f - L_n f\| = 0 \quad \text{for all } f \in \mathscr{F} \tag{2.44}$$

or equivalently, the sequence $\{L_n f; n = 0, 1, 2, \ldots\}$ converges to f for all f in $\mathscr{F}$.

The converse of the above theorem is stated in the theorem below.

Theorem 2.19 (Powell)

Let $\mathscr{F}$ be a Banach space, and let $\{L_n; n = 0, 1, 2, \ldots\}$ be a sequence of linear operators from $\mathscr{F}$ to $\mathscr{F}$. If the sequence of norms $\{\|L_n\|; n = 0, 1, 2, \ldots\}$ is unbounded, then there exists a function $f^* \in \mathscr{F}$ such that the sequence $\{L_n f^*; n = 0, 1, 2, \ldots\}$ diverges.

Powell computed the sequence of norms $\{\|L_n\|; n = 0, 1, 2, \ldots\}$ for the Lagrange interpolating operator, and showed that the sequence of norms became unbounded when interpolating points were chosen with uniform spacing. Thus, a function f^* (Runge's function) could be found to show the divergence of the sequence $\{L_n f^*; n = 0, 1, 2, \ldots\}$. However, when the Lagrange operators were computed with the Chebyshev interpolating points, the sequence of operator-norms remained bounded.

2.6 Minimax Approximation and Exchange Algorithm

The best minimax approximation $Lf\left(A^{*}, x\right)=\sum_{j=0}^{n} a_{j}^{*}\varphi_{j}(x)$, $\varphi_{0}(x)=1$, for a continuous function $f(x)$, $a \le x \le b$, is the element of the subspace $\mathscr{A}$ that minimizes the expression

$$\|f-Lf\|_{\infty}=\max_{a\le x\le b}\left|f(x)-Lf(x)\right|, \quad Lf \in \mathscr{A} \tag{2.45}$$

where $\{\varphi_{j}; j=0,1,\ldots,n\}$ are the known basis functions of $\mathscr{A}$, and $A=\{a_{j}\}$ is the set of $(n+1)$ unknown coefficients.

In Figure 2.4, we show the best minimax approximation of a non-linear function f by a straight line. Note that from the optimum position, if we rotate the line clockwise or anticlockwise, then the maximum error occurring at one of the endpoints will be more than what it is now. On the other hand, if we translate the line upward or downward, then the maximum error occurring at the midpoint or at each endpoint simultaneously will increase from its present value. Note that for the optimum position of the straight line, the maximum errors occurring at these three locations over the interval are all of equal magnitude and alternating signs. This happens to be the characteristics of best minimax approximation as explained in the sequel.

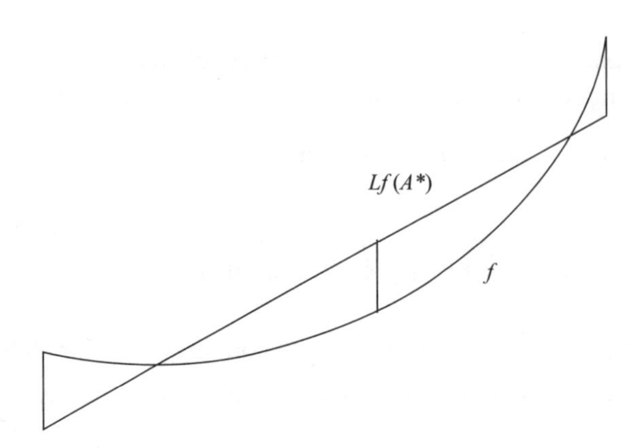

FIGURE 2.4 Minimax approximation by a straight line; The ∞-norm of error shown by the length of the vertical lines

The minimax approximation gets a convenient structure when the linear subspace $\mathscr{A}$ satisfies the Haar conditions as stated below:

1. If $\mathscr{A}$ is a linear subspace of dimension $(n+1)$, and $L(A,x)$, $a \le x \le b$, is a function in $\mathscr{A}$, that is not identically zero, then $L(A,x)$ has at most n zeros in $[a, b]$.
2. If $\{\xi_{i}; i=1,2,\ldots,k\}$, $1 \le k \le n$, is a set of k distinct points from the open interval (a, b), then there exists a function $L(A,x)$, $a \le x \le b$, in $\mathscr{A}$, that changes sign at these k points, and that has no other zeros. Moreover, there is a function in $\mathscr{A}$ that has no zeros in $[a, b]$.

When the Haar conditions are satisfied, the set of basis functions $\{\varphi_j(x); j = 0,1,\ldots,n\}$ is said to form a Chebyshev set, and the linear subspace $\mathscr{A}$ spanned by $\{\varphi_j\}$ is known as the Haar subspace. Some other equivalent definitions of the Chebyshev set and the Haar subspace are given below:

(a) The basis functions $\{\varphi_j; j = 0, 1, \ldots, n\}$ constitute a Chebyshev set in [a, b] if the difference function $L(A_1, x) - L(A_2, x)$ has at most n zeros in [a, b] for $A_1 \neq A_2$.

(b) If $\{\varphi_j; j = 0, 1, \ldots, n\}$ is any basis of a Haar subspace in the interval [a, b], and if $\{\xi_i; i = 0, 1, \ldots, n\}$ is any set of $(n+1)$ distinct points in [a, b], then the $(n+1) \times (n+1)$ matrix whose elements have the values $\left[\varphi_j(\xi_i); i = 0, 1, \ldots, n; j = 0, 1, \ldots, n\right]$ is non-singular.

Two well-known examples of Chebyshev sets and corresponding Haar subspaces are mentioned here:

(i) The set of the first n powers of x together with $1\left(= x^0\right)$, i.e., $\{1, x, x^2, \ldots, x^n\}$ forms a Chebyshev set for any interval [a, b]. The corresponding Haar subspace is $\mathscr{P}_n$, the collection of polynomials of degree at most n.

(ii) The set of the first $2n$ trigonometric functions of x together with $1(= \cos 0)$, i.e., $\{1, \cos x, \sin x, \cos 2x, \sin 2x, \ldots, \cos nx, \sin nx\}$ forms a Chebyshev set for the interval $[0, 2\pi)$. The corresponding Haar subspace is $\mathscr{Q}_n$, the collection of trigonometric series of degree at most n.

Any trigonometric series is 2π-periodic; thus, the points 0 and 2π merge to a single point. Moreover, $\mathscr{Q}_n \subset \mathscr{C}_{2\pi}$, the set of continuous functions that are 2π-periodic, that is, $f(x) = f(x + 2\pi)$ for $f \in \mathscr{C}_{2\pi}$. Note that each of the functions $\cos jx$ and $\sin jx$ has $2j$ zeros in the interval $[0, 2\pi)$. We consider the topic of approximation by trigonometric series in a later section.

We now state the characterization theorem for the best minimax approximation in a Haar subspace in the following theorem.

Theorem 2.20

Let $f \in \mathscr{C}[a, b]$ be a function to be approximated by the function $L(A)$ in the Haar subspace $\mathscr{A} \subset \mathscr{C}[a, b]$, and let $L(A)$ have the form

$$L(A, x) = \sum_{j=0}^{n} a_j \varphi_j(x) \quad \text{with } \varphi_0(x) = 1, \ a \leq x \leq b \tag{2.46}$$

Then, the necessary and sufficient condition that $L\left(A^*\right)$ be the best minimax approximation is that there exist (n + 2) extremal points $a \leq x_0 < x_1 < \cdots < x_{n+1} \leq b$ such that the error magnitudes are

$$\left|f(x_i)-L(A^*,x_i)\right| = \left\|f-L(A^*)\right\|_\infty, \quad i=0,1,\dots,n+1; \tag{2.47}$$

and the error curve alternates $(n + 1)$ times on $[a, b]$, that is,

$$f(x_{i+1})-L(A^*,x_{i+1}) = -\left[f(x_i)-L(A^*,x_i)\right], \quad i=0,1,\dots,n \tag{2.48}$$

First of all, the above theorem says that if the subspace $\mathscr{A}$ is a Haar subspace, then all we need to know to compute the best minimax approximation are the locations of the extremal points $\{x_i; i=0,1,\dots,n+1\}$ and the function values $\{f(x_i)\}$ evaluated at these points. Because in that case, we can solve the system of equations given by

$$f(x_i)-L(A^*,x_i) = (-1)^i d^*, \quad i=0,1,\dots,n+1 \tag{2.49}$$

with the deviation magnitude d^* (discard sign) being the minimax error norm,

$$d^* = \left\|f-L(A^*)\right\|_\infty = \max_{a\le x\le b}\left|f(x)-L(A^*,x)\right|, \tag{2.50}$$

for the unknowns $\{a_j; j=0,1,\dots,n\}$ and d^*. When $\{\varphi_j(x)\}$ is a Chebyshev set, the solution of the linear estimation problem (2.49) is uniquely determined.

However, we know that the direct method as mentioned above is not implementable, because the locations of the extremal points are not known beforehand. In fact, the main effort in computing the best minimax approximation is aimed at finding the locations of extremal points by some technique which may be iterative. To develop the theory in that direction, let us consider the case of minimax approximation on a discrete point set as given in the following theorem.

Theorem 2.21

Let $\mathscr{A}$ be an $(n+1)$-dimensional Haar subspace of $\mathscr{C}[a,b]$, let $a \le \xi_0 < \xi_1 < \cdots < \xi_{n+1} \le b$ be a reference set of points, and let f be any function in $\mathscr{C}[a,b]$.

Then, $L(A,x)=\sum_{j=0}^{n} a_j\varphi_j(x)$ with $\varphi_0(x)=1$, $a\le x\le b$, is the function in $\mathscr{A}$ that minimizes the maximum error magnitude

$$\max_{i=0,1,\dots,n+1}\left|f(\xi_i)-L(A,\xi_i)\right|, \quad L(A)\in\mathscr{A}, \tag{2.51}$$

if and only if the recursive relations

$$f(\xi_{i+1})-L(A,\xi_{i+1}) = -\left[f(\xi_i)-L(A,\xi_i)\right], \quad i=0,1,\dots,n \tag{2.52}$$

are satisfied.

With the chosen reference points as stated in the above theorem, we can solve the linear system of equations

$$\begin{bmatrix} 1 & \varphi_1(x_0) & \cdots & \varphi_n(x_0) & 1 \\ 1 & \varphi_1(x_1) & \cdots & \varphi_n(x_1) & -1 \\ \vdots & \vdots & & \vdots & \vdots \\ 1 & \varphi_1(x_{n+1}) & \cdots & \varphi_n(x_{n+1}) & (-1)^{n+1} \end{bmatrix} \begin{bmatrix} a_0 \\ a_1 \\ \vdots \\ a_n \\ d \end{bmatrix} = \begin{bmatrix} f(\xi_0) \\ f(\xi_1) \\ \vdots \\ f(\xi_{n+1}) \end{bmatrix} \tag{2.53}$$

to determine the approximating function $L(A,x)$ and the deviation $d=(-1)^i\left[f(\xi_i)-L(A,\xi_i)\right]$ for any i. Note that the minimax approximation as obtained on a discrete point set may not be the best minimax approximation of the function f in the interval $[a, b]$, because the set of reference points may not coincide with the set of extremal points as required by Theorem 2.20.

However, the following theorem gives a straightforward clue on how to adjust the reference points iteratively such that they merge with the extremal points at the end.

Theorem 2.22

Let the conditions of Theorem 2.21 be valid. If the error curve $e(x)=f(x)-L(A,x)$ takes on the values $(-1)^i d \ (d>0)$ at $(n+2)$ reference points $a \le \xi_0 < \xi_1 < \cdots < \xi_{n+1} \le b$, then

$$d \le d^* \le \max_{a\le x\le b}\left|f(x)-L(A,x)\right| \tag{2.54}$$

where d^* is the minimax error norm of the best approximation.

Moreover, if the reference $\{\xi_i\}$ is changed to a new reference $\{\xi_i^+\}$ such that the conditions

$$\operatorname{sgn}\left[f\left(\xi_{i+1}^+\right)-L\left(A,\xi_{i+1}^+\right)\right]=-\operatorname{sgn}\left[f\left(\xi_i^+\right)-L\left(A,\xi_i^+\right)\right],\quad i=0,1,\ldots,n \tag{2.55}$$

and

$$\left|f\left(\xi_i^+\right)-L\left(A,\xi_i^+\right)\right|\ge d,\quad i=0,1,\ldots,n+1 \tag{2.56}$$

are satisfied, then

$$d\left(\xi_0^+,\xi_1^+,\ldots,\xi_{n+1}^+\right)>d\left(\xi_0,\xi_1,\ldots,\xi_{n+1}\right) \tag{2.57}$$

provided that there is at least one i for which $\left|f\left(\xi_i^+\right)-L\left(A,\xi_i^+\right)\right|>d$.

The *one-point exchange algorithm* replaces one reference point at a time in between successive iterations, such that the point of maximum approximation error is included as a reference point for the next iteration. In Figure 2.5, the reference set $\{\xi_0,\xi_1,\xi_2,\xi_3,\xi_4\}$ is used to find the minimax approximation which provides the error function as plotted [Powell]. The maximum approximation error is observed at the point η_1. Thus, the new reference set becomes $\{\xi_0,\eta_1,\xi_2,\xi_3,\xi_4\}$, and a new iteration is started. In this way, the minimax approximation of the function f in the given interval is obtained in a finite number of iterations.

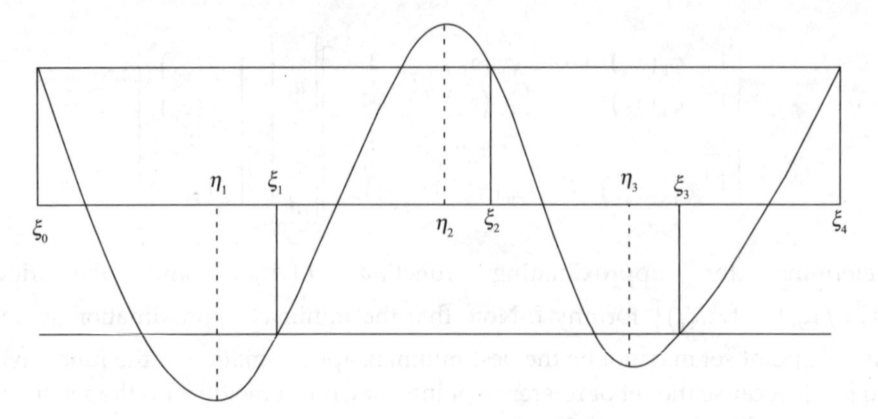

FIGURE 2.5 An error function of the minimax approximation

The following theorem provides an efficient initialization scheme for the iterative exchange algorithm in order to obtain the minimax approximation of a continuous function.

Theorem 2.23

Let $f \in \mathscr{C}[-1,1]$, and let $p \in \mathscr{P}_n$ be the minimax approximation of *f* that is calculated by an exchange algorithm. Then, a suitable initial reference set of points are given by

$$\xi_i = \cos\left[\frac{(n+1-i)\pi}{n+1}\right], \quad i = 0, 1, \ldots, n+1 \tag{2.58}$$

Moreover, if $f \in \mathscr{P}_{n+1}$, then *p* computed with the initial reference is the best minimax approximation of *f*.

When $f \in \mathscr{C}[a, b]$, the initial reference set is given by the following relations

$$\xi_i = \frac{1}{2}(a+b) + \frac{1}{2}(b-a)\cos\left[\frac{(n+1-i)\pi}{n+1}\right], \quad i = 0, 1, \ldots, n+1 \tag{2.59}$$

2.7 Least Absolute Approximation and Linear Programming

The best least-absolute approximation $Lf(A^*, x) = \sum_{j=0}^{n} a_j^* \varphi_j(x)$, $\varphi_0(x) = 1$, for a continuous function $f(x)$, $a \le x \le b$, is the element of the subspace $\mathscr{A}$ that minimizes the expression

$$\|f - Lf\|_1 = \int_a^b |f(x) - Lf(x)|\,dx, \quad Lf \in \mathscr{A} \tag{2.60}$$

where $\{\varphi_j;\ j = 0, 1, \ldots, n\}$ are the known basis functions of $\mathscr{A}$, and $A = \{a_j\}$ is the set of $(n+1)$ unknown coefficients.

In Figure 2.6, we show the best least-absolute approximation of a strictly monotonic function f by a constant value. Note that from the optimum value, if we raise the value as shown, then the area under the absolute error curve increases by the area BCFH and decreases by the area CDEF, so there will be a net increase of twice the area CFG. On the other hand, if we reduce the value as shown in figure, then the area under the absolute error curve will have a net increase of twice the area CST. Note that for the optimum constant value, the error curve changes sign at the midpoint of the interval. Accordingly, the function value evaluated at the midpoint is all that is needed to find the best approximation of a function, provided it is strictly monotonic over the interval. The above observations happen to be the characteristics of best least-absolute approximation as explained in the sequel.

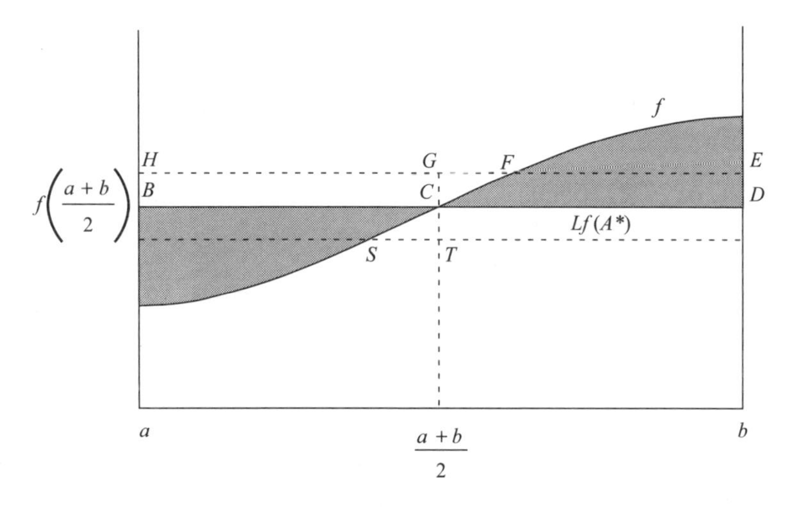

FIGURE 2.6 Least-absolute approximation by a constant value; The 1-norm of error shown by the shaded area

One interesting property of least-absolute approximation is that the approximation depends on the sign function, $\operatorname{sgn}\left[f(x)-Lf\left(A^*, x\right)\right]$, with the values given by

$$\operatorname{sgn}\left[f(x)-Lf\left(A^*,x\right)\right]=\begin{cases}-1, & f(x)<Lf\left(A^*,x\right)\\ 0, & f(x)=Lf\left(A^*,x\right)\\ 1, & f(x)>Lf\left(A^*,x\right)\end{cases} \tag{2.61}$$

rather than on the function $f(x)$ directly. Hence, if the function $f(x)$ changes over some subinterval leaving the above sign function unaltered, then $Lf\left(A^*,x\right)$ remains a best approximation of the changed function. This feature of the 1-norm approximation can reject outliers and provide a robust approximation.

We now state the characterization theorem for a best least-absolute approximation in the following theorem.

Theorem 2.23

Let $f \in \mathscr{C}[a,b]$ be the function to be approximated by the approximating function $L(A) \in \mathscr{A} \subset \mathscr{C}[a,b]$ of the form

$$L(A,x) = \sum_{j=0}^{n} a_j \varphi_j(x) \quad \text{with } \varphi_0(x) = 1, \; a \le x \le b,$$

and let the error function $e(x) = f(x) - L(A,x)$ have a finite number of zeros in $[a, b]$. Then, the necessary and sufficient condition that $L(A^*)$ be a best least-absolute approximation is that

$$\int_a^b \varphi_j(x) \operatorname{sgn}\left[f(x) - L(A^*,x)\right] dx = 0, \quad j = 0,1,\ldots,n \tag{2.62}$$

and there exist $(n + 1)$ points, $a \le x_0 < x_1 < \cdots < x_n \le b$, where the optimum error $e^*(x) = f(x) - L(A^*, x)$ changes sign.

Moreover, the optimum error norm $\|e^*\|_1$ can be obtained as

$$\int_a^b f(x) \operatorname{sgn}\left[f(x) - L(A^*, x)\right] dx = \int_a^b \left|f(x) - L(A^*, x)\right| dx = \|e^*\|_1 \tag{2.63}$$

The least-absolute approximation gets a convenient structure when the basis functions $\{\varphi_j\}$ constitute a Chebyshev set and the corresponding linear space $\mathscr{A}$ becomes a Haar subspace. It helps in two ways. Firstly, it ensures that a continuous function will have a unique best approximation. Secondly, it provides a clue of how the least-absolute approximation can be attempted as a problem of interpolation. The following two theorems are related to the best least-absolute approximation of a continuous function in a Haar subspace.

Theorem 2.24

Let $\mathscr{A}$ be a Haar subspace of $\mathscr{C}[a, b]$. Then, for any function f in $\mathscr{C}[a,b]$, there is a unique best least-absolute approximation in $\mathscr{A}$.

Theorem 2.25

Let $\mathscr{A}$ be an $(n+1)$-dimensional Haar subspace of $\mathscr{C}[a, b]$, and let f be a function in $\mathscr{C}[a, b]$ that is to be approximated.

If $L(A,x) = \sum_{j=0}^{n} a_j \varphi_j(x)$ with $\varphi_0(x) = 1$, $a \le x \le b$, is a best least-absolute approximation of f in $\mathscr{A}$, and if the error function $e^*(x) = f(x) - L(A^*, x)$, $a \le x \le b$, has a finite number of zeros, then $e^*(x)$ has at least $(n + 1)$ zeros, and $L(A^*, x)$ interpolates $f(x)$ at $(n + 1)$ zeros of $e^*(x)$.

The above theorem states that it is possible to find the best approximation provided the positions of zeros of the optimum error function are known. In such case, the approximating function $L(A^*, x)$ can be determined uniquely. The problem becomes more interesting when the optimum error function $e^*(x)$ has exactly $(n + 1)$ zeros. In this case, the positions of zeros of $e^*(x)$ do not depend on $f(x)$.

Theorem 2.26

Let $\mathscr{A}$ be an $(n + 1)$-dimensional Haar subspace of $\mathscr{C}[a,b]$, and let f be a function in $\mathscr{C}[a, b]$ that is to be approximated.

If the optimum error function $e^*(x) = f(x) - L(A^*, x)$, $a \le x \le b$, has exactly $(n + 1)$ zeros, where $L(A^*)$ is the best least-absolute approximation of f in $\mathscr{A}$, then the positions of zeros of $e^*(x)$ do not depend of $f(x)$.

Corollary 2.27

Let the conditions of Theorem 2.26 be satisfied. Let $\mathscr{A}$ be the space $\mathscr{P}_n$ and $[a, b]$ be the interval $[-1, 1]$.

Then, the zeros of the optimum error function $e^*(x) = f(x) - p_n^*(x)$, $-1 \le x \le 1$, where the polynomial $p_n^*(x)$ is the best least-absolute approximation of $f(x)$, are located at

$$\xi_i = \cos\left[\frac{(n+1-i)\pi}{n+2}\right], \quad i = 0, 1, \ldots, n \tag{2.64}$$

Therefore, the approximating polynomial $p_n^*(x)$ can be computed by interpolation satisfying the conditions $\{f(\xi_i) = p_n^*(\xi_i);\ i = 0, 1, \ldots, n\}$, provided the resulting error function $e^*(x)$ changes sign at the points $\{\xi_i\}$ and at no other points.

2.7.1 Linear Programming Method

We now consider the discrete least absolute approximation problem stated as follows:

Let $\{f(x_i);\ i = 1, 2, \ldots, m\}$ be the data-set to be fitted in the least-absolute error sense. Let $\{\varphi_j;\ j = 0, 1, \ldots, n\}$ be a basis of the approximation space $\mathscr{A}$. Then, the least-absolute approximation $Lf(A, x) = \sum_{j=0}^{n} a_j \varphi_j(x)$, $\varphi_0(x) = 1$, for the function $f(x)$ is obtained by minimizing the expression

$$\sum_{i=1}^{m} w_i \left| f(x_i) - Lf(A, x_i) \right| \tag{2.65}$$

where the set of weights $\{w_i; i = 1, 2, \ldots, m\}$ is given, and the set of coefficients $\{a_j;\ j = 0, 1, \ldots, n\}$ is to be determined by linear programming method.

We introduce two new variables u_i and v_i for each data point x_i such that $u_i \geq 0,\ v_i \geq 0$, and $-v_i \leq f(x_i) - Lf(A, x_i) \leq u_i$ for $i = 1, 2, \ldots, m$. It is clear that the new variables are related to the error of approximation as

$$\begin{aligned} u_i &= \max\left(0, f(x_i) - Lf(A, x_i)\right) \\ v_i &= \max\left(0, Lf(A, x_i) - f(x_i)\right) \end{aligned} \quad \text{for } i = 1, 2, \ldots, m \tag{2.66}$$

and

$$u_i + v_i = \left| f(x_i) - \sum_{j=0}^{n} a_j \varphi_j(x_i) \right| \tag{2.67}$$

Therefore, we need to find the least value of the objective function

$$\psi = \sum_{i=1}^{m} w_i \left(u_i + v_i\right) \tag{2.68}$$

subject to the constraints (2.66) and the interpolation conditions

$$f(x_k) = Lf\left(\hat{A}, x_k\right) = \sum_{j=0}^{n} \hat{a}_j \varphi_j(x_k), \quad k = 1, 2, \ldots, n+1 \tag{2.69}$$

where the point-set $\{x_k; k = 1, 2, \ldots, n+1\}$ is a subset of the point-set $\{x_i; i = 1, 2, \ldots, m\}$. Note that we need the $(n + 1)$ equality constraints of interpolation, in addition to the $2m$ constraints of (2.66), to evaluate a total of $(2m + n + 1)$ variables. Moreover, it is to be noted that at most one of u_i and v_i is non-zero and the other is zero, or both are zero at the interpolation points of (2.69).

It follows now the $(2m + n + 1)$ linearly independent constraints define a convex polyhedron of feasible points in the space of $(2m + n + 1)$ variables, and a solution to the problem can be obtained at a vertex of the polyhedron. It is to be pointed out that the approximation is to be determined by the interpolation conditions (2.69). However, the appropriate point-set $\{x_k; k = 1, 2, \ldots, n+1\}$ needs to be found.

In the *simplex method*, an iteration is started by setting the variables to the coordinates of a vertex, and then, by moving along an edge from the vertex, we check whether the objective function can be reduced. A vertex is defined by the interpolation conditions (2.69) in this case, and an edge from the vertex is generated by setting $\left\{a_j = \hat{a}_j + \alpha\, \tilde{a}_j; j = 0, 1, \ldots, n\right\}$, where the coefficients $\left\{\tilde{a}_j\right\}$ are calculated as

$$\sum_{j=0}^{n} \tilde{a}_j \varphi_j(x_l) = 0 \quad \text{with } \tilde{a}_0 = 1, \quad l = 1, 2, \ldots n, \tag{2.70}$$

the point-set $\{x_l;\ l = 1, 2, \ldots, n\}$ being a subset of the point-set $\{x_k; k = 1, 2, \ldots, n+1\}$, and α being a real value.

The objective function on the edge of the polyhedron now becomes

$$\psi(\alpha) = \sum_{i=1}^{m} w_i \left| f(x_i) - \sum_{j=0}^{n} \left(\hat{a}_j + \alpha\, \tilde{a}_j\right) \varphi_j(x_i) \right| \tag{2.71}$$

and an appropriate value of α is found to minimize $\psi(\alpha)$ along the edge. At the point of least value of the objective function, we reach another vertex of the polyhedron. This is indicated by a new point of interpolation, $u_i = v_i = 0$ for some i, and a fresh iteration starts at this juncture.

The above iteration stops when it is not possible to reduce the objective function by moving along one of the edges that meet at the vertex. The coefficients $\{a_j^*;\ j = 0,1,\dots,n\}$ corresponding to this vertex provide the best least-absolute approximation.

2.8 Least Squares Approximation and Orthogonal Polynomials

The best least-squares approximation $Lf(A^*, x) = \sum_{j=0}^{n} a_j^* \varphi_j(x),\ \varphi_0(x) = 1$, for a continuous function $f(x),\ a \le x \le b$, is the element of the subspace $\mathscr{A}$ that minimizes the expression

$$\int_a^b w(x)\left[f(x) - Lf(A,x)\right]^2 dx,\quad Lf \in \mathscr{A} \tag{2.72}$$

where $\{\varphi_j;\ j = 0, 1, \dots, n\}$ are the known basis functions of $\mathscr{A}$, $A = \{a_j\}$ is the set of $(n+1)$ unknown coefficients, and $w(x),\ a \le x \le b$, is the weight function. When $w(x) = 1$, (2.72) is the squared 2-norm of the error function $e(x) = f(x) - Lf(A,x),\ a \le x \le b$.

The least-squares approximation gets many desirable features when the function space is an inner-product linear space. The space of continuous functions $\mathscr{C}[a,b]$ is an inner-product linear space, and for any two functions $f, g \in \mathscr{C}[a,b]$, the inner product is denoted by (f, g) and defined by

$$(f, g) = \int_a^b w(x) f(x) g(x) dx \tag{2.73}$$

A complete inner-product linear space is called a Hilbert space. Note that for $w(x) = 1$, the squared 2-norm of a function f can be computed as

$$(f, f) = \|f\|_2^2 \tag{2.74}$$

Thus, an inner-product space is also a normed space. In other words, we consider some additional properties of the metric function space to get the advantage of least-squares approximation. One such property of the inner-product linear space is the condition of orthogonality, which makes the procedure of least-squares approximation elegantly simple. Two elements of an inner-product space are said to be orthogonal if the inner-product $(f, g) = 0$.

We now state the characterization theorem for the best least-squares approximation in an inner-product space in the following theorem.

Theorem 2.28

Let $f \in \mathscr{C}[a,b]$ be the function to be approximated by the approximating function $L(A) \in \mathscr{A} \subset \mathscr{C}[a,b]$ of the form

$$L(A,x) = \sum_{j=0}^{n} a_j \varphi_j(x) \quad \text{with } \varphi_0(x) = 1, \quad a \le x \le b$$

Then, the necessary and sufficient condition that $L(A^*)$ be the best least-squares approximation is that the error function $e^* = f - L(A^*)$ satisfies the orthogonality conditions

$$\left(e^*, L(A)\right) = 0 \quad \text{for } L(A) \in \mathscr{A}$$

$$\Rightarrow \left(\varphi_i, f - \sum_{j=0}^{n} a_j^* \varphi_j\right) = 0, \quad i = 0, 1, \ldots, n.$$

Thus, according to the above theorem, the solution to the least-squares problem may be regarded as an orthogonal projection of the given function onto the space of approximating functions. The procedure leads to a square system of linear equations to be solved for the unknown coefficients,

$$\sum_{j=0}^{n} (\varphi_i, \varphi_j) a_j^* = (\varphi_i, f), \quad i = 0, 1, \ldots, n \tag{2.75}$$

The above equations are called the normal equations. In general, to calculate the solutions of the normal equations, one needs to compute the inverse of a matrix. Since the inversion of matrix usually causes some loss of accuracy, we look for procedure to get the least-squares approximation without matrix inversion.

Suppose that we can find a set of basis functions $\{\varphi_j; j = 0, 1, \ldots, n\}$ so that the orthogonality conditions

$$(\varphi_i, \varphi_j) = 0, \quad i \ne j \tag{2.76}$$

are satisfied. Then, the matrix of the normal equations becomes diagonal, and accordingly, the solutions of the normal equations can be computed without involving matrix inversion. The following theorem gives the form of the best least-squares approximation when the basis functions are orthogonal.

Theorem 2.29

Let $\mathscr{A}$ be an $(n+1)$-dimensional subspace of $\mathscr{C}[a, b]$, that is spanned by the basis functions $\{\varphi_j; j = 0, 1, \ldots, n\}$. If the orthogonality conditions $(\varphi_i, \varphi_j) = 0, \; i \ne j$, are satisfied, then for any $f \in \mathscr{C}[a, b]$, the best least-squares approximation $L(A^*) \in \mathscr{A}$ is given by

$$L(A^*) = \sum_{j=0}^{n} \frac{(\varphi_j, f)}{\|\varphi_j\|^2} \varphi_j \quad \text{with } a_j^* = \frac{(\varphi_j, f)}{\|\varphi_j\|^2} \tag{2.77}$$

An important case of least-squares approximation arises when the set of approximating functions $\mathscr{A}$ is same as $\mathscr{P}_n$, the space of all polynomials of degree at most n. The basis functions $\{\psi_j = x^j; j = 0, 1, \ldots, n\}$ span the linear space $\mathscr{P}_n$. However, this set of basis functions does not constitute an orthogonal set. A set of orthogonal basis functions $\{\varphi_j; j = 0, 1, \ldots, n\}$ can be constructed by using the recurrence relation for the polynomials $\varphi_j(x)$, $a \le x \le b$ as follows

$$\varphi_j(x) = x\varphi_{j-1}(x) - \sum_{i=0}^{j-1} \frac{(\varphi_i, x\varphi_{j-1})}{\|\varphi_i\|^2} \quad \text{for } j = 1, 2, \ldots, n \quad \text{with } \varphi_0(x) = 1 \tag{2.78}$$

which is orthogonal to all polynomials $\{\varphi_i; i = 0, 1, \ldots, j-1\}$. This procedure is known as Gram-Schmidt orthogonalization.

The following theorem provides the three-term recurrence relation for generating orthogonal polynomials $\{\varphi_j; j = 0, 1, \ldots, n\}$ to span the linear space $\mathscr{P}_n$ for any n.

Theorem 2.30

Let the polynomials $\varphi_j(x)$, $a \le x \le b$ be defined by the recurrence relations

$$\varphi_1(x) = (x - \alpha_1)\varphi_0(x) \quad \text{with } \varphi_0(x) = 1,$$

and

$$\varphi_j(x) = (x - \alpha_j)\varphi_{j-1}(x) - \beta_{j-1}\varphi_{j-2} \quad \text{for } j \ge 2 \tag{2.79}$$

where the real values $\alpha_j = \dfrac{(\varphi_{j-1}, x\varphi_{j-1})}{\|\varphi_{j-1}\|^2}$ and $\beta_j = \dfrac{\|\varphi_j\|^2}{\|\varphi_{j-1}\|^2}$ for $j \ge 1$. Then, for each j, φ_j is a polynomial of degree j, with the coefficient of x^j being unity. Moreover, the polynomials $\{\varphi_j; j = 0, 1, 2, \ldots\}$ constitute an orthogonal set.

One of the main advantages of least-squares approximation in an inner-product space is that the developed theory applies equally well for continuous and discrete functions. In the discrete case, the least-squares approximation to a function $f(x)$ defined on a finite point set $X = \{x_1, x_2, \ldots, x_m\}$ is obtained by minimizing the expression

$$\sum_{i=1}^{m} w_i \left[f(x_i) - Lf(A, x_i) \right]^2, \quad w_i > 0 \tag{2.80}$$

with respect to the set of coefficients A, where $W = \{w_1, w_2, \ldots, w_m\}$ is the set of weights. The set of functions $\{\varphi_j; j = 0, 1, \ldots, n\}$ is said to be orthogonal on the points X with the weights W if the conditions

$$\sum_{i=1}^{m} w_i \varphi_j(x_i) \varphi_k(x_i) = 0 \quad \text{for } j \neq k \tag{2.81}$$

are satisfied. Moreover, the set $\{\varphi_j\}$ is said to be orthonormal if, in addition, we have

$$\sum_{i=1}^{m} w_i \left[\varphi_j(x_i)\right]^2 = 1 \tag{2.82}$$

The following theorem provides the least-squares approximation of a function defined over a finite point set.

Theorem 2.31

Let $\{\varphi_j(x)\}$ be an orthonormal set of functions on points $X = \{x_1, x_2, \ldots, x_m\}$ with weights $W = \{w_1, w_2, \ldots, w_m\}$. Then, a function $f(x)$ defined on X possesses the best least-squares approximation of the form $L(A^*, x) = \sum_{j=0}^{n} a_j^* \varphi_j(x)$ with $\varphi_0(x) = 1$, where the coefficients a_j^* are given by

$$a_j^* = \sum_{i=1}^{m} w_i f(x_i) \varphi_j(x_i) \tag{2.83}$$

A set of orthogonal polynomials $\{P_j(x_i);\ j = 0, 1, \ldots\}$ associated with a finite point-set X and a set of weights W can be generated from the recurrence relation

$$\begin{gathered} P_j(x_i) = (x_i - \alpha_j) P_{j-1}(x_i) - \beta_{j-1} P_{j-2}(x_i) \quad \text{for } j \geq 1 \\ \text{with } \mathrm{P}_0(x_i) = 1,\ \mathrm{P}_{-1}(x_i) = 0 \end{gathered} \tag{2.84}$$

where $\alpha_j = \dfrac{\sum_{i=1}^{m} w_i\, x_i \left[P_{j-1}(x_i)\right]^2}{\sum_{i=1}^{m} w_i \left[P_{j-1}(x_i)\right]^2}$ and $\beta_j = \dfrac{\sum_{i=1}^{m} w_i \left[P_j(x_i)\right]^2}{\sum_{i=1}^{m} w_i \left[P_{j-1}(x_i)\right]^2}$ for $j \geq 1$.

In presence of noise, when the discrete function f is approximated by a series of orthogonal polynomials $\{P_j(x_i);\ j = 0, 1, \ldots\}$ at the point-set X, the approximating series should be truncated at the order of approximation n where the error-variance

$$\sigma_n^2 = \frac{\sum_{i=1}^{m} \left[f(x_i) - \sum_{j=0}^{n} a_j P_j(x_i) \right]^2}{m - n - 1} = \frac{\delta_n^2}{m - n - 1}, \quad n \leq m - 1 \tag{2.85}$$

is either minimum, or does not decrease appreciably any further with the increase of approximation order. This is known as the minimum error-variance criterion, which is a useful tool in reduction of the effect of noise. We assume that the noise added to the signal values is of zero-mean and uncorrelated. Then, it can be shown that the expected value of mean-square error of estimate is the ratio of the sum of square deviations between the estimated and sampled values and the residual degrees of freedom. More on this topic can be found in [Wilks]. Note that under the interpolation condition with $n = m - 1$, the sum of square deviations δ_n^2 is zero and the residual degrees of freedom $(m - n - 1)$ is also zero.

2.8.1 Least Squares Approximation of Periodic Functions

A real-valued continuous function $f(x)$ is 2π-periodic if the periodicity condition

$$f(x+2\pi) = f(x), \quad -\infty < x < \infty \tag{2.86}$$

is satisfied. Since the variable x can be scaled, if necessary, we denote a periodic continuous function f as $f \in \mathscr{C}_{2\pi}$, which is the set of all continuous functions that are 2π-periodic. The set of approximating functions is usually composed of trigonometric series of the form

$$q(x) = \frac{1}{2}a_0 + \sum_{j=1}^{n}\left[a_j \cos(jx) + b_j \sin(jx)\right], \quad -\infty < x < \infty \tag{2.87}$$

with real parameters $\{a_j;\ j = 0, 1, \ldots, n\}$ and $\{b_j;\ j = 1, 2, \ldots, n\}$. For a fixed n, we write the approximating function $q \in \mathscr{Q}_n$, which is the set of all trigonometric series of degree at most n. Note that a trigonometric function of the form $\cos^j x \ \sin^k x \in \mathscr{Q}_n$ if $j + k \le n$ for non-negative integers j and k. The dimension of $\mathscr{Q}_n$ is $(2n + 1)$ and $\mathscr{Q}_n \subset \mathscr{C}_{2\pi}$.

A periodic continuous function can be approximated by a trigonometric series with an arbitrary high accuracy provided the order of approximation n is chosen to be sufficiently large. This is the statement of the following theorem.

Theorem 2.32

For a function $f \in \mathscr{C}_{2\pi}$ and for any $\varepsilon > 0$, there exists a trigonometric series

$$q(x) = \frac{1}{2}a_0 + \sum_{j=1}^{n}\left[a_j \cos(jx) + b_j \sin(jx)\right], \quad -\infty < x < \infty,$$

such that the bound on the error-norm

$$\|f - q\|_\infty \le \varepsilon$$

is satisfied.

The least-squares trigonometric series approximation $q(x)$, $-\infty < x < \infty$, for a function $f \in \mathscr{C}_{2\pi}$, is calculated by minimizing the expression

$$\int_{-\pi}^{\pi} \left[f(x) - q(x)\right]^2 dx, \quad q \in \mathscr{Q}_n \tag{2.88}$$

The above procedure can be conveniently carried out using the orthogonality conditions of the trigonometric functions as given below

$$\int_{-\pi}^{\pi} \cos(jx)\cos(kx)\,dx = 0, \quad j \neq k$$

$$\int_{-\pi}^{\pi} \sin(jx)\sin(kx)\,dx = 0, \quad j \neq k \tag{2.89}$$

$$\int_{-\pi}^{\pi} \cos(jx)\sin(kx)\,dx = 0$$

The desired trigonometric series approximation can be computed as stated in the following theorem.

Theorem 2.33

Let $f \in \mathscr{C}_{2\pi}$ be the function to be approximated by the trigonometric series $q \in \mathscr{Q}_n \subset \mathscr{C}_{2\pi}$ of the form

$$q(x) = \frac{1}{2}a_0 + \sum_{j=1}^{n}\left[a_j \cos(jx) + b_j \sin(jx)\right], \quad -\infty < x < \infty.$$

Then, the necessary and sufficient condition that q be the best least-squares approximation is that the coefficients have the values

$$a_j = \frac{1}{\pi}\int_{-\pi}^{\pi} f(\theta)\cos(j\theta)\,d\theta, \quad j = 0, 1, \ldots, n$$

$$b_j = \frac{1}{\pi}\int_{-\pi}^{\pi} f(\theta)\sin(j\theta)\,d\theta, \quad j = 1, 2, \ldots, n \tag{2.90}$$

The approximation of a periodic function $f(x)$ by the trigonometric series $q(x)$ as above is known as the finite Fourier series representation.

In the discrete function case, we find an approximation for the function using the equally spaced function values $\left\{f\left(\frac{2\pi i}{m}\right);\ i = 1, 2, \ldots, m\right\}$. The approximating trigonometric series now becomes

$$q\left(\frac{2\pi i}{m}\right) = \frac{1}{2}a_0 + \sum_{j=1}^{n}\left[a_j \cos\left(\frac{2\pi ij}{m}\right) + b_j \sin\left(\frac{2\pi ij}{m}\right)\right], \quad i = 1, 2, \ldots, m \tag{2.91}$$

where the coefficients have the values

$$a_j = \frac{2}{m}\sum_{i=1}^{m} f\left(\frac{2\pi i}{m}\right)\cos\left(\frac{2\pi ij}{m}\right), \quad j = 0, 1, \ldots, n$$

$$b_j = \frac{2}{m}\sum_{i=1}^{m} f\left(\frac{2\pi i}{m}\right)\sin\left(\frac{2\pi ij}{m}\right), \quad j = 1, 2, \ldots, n \tag{2.92}$$

The above approximation of a periodic discrete function is known as the discrete Fourier series representation. Note that in deriving the expressions for the coefficients $\{a_j\}$ and $\{b_j\}$, the following orthogonality conditions are used

$$\begin{aligned}
&\sum_{i=1}^{m} \cos\left(\frac{2\pi ij}{m}\right)\cos\left(\frac{2\pi ik}{m}\right) = 0, \quad j \neq k \\
&\sum_{i=1}^{m} \sin\left(\frac{2\pi ij}{m}\right)\sin\left(\frac{2\pi ik}{m}\right) = 0, \quad j \neq k \\
&\sum_{i=1}^{m} \cos\left(\frac{2\pi ij}{m}\right)\sin\left(\frac{2\pi ik}{m}\right) = 0
\end{aligned} \tag{2.93}$$

2.9 Signal Processing Applications I: Complex Exponential Signal Representation

A complex-valued signal consisting of M weighted complex exponentials is expressed as

$$g(t) = \sum_{i=1}^{M} b_{ci} \exp(s_i t), \quad t \geq 0 \tag{2.94}$$

where $s_i = \alpha_i + j\omega_i$, with α_i negative, and $b_{ci} = b_i \exp(j\varphi_i)$ is complex in general. A real function of the form of damped sinusoids or cosinusoids,

$$g(t) = \sum_{i=1}^{\Lambda} \beta_i \exp(\alpha_i t) \begin{cases} \sin(\omega_i t + \varphi_i) \\ \cos(\omega_i t + \varphi_i) \end{cases} \tag{2.95}$$

can be transformed into (2.94) where the $M = 2\Lambda$ values for b_{ci} or s_i will occur in complex conjugate pairs.

The problem of parameter estimation of a sum of complex exponentials is closely related to the problem of system identification. Indeed, the impulse response of a linear system can be characterized by a sum of weighted complex exponentials where the complex frequencies s_i are the distinct poles of the system and the complex amplitudes b_{ci} are the residues at the poles.

The parameter estimation problem for the complex exponential signals can be solved in two steps. First, the s_i-values are determined. Then substituting the known values of s_i, the b_{ci}-values are computed by solving a set of linear equations. It is assumed here that the exact order of the model M is known *a priori*.

The sampled values of the modelled signal satisfy a difference equation with constant coefficients. Thus, the direct problem of estimation of s_i-parameters, which is a non-linear problem, is converted into an indirect and linear problem of estimation of coefficients of the difference equation. Once the coefficients are estimated, a polynomial equation is formed and its roots computed to determine the s_i-parameters.

2.9.1 Prony's Method

Let the sampled values of the signal represented by (2.94) be given at $2M$ uniformly spaced points of the independent variable $t = nT$,

$$g_n = \sum_{i=1}^{M} b_{ci} z_i^n, \quad n = 0, 1, \ldots, 2M-1 \tag{2.96}$$

where $z_i = \exp(s_i T)$, T being the sampling interval.

Let us assume that the M solutions for z_i are the roots of the polynomial equation

$$a_M z^M + a_{M-1} z^{M-1} + \cdots + a_1 z + a_0 = 0 \tag{2.97}$$

Note that the solutions for z_i or s_i can be obtained provided the values for a_i are known.

By utilizing (2.96) and (2.97), a set of linear equations can be written in matrix form as follows,

$$\begin{bmatrix} g_0 & g_1 & \cdots & g_{M-1} & g_M \\ g_1 & g_2 & \cdots & g_M & g_{M+1} \\ \vdots & \vdots & & \vdots & \vdots \\ g_{M-1} & g_M & \cdots & g_{2M-2} & g_{2M-1} \end{bmatrix} \begin{bmatrix} a_0 \\ \vdots \\ a_{M-1} \\ a_M \end{bmatrix} = \mathbf{0} \tag{2.98}$$

Setting $a_M = 1$, (2.98) can be written as

$$\begin{bmatrix} g_0 & g_1 & \cdots & g_{M-1} \\ g_1 & g_2 & \cdots & g_M \\ \vdots & \vdots & & \vdots \\ g_{M-1} & g_M & \cdots & g_{2M-2} \end{bmatrix} \begin{bmatrix} a_0 \\ a_1 \\ \vdots \\ a_{M-1} \end{bmatrix} = - \begin{bmatrix} g_M \\ g_{M+1} \\ \vdots \\ g_{2M-1} \end{bmatrix} \tag{2.99}$$

which can be solved for a_i. According to (2.99), each sampled value of the signal can be 'predicted' by a linear combination of the M past values of the signal. The a_i-values are called the linear prediction coefficients. An alternative matrix equation can be found from (2.98) with $a_0 = 1$. The resulting equations provide backward prediction, whereas in (2.99) we get forward prediction.

After computing the z_i-roots of (2.97), the solutions for s_i are determined by the following relations:

$$s_i = \frac{\ln|z_i|}{T} \pm j \frac{\arg(z_i)}{T} \tag{2.100}$$

where $|.|$ and $\arg(.)$ denote the magnitude and argument of the complex variable, respectively.

2.9.2 The Derivative Methods

It may be anticipated that if the signal and its delayed values satisfy a difference equation whose coefficients form a polynomial equation in z-variable, then the signal and its derivatives will also satisfy a difference equation whose coefficients will now form a polynomial equation in s-variable.

Assume that the signal and all its derivatives up to order $2M-1$, evaluated at a single point, t_0 of the t-axis are available. Differentiating (2.94) n times and substituting $t = t_0$, these values are expressed as follows,

$$g^k = \left. \frac{d^k g(t)}{dt^k} \right|_{t=t_0} = \sum_{i=1}^{M} s_i^k b_{ci} \exp(s_i t_0); \quad k = 0, 1, \ldots, 2M-1; \tag{2.101}$$

where g^0 is the signal value.

Assume now that the M solutions for s_i are the roots of the polynomial equation

$$a_M s^M + a_{M-1} s^{M-1} + \cdots + a_1 s + a_0 = 0. \tag{2.102}$$

Then utilizing (2.101) and (2.102), a matrix equation similar to (2.98) can be written,

$$\begin{bmatrix} g^0 & g^1 & \cdots & g^{M-1} & g^M \\ g^1 & g^2 & \cdots & g^M & g^{M+1} \\ \vdots & \vdots & & \vdots & \vdots \\ g^{M-1} & g^M & \cdots & g^{2M-2} & g^{2M-1} \end{bmatrix} \begin{bmatrix} a_0 \\ \vdots \\ a_{M-1} \\ a_M \end{bmatrix} = \mathbf{0} \tag{2.103}$$

which yields to

$$\begin{bmatrix} g^0 & g^1 & \cdots & g^{M-1} \\ g^1 & g^2 & \cdots & g^M \\ \vdots & \vdots & & \vdots \\ g^{M-1} & g^M & \cdots & g^{2M-2} \end{bmatrix} \begin{bmatrix} a_0 \\ a_1 \\ \vdots \\ a_{M-1} \end{bmatrix} = - \begin{bmatrix} g^M \\ g^{M+1} \\ \vdots \\ g^{2M-1} \end{bmatrix} \tag{2.104}$$

with $a_M = 1$.

Solving (2.104) for a_i, and then forming the polynomial equation (2.102), the values for s_i can be obtained by extraction of roots.

We shall now develop an alternative Derivative method. Let us assume that the signal values and all its derivatives up to order M are known at arbitrarily chosen M distinct points $\{t_1, t_2, \ldots, t_M\}$ of the t-axis. Then similar to the first row of (2.103), a set of linear equations can be written in matrix form as follows,

$$\begin{bmatrix} g_1 & g_1^1 & \cdots & g_1^{M-1} & g_1^M \\ g_2 & g_2^1 & \cdots & g_2^{M-1} & g_2^M \\ \vdots & \vdots & & \vdots & \vdots \\ g_M & g_M^1 & \cdots & g_M^{M-1} & g_M^M \end{bmatrix} \begin{bmatrix} a_0 \\ \vdots \\ a_{M-1} \\ a_M \end{bmatrix} = \mathbf{0} \tag{2.105}$$

where $g_n^k = \sum_{i=1}^{M} s_i^k b_{ci} \exp(s_i t_n)$ is the kth order derivative of the signal evaluated at $t = t_n$ which are not necessarily at uniform spacing.

Rewriting (2.105) with $a_M = 1$, we get

$$\begin{bmatrix} g_1 & g_1^1 & \cdots & g_1^{M-1} \\ g_2 & g_2^1 & \cdots & g_2^{M-1} \\ \vdots & \vdots & & \vdots \\ g_M & g_M^1 & \cdots & g_M^{M-1} \end{bmatrix} \begin{bmatrix} a_0 \\ a_1 \\ \vdots \\ a_{M-1} \end{bmatrix} = - \begin{bmatrix} g_1^M \\ g_2^M \\ \vdots \\ g_M^M \end{bmatrix} \tag{2.106}$$

which can be solved for a_i. Requirement of all derivatives up to order $2M-1$, for a system of order M, does not make the Derivative method of (2.104) suitable in practice. For the alternative Derivative method of (2.106), we employ signal derivatives up to order M computed at M distinct points of the t-axis. Unless stated otherwise, we shall refer to the alternative Derivative method as the Derivative method for estimation of signal parameters.

2.9.3 The Integral Method

The difference equation can also be written in terms of the signal value $g(t)$ and its anti-derivatives $\{g^{-k}(t); k=1,2,\dots,M\}$ for any t. The anti-derivatives are given as follows,

$$g^{-k}(t) = g^{-k}(t_0) + \int_{t_0}^{t} g^{-(k-1)}(t)dt \tag{2.107}$$

where t_0 is the initial time and $g^{-k}(t_0)$ is the unknown initial condition.

It is left as an exercise here that with $a_M = 1$, the matrix equation in this case becomes

$$[\mathbf{A}\,|\,\mathbf{B}]\begin{bmatrix} a_{M-1} \\ \vdots \\ a_0 \\ \cdots \\ \mathbf{c} \end{bmatrix} = -\begin{bmatrix} g(t_1) \\ g(t_2) \\ \vdots \\ g(t_{2M}) \end{bmatrix} \tag{2.108}$$

where

$$\mathbf{A} = \begin{bmatrix} g_0^{-1}(t_1) & g_0^{-2}(t_1) & \cdots & g_0^{-M}(t_1) \\ g_0^{-1}(t_2) & g_0^{-2}(t_2) & \cdots & g_0^{-M}(t_2) \\ \vdots & \vdots & & \vdots \\ g_0^{-1}(t_{2M}) & g_0^{-2}(t_{2M}) & \cdots & g_0^{-M}(t_{2M}) \end{bmatrix},$$

$$\mathbf{B} = \begin{bmatrix} 1 & \frac{(t_1-t_0)}{1!} & \frac{(t_1-t_0)^2}{2!} & \cdots & \frac{(t_1-t_0)^{M-1}}{(M-1)!} \\ 1 & \frac{(t_2-t_0)}{1!} & \frac{(t_2-t_0)^2}{2!} & \cdots & \frac{(t_2-t_0)^{M-1}}{(M-1)!} \\ \vdots & \vdots & \vdots & & \vdots \\ 1 & \frac{(t_{2M}-t_0)}{1!} & \frac{(t_{2M}-t_0)^2}{2!} & \cdots & \frac{(t_{2M}-t_0)^{M-1}}{(M-1)!} \end{bmatrix},$$

$$\mathbf{c} = \begin{bmatrix} a_{M-1}g^{-1}(t_0) + a_{M-2}g^{-2}(t_0) + \cdots + a_0 g^{-M}(t_0) \\ a_{M-2}g^{-1}(t_0) + a_{M-3}g^{-2}(t_0) + \cdots + a_0 g^{-(M-1)}(t_0) \\ \vdots \\ a_1 g^{-1}(t_0) + a_0 g^{-2}(t_0) \\ a_0 g^{-1}(t_0) \end{bmatrix},$$

and, $g_0^{-k}(t_n) = \int_{t_0}^{t_n} \cdots \int_{k \text{ times}} g(t)dt^k$ are the zero-initial conditioned integrals of the signal. The integrals are assumed to be known at arbitrarily chosen 2*M* distinct points $\{t_n; n = 1, 2, \ldots, 2M\}$ of the independent variable axis.

2.9.4 Linear Prediction Model for Noisy Data and Extended Order Modelling

We will consider now the problem of estimation of signal parameters from a given set of signal values at uniform spacing when the signal samples are corrupted with noise.

The discrete-time signal $g[n] = g(nT)$ of (2.94) given by $g[n] = \sum_{i=1}^{M} b_{ci} z_i^n$ for $n = 0, 1, \ldots, N-1$, with $z_i = \exp(s_i T)$, satisfies the linear prediction model as follows,

$$g[n] = -\sum_{i=1}^{M} a_{M,M-i}\, g[n-i], \quad n \geq M \tag{2.109}$$

with proper initial conditions $\{g[0], g[1], \ldots, g[M-1]\}$. The coefficients $a_{M,M-i}$ form the characteristic polynomial of the prediction-error filter $A_M(z)$,

$$\begin{aligned} A_M(z) &= 1 + \sum_{i=1}^{M} a_{M,M-i} z^{-i} \\ &= \prod_{i=1}^{M} \left(1 - z_i z^{-1}\right) \end{aligned} \tag{2.110}$$

whose M roots are located at $z_i = \exp(s_i T)$.

Let us now form an extended-order prediction-error filter whose characteristic polynomial $A_L(z)$ is given by

$$\begin{aligned} A_L(z) &= A_M(z) B_{L-M}(z) \\ &= \prod_{i=1}^{M} \left(1 - z_i z^{-1}\right)\left[1 + b_{L-M-1} z^{-1} + \cdots + b_1 z^{-(L-M-1)} + b_0 z^{-(L-M)}\right] \\ &= \prod_{i=1}^{L} \left(1 - z_i z^{-1}\right), \quad L \geq M \end{aligned} \tag{2.111}$$

where $B_{L-M}(z)$ is an arbitrary polynomial of degree $(L-M)$. Note that all of the signal modes z_i are present in the extended-order model, and the additional modes may be used to model additive noise. It then follows directly that the discrete-time signal $g[n]$ satisfies the extended-order linear prediction model given by

$$g[n] = -\sum_{i=1}^{L} a_{L,L-i}\, g[n-i], \quad n \geq L \tag{2.112}$$

with proper initial conditions $\{g[0], g[1], \ldots, g[L-1]\}$.

Applying Prony's method based on the linear prediction model with extended-order modelling, we now estimate M signal and L–M noise z_i or s_i modes. In order to distinguish the signal modes from the noise modes, we compute the complex amplitudes b_{ci} for all components.

With the known values of the modes $\{z_i; i = 1, \ldots, L\}$, the sampled data with additive noise $w[n]$ represented by $y[n] = g[n] + w[n] = \sum_{i=1}^{L} b_{ci} z_i^n$ for $n = 0, 1, \ldots, N-1$ can be written in matrix form as

$$\begin{bmatrix} 1 & 1 & \cdots & 1 \\ z_1 & z_2 & \cdots & z_L \\ \vdots & \vdots & & \vdots \\ z_1^{N-1} & z_2^{N-1} & \cdots & z_L^{N-1} \end{bmatrix} \begin{bmatrix} b_{c1} \\ b_{c2} \\ \vdots \\ b_{cL} \end{bmatrix} = \begin{bmatrix} y[0] \\ y[1] \\ \vdots \\ y[N-1] \end{bmatrix} \tag{2.113}$$

or,

$$\mathbf{Z}_V \mathbf{b}_c = \mathbf{y} \quad .$$

where $\mathbf{Z}_V$ is the Vandermonde matrix, $\mathbf{b}_c$ is the complex amplitude vector, and $\mathbf{y}$ is the data vector. We find the least squares solution of $\mathbf{b}_c$ as

$$\mathbf{b}_c = \left[\mathbf{Z}_V{}^H \mathbf{Z}_V \right]^{-1} \mathbf{Z}_V{}^H \mathbf{y} \tag{2.114}$$

and arrange the computed amplitudes (residue-magnitudes) $|b_{ci}| = b_i$ in descending order. Typically, there exist M larger amplitudes $(b_M \gg b_{M+1})$. Therefore, it is easy to distinguish the signal modes from the noise modes.

2.9.5 Transient Signal Analysis I

Parametric modelling and parameter estimation are important steps in model-based signal processing. For instance, in electromagnetic pulse (EMP) situations, the EMP pickup can be characterized by a sum of weighted complex exponentials whose parameters are to be determined. The parameters are a means of coding the various pulse waveforms, and the signal representation thus obtained can be readily employed to analyse responses in various subsystems under EMP environment. A typical EMP signal has the characteristics of having a short rise time and rather slow decay rate. This situation provides a strong motivation to use a non-uniform sampling strategy over the uniform sampling strategy. An approach to process signal samples on a non-uniform grid will be demonstrated in the following examples which are taken from [Sircar].

Non-uniform sampling may be chosen over uniform sampling because of various reasons. Sampling efficiency and noise consideration are two such reasons. An efficient sampling scheme should sample a signal with adjustable sampling rate. The idea is to sample the signal more densely where its variation is relatively rapid. Maintaining the same sampling rate in the region where the signal varies slowly is inefficient. For complete reconstruction of a signal in an efficient way, the instantaneous sampling rate must be determined by local estimate of the bandwidth of the signal. Obviously, when the local bandwidth of a signal changes from region to region, any efficient sampling scheme will not sample the signal at uniform spacing of the independent variable axis.

Furthermore, when zero-mean random noise is superimposed on the signal samples, it is advisable to sample more points of signal where the signal values are far away from zero-crossing. In this way, we can achieve a higher signal-to-noise ratio over the sampled set of values. Once again, any sampling technique to implement the above scheme will sample the signal at a set of non-uniformly distributed points of the independent variable axis.

In the following examples, we will consider arbitrarily distributed sampled points, except that the condition for complete reconstruction of the signal must be satisfied, as mentioned above. Note that the uniform sampling scheme is a special case of our generalized approach.

Given the sampled set of signal values at non-uniformly distributed points of the independent variable axis, we essentially go through three stages of procedure:

(i) Find a closed-form expression for the signal over the sampling span, which in turn can be used to compute the signal values at uniform spacing, or the signal derivatives or zero-initial conditioned signal integrals at the sampled points.

(ii) Estimate the complex frequency $s_i = \alpha_i + j\omega_i$ parameters of the complex exponential signals $\sum_i b_{ci} \exp(s_i t)$, by first employing the Prony's method or the Derivative method or the Integral method to find the linear prediction coefficients, and then, evaluating the roots of the characteristic polynomial of the prediction error filter.

(iii) Estimate the complex amplitude $b_{ci} = b_i \exp(j\varphi_i)$ parameters of the complex exponential signals by solving a set of linear equations.

2.9.6 Simulation Study: Example 1 (Non-uniform Sampling)

The signal $g(t) = e^{-t}\cos(2t) + e^{-3t}\cos(5t)$ is sampled at 9 non-uniformly spaced points $\{0, 0.08, 0.12, 0.2, 0.24, 0.32, 0.4, 0.44, 0.48\}$ in single and double precisions.

By utilizing the Gregory-Newton interpolating polynomial $p_8(t)$, we compute the signal values at uniform spacing of $\Delta t = 0.06$ over the sampling span. These values are used in the Prony's method.

We also compute all signal derivatives up to order 8 at $t = 0.24$, and use these values in the Derivative method. For the alternative Derivative method, the signal derivatives up to order 4 are computed at the sampled points $\{0.12, 0.2, 0.24, 0.32, 0.4\}$.

In each case, the conjugate gradient (CG) method is employed to solve the least squares problem and find the coefficients of the polynomial in z (or s). The CG method is described in Chapter 3. A standard utility subroutine is then used to evaluate the roots of the polynomial equation. The subroutine uses the method of factorization of the polynomial in quadratic terms iteratively. From which, the complex roots are evaluated. The results are shown in Table 2.1.

In the above example, we carefully observe the large deviation of the computed s_i -roots caused by quantization error, while utilizing the Prony's method.

Note that the quantization error gets magnified in the successive order divided-differences, and thus, the higher order derivatives used in the two Derivative methods are affected more by quantization error as shown in Tables 2.2 and 2.3. Yet these values yield results with reasonable accuracy.

TABLE 2.1 Estimated signal parameters

Signal samples		Computed s_i-values: The Prony's method	Computed s_i-values: The Derivative method	Computed s_i-values: The alternative Derivative method	True values
Single precision		$-0.44 \pm j0.58$	$-0.80 \pm j1.70$	$-0.72 \pm j1.41$	
		$-3.26 \pm j4.91$	$-3.06 \pm j4.95$	$-3.12 \pm j4.95$	$-1 \pm j2$
Double precision		$-1.12 \pm j2.58$	$-1.12 \pm j2.22$	$-1.04 \pm j2.08$	$-3 \pm j5$
		$-2.90 \pm j4.94$	$-2.93 \pm j5.01$	$-2.97 \pm j5.00$	

TABLE 2.2 Signal values computed at uniform spacing

t_i	$p(t_i)$	$p^*(t_i)$	$g(t_i)$
0.00	2.0000000	2.0000000	2.0000000
0.06	1.7329550	1.7329560	1.7329561
0.12	1.4373159	1.4373166	1.4373166
0.18	1.1439686	1.1439689	1.1439689
0.24	0.87411499	0.87411340	0.87411340
0.30	0.64018536	0.64018328	0.64018329
0.36	0.44736290	0.44736028	0.44736027
0.42	0.29535484	0.29535271	0.29535272
0.48	0.18017673	0.18017562	0.18017562

$p(t_i)$, $p^*(t_i)$ - Computed values fitting the signal samples in single and double precisions, respectively.
$g(t_i)$ - Analytically computed values of the signal.

TABLE 2.3 Higher order signal derivatives computed at $t = 0.24$

Order n	$\frac{d^n}{dt^n}[p(t)]_{t=0.24}$	$\frac{d^n}{dt^n}[p^*(t)]_{t=0.24}$	$\frac{d^n}{dt^n}[g(t)]_{t=0.24}$
0	0.87411499	0.87411340	0.87411340
1	-4.2217178	-4.2217278	-4.2217277
2	10.147705	10.147892	10.147897
3	38.785904	38.788317	38.787794
4	-562.51099	-562.70809	-562.71507
5	2246.9580	2245.7721	2246.7429
6	5031.9727	5197.4699	5198.1794
7	-106771.12	-106335.78	-107617.17
8	537626.25	455659.32	471310.97

The example demonstrates the danger of solving an ill-conditioned problem, and instability associated with it. This topic is discussed in Chapter 3.

2.9.7 Smoothing

When random additive noise is superimposed on the signal values, the difference table shows a completely erratic pattern. This difference table cannot be used reliably to find the Gregory-Newton interpolating polynomial for the signal. Thus, before forming the difference table, it is necessary to reduce the effect of noise on the signal values by some smoothing technique.

A heuristic approach whose effectiveness is no better than the other neighbouring techniques described in literature is presented. The computations involved in this approach, however, are much simpler.

Let us assume that the mean value of the noise superimposed on the signal values at the sampled points is zero. Therefore, any averaging technique to redistribute the noise component of one point to its neighbouring points will effectively reduce the noise portion of the composite signal. The redistribution can be effected by linear interpolation, provided the signal is sampled densely enough along the t-axis to assume that linear interpolation is quite satisfactory between successive sampled points.

In Figure 2.7, y_i^* are the actual signal values, and y_i are the values of the signal when random noise is mixed with it. After one iteration, the corrected value of y_2 will be given as,

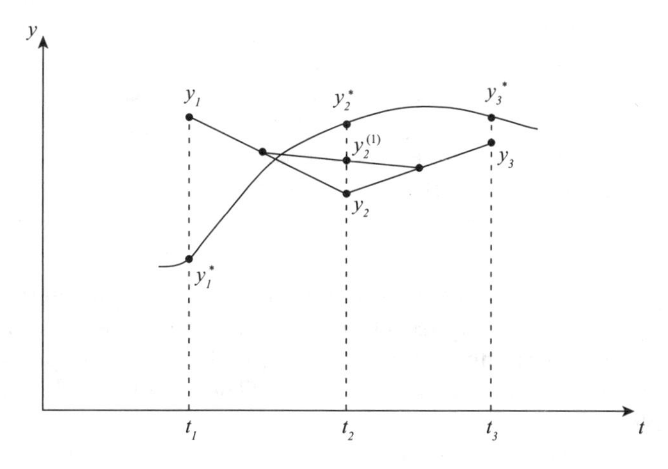

FIGURE 2.7 Averaging technique

$$y_2^{\langle 1 \rangle} = \frac{(y_1 + y_2)(t_3 - t_2) + (y_2 + y_3)(t_2 - t_1)}{2(t_3 - t_1)} \tag{2.115}$$

which is shown to be closer to the actual value of the signal at that abscissa.

The effectiveness of the method described depends on how judiciously the iterations are applied. Because continued iterations on data will not only smooth the noisy signal values, but eventually merge the data set onto a straight line.

2.9.8 Simulation Study: Example 2 (Noisy Signal)

The signal $g(t)=e^{-t}\cos(2t)$ is sampled at uniform spacing, and uniform zero-mean noise r_i, generated by a table of random digits is mixed with the signal setting the signal-to-noise (SNR) level at 9.663 dB.

The noise corrupted values of the signal $y_i = g_i + r_i$ are smoothed by the averaging technique. The up or down arrow indicates whether the value is higher or lower, respectively, than its previous entry. Iterations are continued until a fairly smooth pattern in the signal values has emerged as shown in Table 2.4.

TABLE 2.4 Smoothing of noisy signal

t_i	g_i	r_i	y_i	$y_i^{\langle 1\rangle}$	$y_i^{\langle 2\rangle}$
0.0	1.0000	−0.1825	0.9114		
0.2	0.7541 ↓	0.3849	0.9410 ↑	0.7953	
0.3	0.6114 ↓	−0.1907	0.5188 ↓	0.6058 ↓	0.6174
0.5	0.3277 ↓	−0.2711	0.1961 ↓	0.2965 ↓	0.3241 ↓
0.6	0.1989 ↓	0.2821	0.3359 ↑	0.2245 ↓	0.2136 ↓
0.8	−0.0131 ↓	−0.0815	−0.0527 ↓	0.0149 ↓	0.0080 ↓
1.0	−0.1531 ↓	−0.0365	−0.1708 ↓	−0.2225 ↓	−0.2046 ↓
1.1	−0.1959 ↓	−0.3891	−0.3848 ↓	−0.2874 ↓	−0.2657 ↓
1.2	−0.2221 ↓	0.0270	−0.2090 ↑	−0.2656 ↑	−0.2710 ↓
1.4	−0.2324 ↓	0.0728	−0.1971 ↑	−0.2542 ↑	−0.2728 ↓
1.5	−0.2209 ↑	−0.2914	−0.3624 ↓	−0.3042 ↓	−0.2769 ↓
1.7	−0.1766 ↑	−0.3437	−0.3435 ↑	−0.2401 ↑	−0.1955 ↑
1.9	−0.1183 ↑	0.4270	0.0890 ↑	0.0025 ↑	−0.0237 ↑
2.0	−0.0885 ↑	0.2761	0.0456 ↓	0.0453 ↑	
2.2	−0.1965 ↓	0.3166	−0.0428 ↓		

(a) We now obtain the derivative value $\dot{y}_i$ (or $\ddot{y}_i$) at the centre of an interval by the divided difference. The derivative values at the sampled points are computed by linear interpolation. The averaging technique is applied on the computed values of the derivatives when the values lack smoothness. The computed derivative values are shown in Table 2.5.

By utilizing the alternative Derivative method, the matrix equation is formed in five rows with $a_2 = 1$. The least squares solution provides,

$$a_1 = 1.71 \text{ and } a_0 = 7.75$$

The s_i -values computed are:

$$s_{1,2} = -0.85 \pm j2.65,$$

whereas the actual values are $-1 \pm j2$.

TABLE 2.5 Smoothed signal and its derivatives

t_i	$y_i^{\langle 2 \rangle}$	Dy_i	$\dot{y}_i$	$D\dot{y}_i$	$\ddot{y}_i$	$\ddot{y}_i^{\langle 1 \rangle}$
0.3	0.6174					
		−1.4665				
0.5	0.3241 ↓		−1.2255			
		−1.1050		1.4620		
0.6	0.2136 ↓		−1.0793 ↑		1.0310	
		−1.0280		0.1690		
0.8	0.0080 ↓		−1.0455 ↑		0.7940 ↓	1.4892
		−1.0630		1.4190		
1.0	−0.2046 ↓		−0.7617 ↑		3.3377 ↑	3.0069 ↑
		−0.6110		4.2970		
1.1	−0.2657 ↓		−0.3320 ↑		3.6170 ↑	3.1358 ↑
		−0.0530		2.9370		
1.2	−0.2710 ↓		−0.0383 ↑		1.9713 ↓	2.3475 ↓
		−0.0090		0.0400		
1.4	−0.2728 ↓		−0.0303 ↑		0.9373 ↓	1.3967 ↓
		−0.0410		1.3860		
1.5	−0.2769 ↓		0.1083 ↑		1.7985 ↑	
		0.4070		2.6235		
1.7	−0.1955 ↑		0.6330 ↑			
		0.8590				
1.9	−0.0237 ↑					

(b) At the next stage, the smoothed signal values $y_i^{\langle 2 \rangle}$ are used to calculate the higher order integrals of the signal by the trapezoidal rule.

The initial values of the integrals α and β, as entered in Table 2.6, are unknowns.

TABLE 2.6 Signal and its anti-derivatives

t_i	y_i	y_i^{-1}	y_i^{-2}
0.3	0.6174	$\alpha + 0.0000$	$\beta + 0.0\alpha + 0.0000$
0.5	0.3241	$\alpha + 0.0942$	$\beta + 0.2\alpha + 0.0094$
0.6	0.2136	$\alpha + 0.1211$	$\beta + 0.3\alpha + 0.0202$
0.8	0.0080	$\alpha + 0.1433$	$\beta + 0.5\alpha + 0.0466$
1.0	−0.2046	$\alpha + 0.1236$	$\beta + 0.7\alpha + 0.0733$
1.1	−0.2657	$\alpha + 0.1001$	$\beta + 0.8\alpha + 0.0845$
1.2	−0.2710	$\alpha + 0.0733$	$\beta + 0.9\alpha + 0.0932$
1.4	−0.2728	$\alpha + 0.0458$	$\beta + 1.1\alpha + 0.1051$
1.5	−0.2769	$\alpha + 0.0183$	$\beta + 1.2\alpha + 0.1083$
1.7	−0.1955	$\alpha - 0.0289$	$\beta + 1.4\alpha + 0.1072$
1.9	−0.0237	$\alpha - 0.0508$	$\beta + 1.6\alpha + 0.0992$

Setting $a_2 = 1$, we write the set of linear equations according to the Integral method as,

$$a_0 y_i^{-2} + a_1 y_i^{-1} = -y_i$$

By utilizing the sum equations, we then eliminate the terms $(a_1\alpha + a_0\beta)$ and $a_0\alpha$ from the set of equations in two steps.

The matrix equations in nine rows are solved for a_0, a_1 in the least squares sense,

$$a_1 = 2.06 \text{ and } a_0 = 7.73$$

The s_i-values computed are:

$$s_{1,2} = -1.03 \pm j2.58,$$

whereas the actual values are $-1 \pm j2$.

The example presented above demonstrates how an appropriate smoothing technique can be utilized when the signal samples are corrupted with zero-mean noise. However, it should be pointed out that with increase of the number of sinusoidal components of a test signal, the detection of a 'fairly smooth pattern' in the signal values can be difficult. Consequently, the implementation of the averaging technique presented above may become complicated in those cases.

2.9.9 Simulation Study: Example 3 (Parameter Estimation – Non-uniform Sampling)

The signal $g(t) = e^{-t}\sin(2t) + e^{-1.5t}\sin(5t)$ is sampled at 20 non-uniformly distributed abscissas. The sampled points are chosen completely arbitrarily, except that no sampling interval is allowed to exceed $\frac{1}{8f}$, where f is the highest frequency present in the signal.

A closed-form expression for the signal is found by utilizing the orthogonal polynomial approximation and minimum error-variance criterion. The polynomial series is truncated at the order of approximation 13, where the error-variance gets the minimum value of 0.269×10^{-11}. The truncated polynomial series then provides all the values needed in various estimation techniques. The least squares problems are solved by the conjugate gradient method. The computed values of the s_i-roots are tabulated below. The true values are shown for comparison.

TABLE 2.7 Estimated signal parameters by different methods

Methods employed	The s_i-roots computed	True values
The Prony's method	$-0.51 \pm j1.93$, $-1.58 \pm j4.87$	
The Derivative method	$-0.99 \pm j1.98$, $-1.54 \pm j5.02$	$-1.0 \pm j2.0$, $-1.5 \pm j5.0$
The Integral method	$-1.05 \pm j1.98$, $-1.50 \pm j5.04$	

To obtain the results by the Prony's method, the signal values are computed at equispaced points, using the polynomial expression for the signal. The s_i-roots computed by the Prony's method are found to be less accurate than those computed by the other two estimation methods.

2.9.10 Simulation Study: Example 4 (Parameter Estimation – Noisy Signal)

In this example, we show the effect of noise on the computed s_i-values. The signal $g(t) = e^{-0.3t}\sin(2t) + e^{-0.5t}\sin(5t)$ is sampled at 60 non-uniformly distributed points of the independent variable axis. The uniform zero-mean computer-synthesized noise is then mixed with the sampled signal values, setting the SNR levels at 40, 20 and 10 dB.

In each noise-level case, the Derivative and the Integral methods are employed to compute the s_i-values. These results are tabulated below together with the true values.

The values of derivatives computed at the middle 40 points are used in the Derivative method, whereas the Integral method uses the zero-initial conditioned integrals evaluated at the first 40 points. The error-variance plots at no noise and when SNR=10 dB, are shown in Figure 2.8. In each case, the vertical line is drawn at the approximation order where the polynomial series is truncated.

We observe that the computation of higher order derivatives of the signal with high accuracy is a matter of concern. Compared to that, the higher order integrals of the signal can be evaluated with higher accuracy. This advantage of the Integral method over the Derivative method, however, is not apparent from the computed results of Table 2.8. The increased difficulty of solving the system equation in the Integral method, because of its higher dimensionality, may be responsible for the comparable performance of the two systems. More on this issue has been explored in the next chapter.

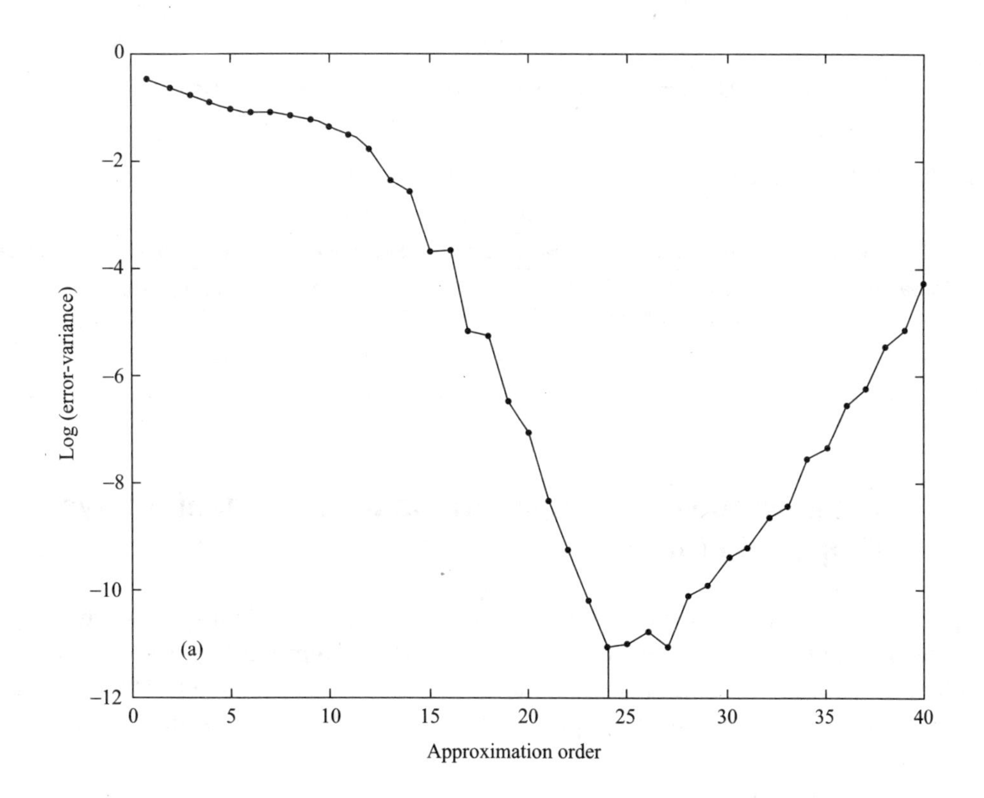

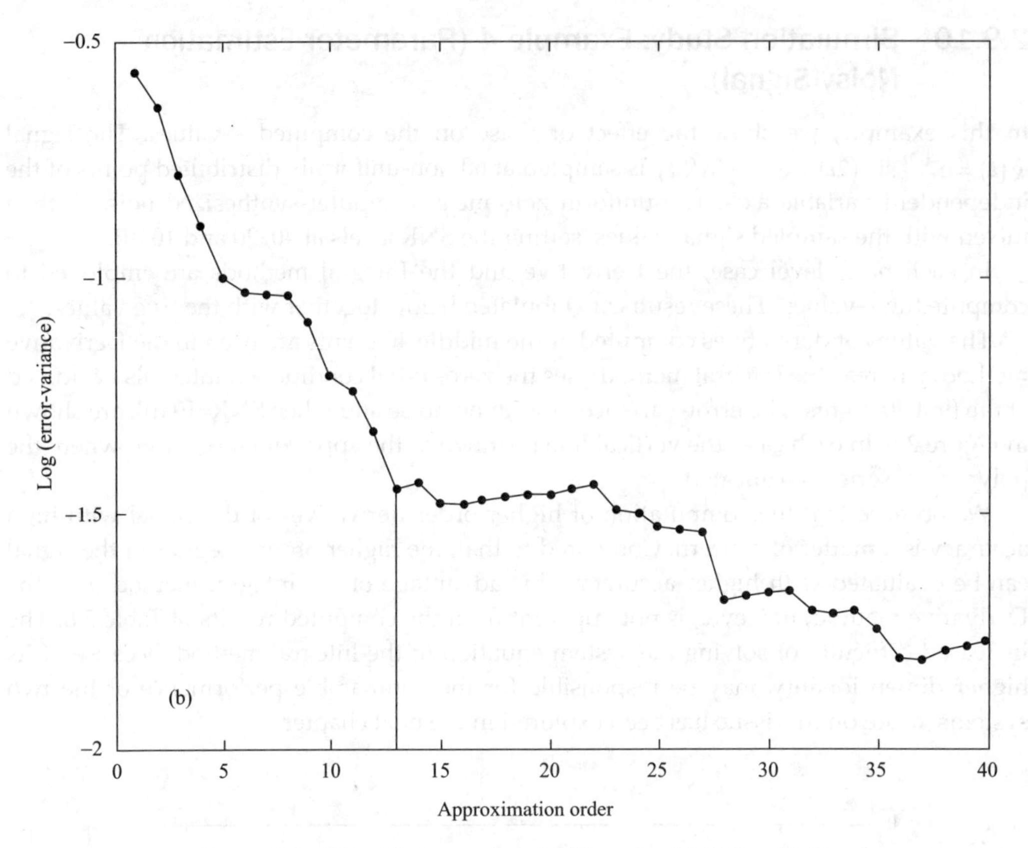

FIGURE 2.8 Error-Variance plots (a) No noise (b) SNR = 10 dB

TABLE 2.8 Estimated signal parameters at various SNR levels

Methods employed	The s_i-values computed				True values
	No noise	SNR = 40 dB	SNR = 20 dB	SNR = 10 dB	
The Derivative method	$-0.3001 \pm j1.9998$, $-0.4999 \pm j4.9999$	$-0.313 \pm j1.92$, $-0.484 \pm j4.96$	$-0.281 \pm j2.03$, $-0.471 \pm j4.96$	$-0.22 \pm j2.16$, $-0.81 \pm j5.30$	$-0.3 \pm j2.0$, $-0.5 \pm j5.0$
The Integral method	$-0.2993 \pm j1.9997$, $-0.5009 \pm j4.9997$	$-0.299 \pm j1.99$, $-0.484 \pm j5.03$	$-0.292 \pm j1.98$, $-0.458 \pm j5.06$	$-0.17 \pm j2.14$, $-0.55 \pm j5.07$	

2.10 Signal Processing Applications II: Numerical Analysis of Optical Fibres

In optical fibre communications, various fibre designs are proposed for specific advantages in different applications. However, the wave equation representing the propagation problem cannot be solved analytically for any refractive index profile other than step and parabolic functions. Therefore, an improved numerical technique is needed for analysing fibres with arbitrary profiles.

2.10.1 Numerical Analysis of Optical Fibres with Arbitrary Refractive Index Profile

The optical fibre under analysis is shown in Figure 2.9. By using the method of separation of variables for a cylindrically symmetric graded-index fibre, the radial variation $F(r)$ of the electric/magnetic field can be shown to satisfy the scalar wave equation

$$\frac{d^2F}{dr^2}+\frac{1}{r}\frac{dF}{dr}+\left[\left(\frac{2\pi}{\lambda}\right)^2 n^2(r)-\beta^2-\frac{m^2}{r^2}\right]F=0 \tag{2.116}$$

where λ is the free-space wavelength, $n(r)$ is the index of refraction, β is the propagation constant, and m is the azimuthal mode number.

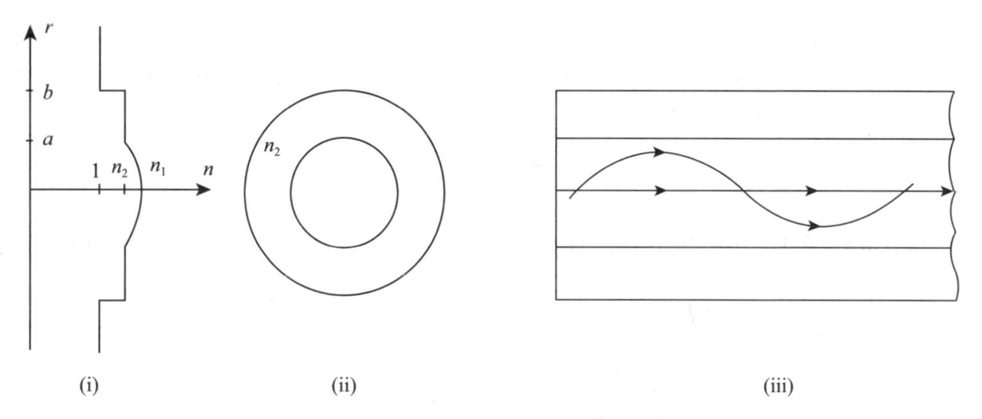

FIGURE 2.9 Graded index fibre (i) Refractive index profile (ii) Transverse section showing core-cladding boundary (iii) Longitudinal section showing ray trajectories

Let the index of refraction be expressed as

$$n(r)=\begin{cases} n_1\left[1-2\Delta\, f(r)\right]^{1/2} & \text{for } r\le a \\ n_1\left[1-2\Delta\right]^{1/2}=n_2 & \text{for } r>a \end{cases} \tag{2.117}$$

where $\Delta=\dfrac{n_1^2-n_2^2}{2n_1^2}$, and the profile function $f(r)$ is chosen such that $f(0)=0$ and $f(a)=1$.

Introducing the normalized parameters

$$x=\frac{r}{a},$$

$$\eta=\frac{\beta^2-\beta_2^2}{\beta_1^2-\beta_2^2},\quad \eta_1=1-\eta=\frac{\beta_1^2-\beta^2}{2\Delta\,\beta_1^2}, \tag{2.118}$$

$$u=(2\Delta)^{1/2}a\beta_1$$

with $\beta_1=2\pi n_1/\lambda$ and $\beta_2=2\pi n_2/\lambda$, (2.116) can be rewritten as

$$\frac{d^2F}{dx^2}+\frac{1}{x}\frac{dF}{dx}+\left\{u^2\left[\eta_1-f(x)\right]-\frac{m^2}{x^2}\right\}F=0 \tag{2.119}$$

where (2.117) is substituted for $n(r)$.

Let the solution of (2.119) have the general form

$$F(x) = x^m R(x) \tag{2.120}$$

Substituting (2.120) into (2.119) and rearranging, we get in the core ($x \le 1$),

$$\frac{d^2R}{dx^2} + \frac{2m+1}{x}\frac{dR}{dx} + u^2\left[\eta_1 - f(x)\right]R = 0 \tag{2.121}$$

Now, suppose that the profile function $f(x)$ for $x \le 1$ has the following form:

$$f(x) = a_0T_0(x) + a_2T_2(x) + \cdots + a_pT_p(x), \quad p \text{ even} \tag{2.122}$$

where $T_j(x)$ are the Chebyshev polynomials of even degree j, and a_j denote the coefficients of the polynomial series. Note that only even degree polynomials are included in (2.122) to ensure that the following condition

$$\frac{dn(x)}{dx} = 0 \tag{2.123}$$

is satisfied at $x = 0$. We use the property that $T_j(-x) = T_j(x)$ for j even, and $T_j(-x) = -T_j(x)$ for j odd.

The coefficients a_j are calculated as shown below:

$$\begin{aligned} a_0 &= \frac{2}{\pi}\int_0^1 \frac{f(x)T_0(x)}{\sqrt{1-x^2}}dx \\ a_j &= \frac{4}{\pi}\int_0^1 \frac{f(x)T_j(x)}{\sqrt{1-x^2}}dx \quad \text{for } j = 2, 4, \ldots, p \end{aligned} \tag{2.124}$$

Now, we express the solution $R(x)$ of (2.121) for $x \le 1$ in terms of the Chebyshev polynomials as follows:

$$R(x) = b_0T_0(x) + b_2T_2(x) + \cdots + b_qT_q(x), \quad q \text{ even} \tag{2.125}$$

where b_j are the coefficients of the Chebyshev polynomials of even degree j. Only even degree polynomials are included in (2.125) to satisfy the condition of radial symmetry for $R(x)$.

We now consider some properties of the Chebyshev polynomials of even degree,

$$\frac{d^2T_k(x)}{dx^2} = 4k\left[1(k-1)T_{k-2}(x) + 2(k-2)T_{k-4}(x) + \cdots + j(k-j)T_{k-2j} + \cdots + \frac{k^2}{8}T_0(x)\right]$$

$$\frac{1}{x}\frac{dT_k(x)}{dx} = 4k\left[T_{k-2}(x) + \cdots + T_{k+2-4j}(x) + \cdots + T_e(x)\right] \quad \text{with } T_e(x) = T_2(x) \text{ or } T_0(x)/2$$

$$T_l(x)T_k(x) = \left[T_{l+k}(x) + T_{l-k}(x)\right]/2 \quad \text{for } l \ge k \tag{2.126}$$

In order to solve (2.121), the expansions of $f(x)$ from (2.122) and $R(x)$ from (2.125) are introduced, and the equation is simplified by using the identities (2.126). We then arrange the terms of (2.121) with respect to the Chebyshev polynomials of ascending degrees, and equate the resulting coefficient of each Chebyshev polynomial to zero. From the above procedure, we get the following matrix equation,

$$\mathbf{AX}+(2m+1)\mathbf{BX}+\mathbf{CX}=\mathbf{0} \tag{2.127}$$

where $\mathbf{X}=\left[b_0, b_2, b_4, \ldots, b_q\right]^T$

$$\mathbf{A}=4\begin{bmatrix} 0, & \frac{1}{2}\cdot 2, & \frac{1}{2}\cdot 4\cdot 2\cdot(4-2), & \frac{1}{2}\cdot 6\cdot 3\cdot(6-3), & \cdots \\ 0, & 0, & 4\cdot 1\cdot(4-1), & 6\cdot 2\cdot(6-2), & \cdots \\ 0, & 0, & 0, & 6\cdot 1\cdot(6-1), & \cdots \\ \vdots & \vdots & \vdots & \vdots & \ddots \end{bmatrix}$$

$$\mathbf{B}=4\begin{bmatrix} 0, & \frac{1}{2}\cdot 2, & 0, & \frac{1}{2}\cdot 6, & 0, & \frac{1}{2}\cdot 10, & \cdots \\ 0, & 0, & 4, & 0, & 8, & 0, & \cdots \\ 0, & 0, & 0, & 6, & 0, & 10, & \cdots \\ \vdots & \vdots & \vdots & \vdots & \vdots & \vdots & \ddots \end{bmatrix}$$

$$\mathbf{C}=\frac{1}{2}\begin{bmatrix} 2c_0, & c_2, & c_4, & c_6, & \cdots \\ 2c_2, & 2c_0+c_4, & c_2+c_6, & c_4+c_8, & \cdots \\ 2c_4, & c_2+c_6, & 2c_0+c_8, & c_2+c_{10}, & \cdots \\ \vdots & \vdots & \vdots & \vdots & \ddots \end{bmatrix}$$

The elements of matrix c are related to the coefficients a_j of (2.122) as shown below:

$$\begin{aligned} c_0 &= u^2(\eta_1 - a_0) \\ c_2 &= -u^2 a_2 \\ c_4 &= -u^2 a_4 \\ &\vdots \end{aligned} \tag{2.128}$$

The homogenous matrix equation (2.127) can be solved by using the conventional numerical technique. Once the solution of $R(x)$ is computed from (2.125), the field solution $F(x)$ in the core for $x \le 1$ can be obtained from (2.120). In the cladding for $x > 1$, the refractive index is constant, and $f(x) = 1$. In this case, the general solution of (2.119) is of the form $A_m K_m\left(u\eta^{1/2}x\right)$, where K_m is the modified Bessel function of the second kind and A_m is an arbitrary constant. This field tends to zero as $x \to \infty$.

For the condition of continuity of the field and its first derivative at the core-cladding boundary, the following equation

$$\frac{1}{F(x)}\frac{dF(x)}{dx}=\frac{1}{K_m\left(u\eta^{1/2}x\right)}\frac{dK_m\left(u\eta^{1/2}x\right)}{dx} \tag{2.129}$$

is to be satisfied at $x=1$. Using the properties of the Bessel functions and the relation (2.120), we get

$$\left.\frac{1}{R(x)}\frac{dR(x)}{dx}\right|_{x=1}=-u\eta^{1/2}\left.\frac{K_{m+1}\left(u\eta^{1/2}x\right)}{K_m\left(u\eta^{1/2}x\right)}\right|_{x=1} \tag{2.130}$$

Next, introducing the identities

$\left.\frac{dT_k(x)}{dx}\right|_{x=1}=k^2$ and $T_k(x=1)=1$ together with (2.125) in (2.130), we obtain

$$\frac{4b_2+16b_4+\cdots+q^2b_q}{b_0+b_2+b_4+\cdots+b_q}=-u\eta^{1/2}\frac{K_{m+1}\left(u\eta^{1/2}\right)}{K_m\left(u\eta^{1/2}\right)} \tag{2.131}$$

Finally, the solution of (2.121) is to be found with some appropriate value(s) of the normalized propagation constant $\eta_1=1-\eta$ for which the condition (2.131) will be satisfied. Accordingly, several iterations of solution will be needed to evaluate the permissible value(s) of $\eta_1=\eta_{1,mv}$, where v is the radial mode number representing the number of half periods of the oscillatory field in the core. Thus, the values of the normalized propagation constant relate to various modes of propagation in the optical fibre under analysis.

We now present an alternative numerical technique to solve the wave equation given by (2.121). Let the profile function $f(x)$ in the core for $x\le 1$ be expressed as follows:

$$f(x)=s_0P_0(x)+s_2P_2(x)+\cdots+s_tP_t(x),\quad t \text{ even} \tag{2.132}$$

where $p_j(x)$ are the Ralston polynomials of even degree j, and s_j denote the coefficients of the polynomial series. The Ralston polynomials $P_j(x)$ constitute an orthogonal set over the collection of sampled points $\{x_i\}$. Accordingly, the coefficients s_j of (2.132) are given by

$$s_j=\frac{\mu_j}{\delta_j},\quad \mu_j=\sum_{i=1}^{k}f_iP_j(x_i),\quad \delta_j=\sum_{i=1}^{k}\left[P_j(x_i)\right]^2 \tag{2.133}$$

where f_i are the sampled values of the function at x_i, and k is the number of samples.

The polynomials $P_j(x)$ can be evaluated by the recurrence relation

$$P_j(x_i)=(x_i-\alpha_j)P_{j-1}(x_i)-\gamma_{j-1}P_{j-2}(x_i),\quad j\ge 1$$
$$\text{with } P_0(x_i)=1,\ P_{-1}(x_i)=0$$

and

$$\alpha_j=\frac{1}{\delta_{j-1}}\sum_{i=1}^{k}x_i\left[P_{j-1}(x_i)\right]^2,\quad \gamma_j=\frac{\delta_j}{\delta_{j-1}} \tag{2.134}$$

In (2.132), the order of approximation t is chosen such that the error-variance

$$\sigma_t^2 = \frac{\sum_{i=1}^{k}\left[f_i - \{s_0P_0(x_i) + s_2P_2(x_i) + \cdots + s_tP_t(x_i)\}\right]^2}{k-t-1} \tag{2.135}$$

is either minimum or does not decrease appreciably any further with increasing t. Note that only even degree polynomials are included in (2.132) to satisfy the condition of radial symmetry (2.123) for $n(x)$ at $x = 0$.

Now, let the solution of (2.121) for $x \le 1$ be given by

$$R(x) = d_0P_0(x) + d_2P_2(x) + \cdots + d_lP_l(x), \quad l \text{ even} \tag{2.136}$$

and hence, its first and second order derivatives are

$$\frac{dR(x)}{dx} = d_2P_2^1(x) + d_4P_4^1(x) + \cdots + d_lP_l^1(x), \quad P_0^1(x) = 0$$

and

$$\frac{d^2R(x)}{dx^2} = d_2P_2^2(x) + d_4P_4^2(x) + \cdots + d_lP_l^2(x), \quad P_0^2(x) = 0$$

respectively. The generating polynomials for the first and second order derivatives can be evaluated by the recurrence relations

$$P_j^1(x_i) = P_{j-1}(x_i) + (x_i - \alpha_j)P_{j-1}^1(x_i) - \gamma_{j-1}P_{j-2}^1(x_i), \quad j \ge 2$$
$$\text{with } P_1^1(x_i) = 1, \; P_0^1(x_i) = 0$$

and (2.137)

$$P_j^2(x_i) = 2P_{j-1}^1(x_i) + (x_i - \alpha_j)P_{j-1}^2(x_i) - \gamma_{j-1}P_{j-2}^2(x_i), \quad j \ge 2$$
$$\text{with } P_1^2(x_i) = 0, \; P_0^2(x_i) = 0$$

where α_j and γ_j are as defined in (2.134). Note that for radial symmetry of profile function, the field $R(x)$ will also have radial symmetry, and hence, only even degree polynomials are included in (2.136). The even degree Ralston polynomials are symmetric about $x = 0$, and the odd degree polynomials are anti-symmetric about $x = 0$.

Now, substituting (2.132) and (2.136) in (2.121), and evaluating the equation at the points of sampling $\{x_i; i = 1, 2, \ldots, k\}$, we get the following matrix equation

$$\begin{bmatrix} g[1,0] & g[1,2] & \cdots & g[1,l] \\ g[2,0] & g[2,2] & \cdots & g[2,l] \\ \vdots & \vdots & \ddots & \vdots \\ g[k,0] & g[k,2] & \cdots & g[k,l] \end{bmatrix} \begin{bmatrix} d_0 \\ d_2 \\ \vdots \\ d_l \end{bmatrix} = \mathbf{0} \tag{2.138}$$

where

$$g[i,j] = P_j^2(x_i) + \frac{2m+1}{x_i}P_j^1(x_i) + u^2\left[\eta_1 - \{s_0P_0(x_i) + s_2P_2(x_i) + \cdots + s_tP_t(x_i)\}\right]P_j(x_i)$$
$$i = 1, 2, \ldots, k; \; j = 0, 2, \ldots, l$$

The homogeneous matrix equation (2.138) can be solved by using the conventional numerical technique when $k \geq l/2$. Once the solution $R(x)$ is computed from (2.136), the boundary condition (2.130) at $x = 1$ can be enforced to obtain the following relation

$$\frac{d_2 P_2^1(1) + d_4 P_4^1(1) + \cdots + d_l P_l^1(1)}{d_0 P_0(1) + d_2 P_2(1) + d_4 P_4(1) + \cdots + d_l P(1)} = -u\eta^{1/2} \frac{K_{m+1}\left(u\eta^{1/2}\right)}{K_m\left(u\eta^{1/2}\right)} \tag{2.139}$$

Finally, we have to find the solution of (2.121) with some appropriate value(s) of the normalized propagation constant $\eta_1 = 1 - \eta$ for which (2.139) will be satisfied. Accordingly, several iterations of solution will be needed to get the permissible value(s) of η_1.

2.10.2 Simulation Study: Example 5 (α-law Profile Fibre)

For the purpose of analysis, we consider an α-law profile fibre with the profile function $f(x)$ given by $f(x) = x^{\alpha}, \quad \alpha = 1.885$.

For the method based on Chebyshev polynomial interpolation (CPI), a total of 236 points of $f(x)$ are sampled over the interval $0 < x < 1$, and the integral (2.124) is evaluated by using a quadrature formula for high accuracy. The point $x = 1$ is excluded from the quadrature formula, where the integrand becomes infinite. We compute the field solution $R(x)$ for the LP_{01} and LP_{11} modes corresponding to $m = 0$ and $m = 1$. It turns out the field solution $R(x)$ in each case depends crucially on the highest degree of polynomial (q) included in (2.125) as shown in Figures 2.10 and 2.13.

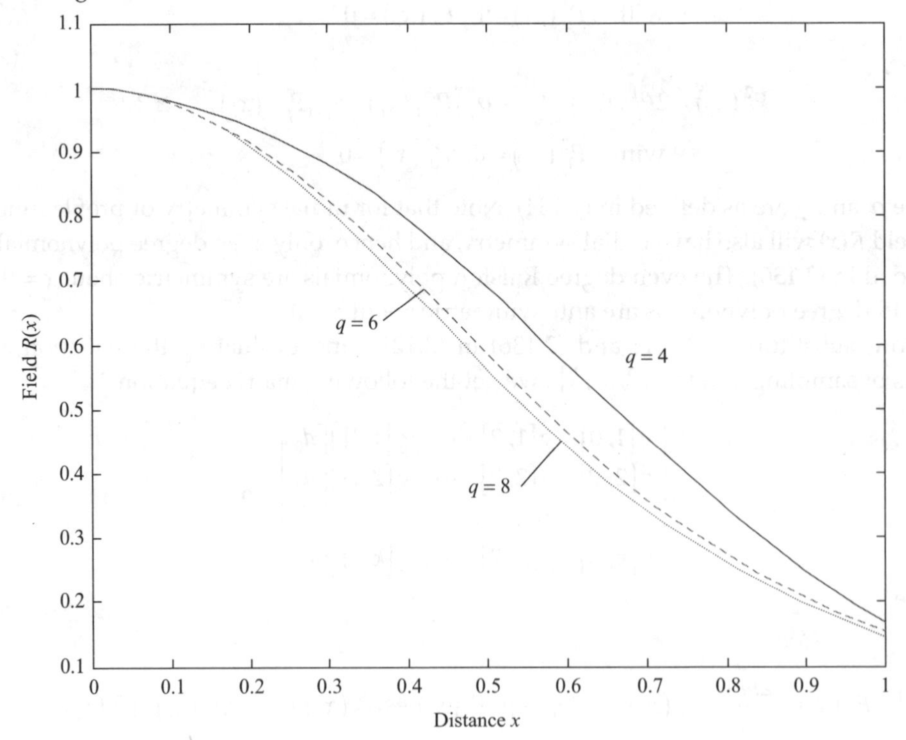

FIGURE 2.10 Plot for the field solution of LP_{01} mode computed by the CPI-based method

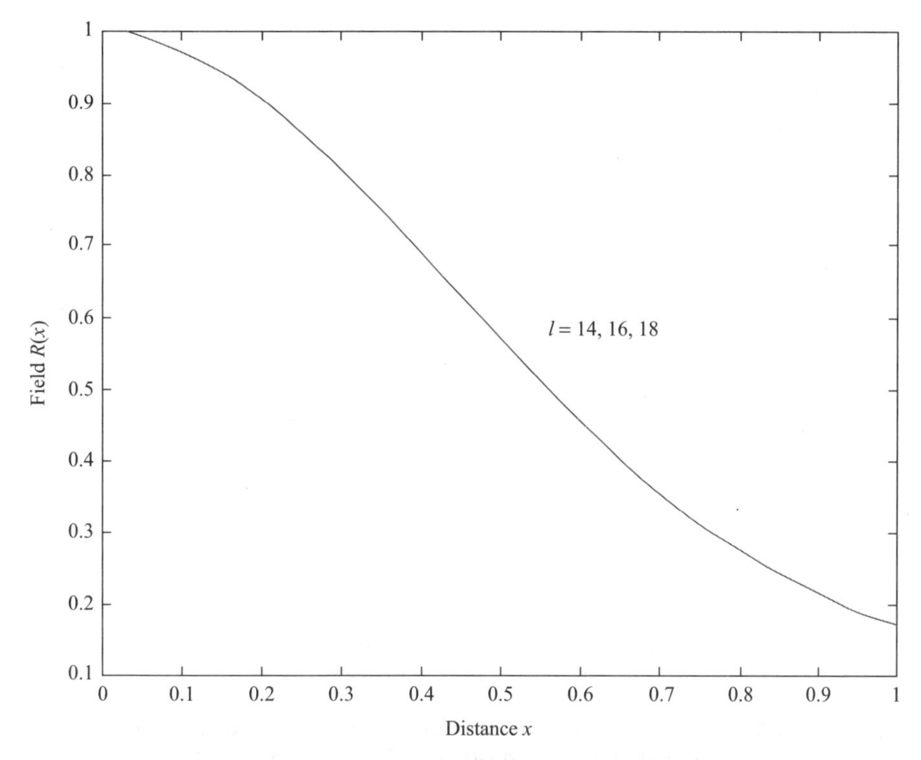

FIGURE 2.11 Plot for the field solution of LP_{01} mode computed by the OPA-based method

For the method based on orthogonal polynomial approximation (OPA), the profile function is sampled at k points in the interval $-1 \le x \le 1$. the number of points k is increased as the normalized frequency u is increased. Note that the highest degree of polynomial (l) to be included in (2.136) can be chosen by the criterion of minimum error-variance. Moreover, it is noticed that the computed field solutions $R(x)$ for the LP_{01} and LP_{11} modes are quite stable when the order of approximation is varied from the optimal order as shown in Figures 2.11 and 2.14.

The field solutions computed by the above two methods for the LP_{01} and LP_{11} modes are plotted in Figures 2.12 and 2.15 for comparison. The results of analysis for the fibre employing the two methods are presented in Table 2.9, where the k-point quadrature formula is used for evaluating the integrals for determining the coefficients.

We now investigate the probable sources of the numerical instability while using the CPI-based method as shown in Figures 2.10 and 2.13. It is to be pointed out that the Chebyshev polynomial series is an infinite-order interpolatory (exact) representation for a continuous function. When this infinite series is truncated at some finite order, the resulting finite series shows the equi-ripple characteristics of the Chebyshev polynomials. This situation may be desirable when we are interested only in the representation for the signal. However, the equi-ripple characteristics can provide very inaccurate derivative values. Therefore, the CPI-based methods may not provide accurate results for the solutions of differential equations. It is to be emphasized that the order of the Chebyshev series cannot be increased arbitrarily because the higher-order coefficients of the series may not be accurate.

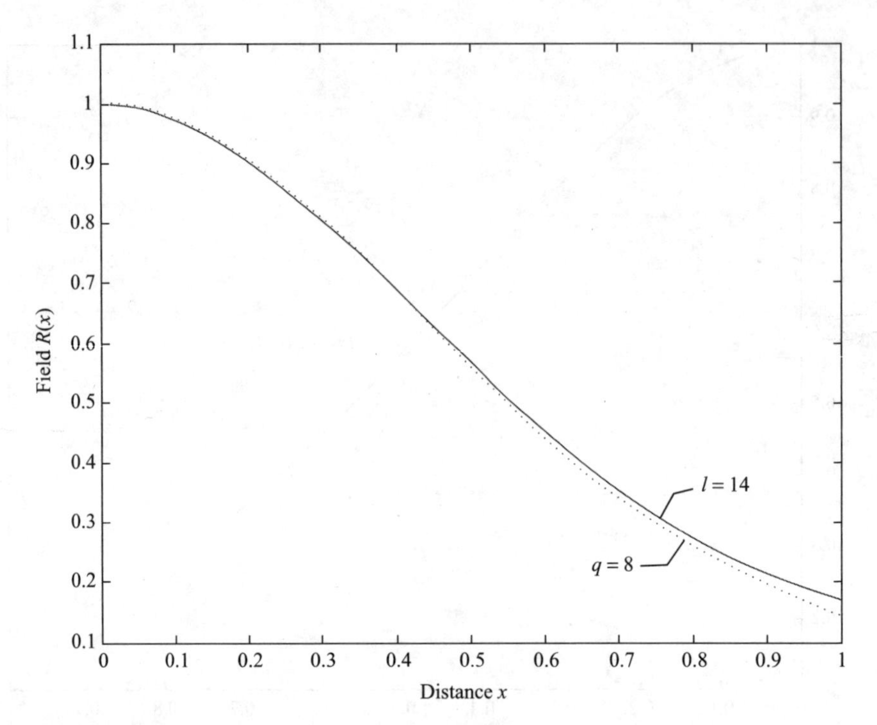

FIGURE 2.12 Plots for the field solutions of LP_{01} mode computed by the two methods: OPA-based (solid); CPI-based (dotted)

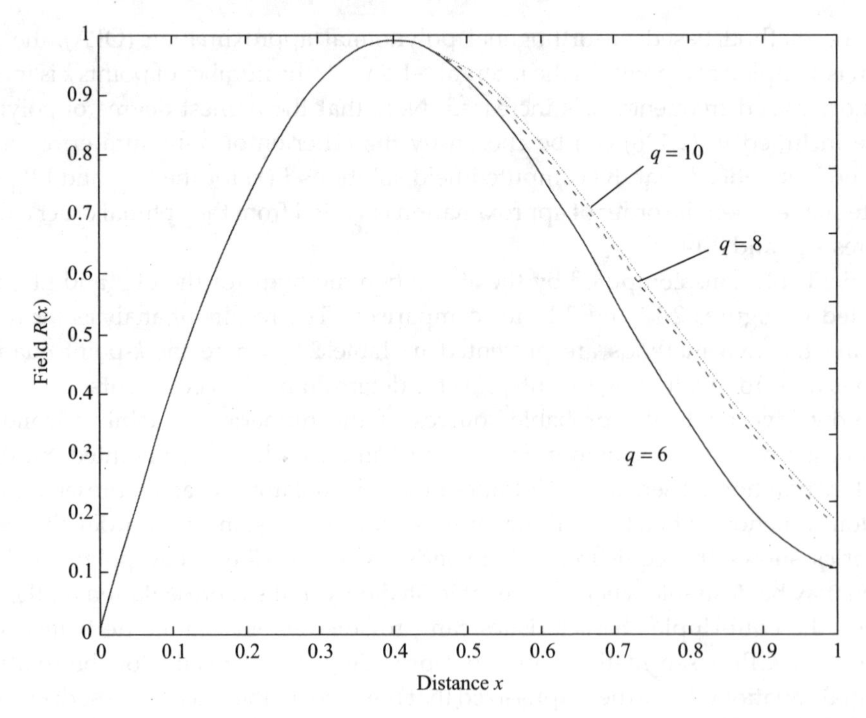

FIGURE 2.13 Plot for the field solution of LP_{11} mode computed by the CPI-based method

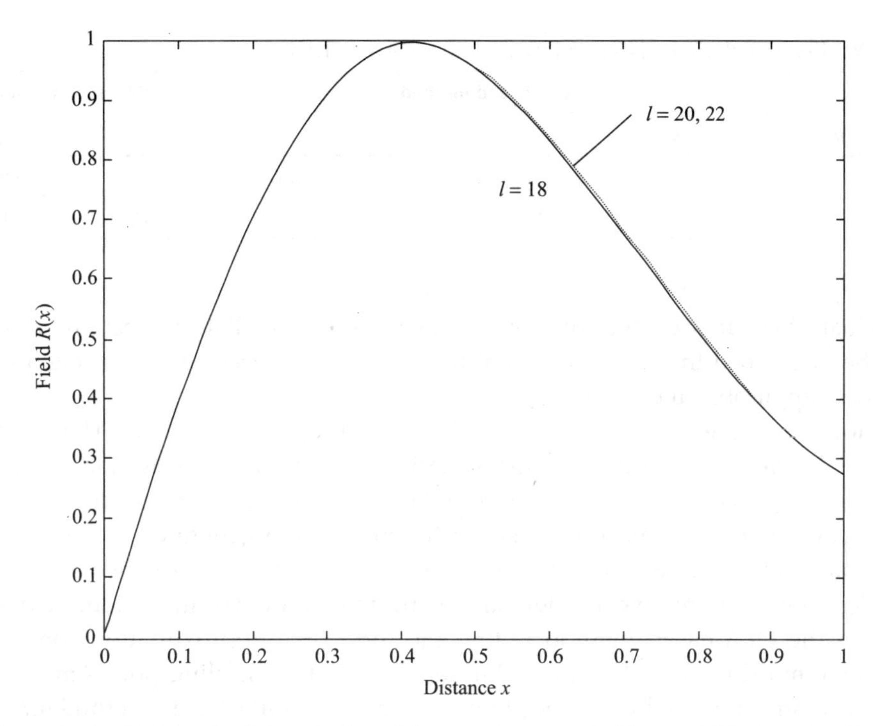

FIGURE 2.14 Plot for the field solution of LP_{11} mode computed by the OPA-based method

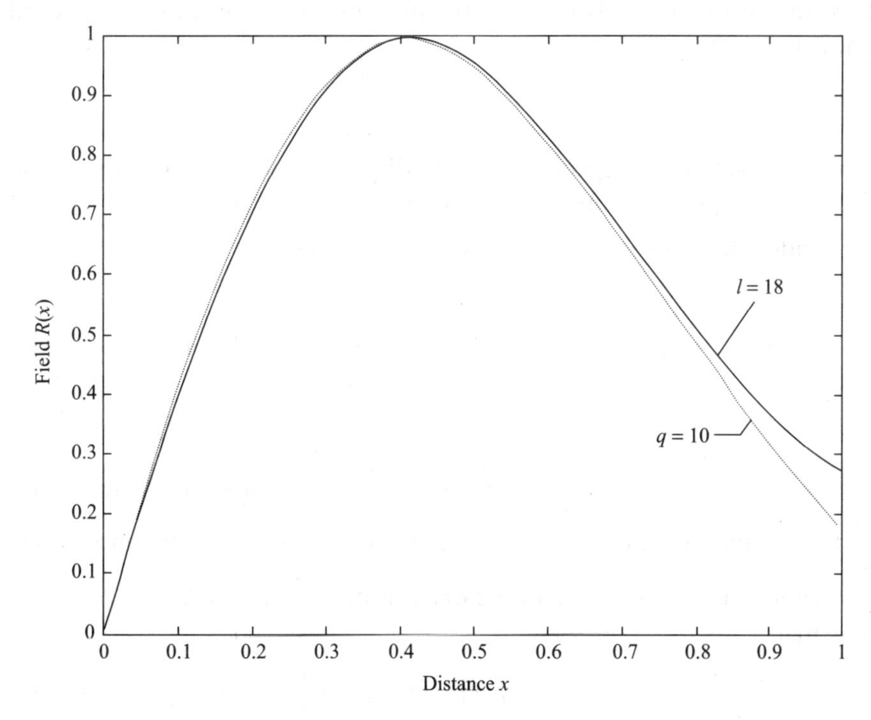

FIGURE 2.15 Plots for the field solutions of LP_{11} mode computed by the two methods: OPA-based (solid); CPI-based (dotted)

TABLE 2.9 Results of analysis for α-law profile fibre, $\alpha = 1.885$

Mode of propagation	u	CPI-based method			OPA-based method		
		k	q	η_1	k	l	η_1
LP_{01}	4	236	8	0.505	21	14	0.505
LP_{11}	6	236	10	0.675	41	18	0.67

The evaluations of these coefficients involve integrations of oscillatory functions. Moreover, it is to be mentioned that the criterion of minimum error-variance for finding a suitable order is not applicable in this case.

We now assess the OPA-based method for analysis of an optical fibre. The orthogonal polynomial series whose order of approximation is chosen by the criterion of minimum error-variance provides an accurate representation for the signal and its derivatives. The Ralston polynomials chosen for signal representation are generated to be orthogonal over an arbitrary collection of sampled points. The evaluation of the coefficients of the series does not involve numerical integration or matrix inversion. A desirable feature of the OPA-based method is that the functional approximation conveniently leads to the multipoint evaluation technique, where the sampling points may be non-uniformly distributed for better sampling efficiency. The OPA-based method for solving differential equations is easy to implement on a computer, and the method provides accurate results, provided sufficient number of sampling points are used in digitalizing the unknown solution.

Problems

Some sample problems are given below. Additional problems can be obtained from standard texts on approximation theory, and numerical analysis.

P2.1 The divided difference of order $k > 1$ of $f(x)$ is defined by

$$f[a_1, \dots, a_k] = \frac{f[a_2, \dots, a_k] - f[a_1, \dots, a_{k-1}]}{a_k - a_1} \text{ with } f[a_i] = f(a_i).$$

Prove that

$$f[a_1, \dots, a_k] = \sum_{i=1}^{k} \frac{f(a_i)}{(a_i - a_1)\cdots(a_i - a_{i-1})(a_i - a_{i+1})\cdots(a_i - a_k)}$$

and thus deduce that the order of the arguments in a divided difference is immaterial.

P2.2 The divided difference of order $k > 1$ of $f(x)$ is defined as in P2.1.
Show that

$$f[a_1, \dots, a_{k-1}, x] = f[a_1, \dots, a_k] + (x - a_k) f[a_1, \dots, a_k, x]$$

and use this result to derive the formula

$$f(x) = f[a_1] + (x - a_1) f[a_1, a_2] + (x - a_1)(x - a_2) f[a_1, a_2, a_3] + \cdots$$
$$+ (x - a_1)(x - a_2) \cdots (x - a_{n-1}) f[a_1, \ldots, a_n] + e(x)$$

where $e(x) = p_n(x) f[a_1, \ldots, a_n, x]$.

P2.3 If the functions $f(\theta) = \sum_{k=0}^{n} (a_k \cos k\theta + b_k \sin k\theta)$ and $g(\theta) = \sum_{k=0}^{n} (\alpha_k \cos k\theta + \beta_k \sin k\theta)$ agree at $(2n + 1)$ points of the interval $[0, 2\pi)$, then show that $a_k = \alpha_k$ for $k = 0, 1, \ldots, n$, and $b_k = \beta_k$ for $k = 1, \ldots, n$.

P2.4 The Tchebycheff polynomials are defined on $[-1, 1]$ as follows:

$$T_n(x) = \cos(n \cos^{-1} x) = \cos(n\theta), \text{ and } U_n(x) = \frac{\sin(n+1)\theta}{\sin\theta}, \quad x = \cos\theta.$$

(i) Prove that $T_0(x) = 1$, $T_1(x) = x$, $T_2(x) = 2x^2 - 1$, $T_3(x) = 4x^3 - 3x$, $T_4(x) = 8x^4 - 8x^2 + 1$, $T_5(x) = 16x^5 - 20x^3 + 5x$, $T_6(x) = 32x^6 - 48x^4 + 18x^2 - 1$.

(ii) Prove by induction that

$$T_n(\cos\theta) = \sum_{k=0}^{n} a_{nk} \cos^k \theta, \text{ and}$$

$$T_{n+1}(x) = 2xT_n(x) - T_{n-1}(x), \quad x = \cos\theta.$$

Similarly, $U_{n+1}(x) = 2xU_n(x) - U_{n-1}(x)$.

(iii) Show that the sequence of Tchebycheff polynomials of first kind $\left\{ \frac{T_0}{\sqrt{2}}, T_1, T_2, \ldots \right\}$ is orthonormal with respect to the inner product $\langle f, g \rangle = \frac{2}{\pi} \int_{-1}^{1} f(x) g(x) \frac{dx}{\sqrt{1 - x^2}}$, and the polynomials of the second kind $\left\{ \sqrt{\frac{2}{\pi}} U_n \right\}$ are orthonormal with respect to the inner product $\langle f, g \rangle = \int_{-1}^{1} f(x) g(x) \sqrt{1 - x^2}\, dx$.

P2.5 Find the n zeros of the Chebyshev polynomial $T_n(x)$ in the interval $[-1, 1]$. Find the $(n + 1)$ points of extrema. Prove that of all polynomials of degree n with coefficients of x^n equal to 1, the Chebyshev polynomial of degree n multiplied by $\frac{1}{2^{n-1}}$ oscillates with minimum maximum amplitude on the interval $[-1, 1]$.

P2.6 The n-order all-pole Chebyshev low-pass filter is designed with the squared-magnitude frequency response given by

$$|H(\Omega)|^2 = \frac{1}{1 + \varepsilon^2 T_n^2(\Omega)}$$

where ε^2 is a small positive number, and $T_n(\Omega)$ is the Chebyshev polynomial of degree n expressed as

$$T_n(\Omega) = \begin{cases} \cos\left(n\cos^{-1}\Omega\right), & |\Omega| \le 1 \\ \cosh\left(n\cosh^{-1}\Omega\right), & |\Omega| > 1 \end{cases}$$

Show that the filter is the implementation of minimization of maximum error between the frequency response of an ideal low-pass filter and the frequency response of a realizable n-order all-pole filter. Some Chebyshev polynomials are given in P2.4 (i). Explain with a plot of the frequency response.

P2.7 Find the linear polynomial $p_1(x) = a_1 + b_1x$ and quadratic polynomial $p_2(x) = a_2 + b_2x + c_2x^2$ that best approximate the function $f(x) = x^3$ on the interval [0,1] in the least squares sense. The inner product between two functions is defined to be $(f, p) = \int_0^1 f(x)p(x)dx$. Comment on why the values of a_1, a_2 and b_1, b_2 are not same.

P2.8 The Rademacher functions R_n are defined in the following way: Given n, divide [0,1) into 2^n subintervals $\left[0, \frac{1}{2^n}\right), \left[\frac{1}{2^n}, \frac{2}{2^n}\right), \ldots$, and define R_n to be alternatively +1 and −1 in these successive intervals. Prove that these functions form an orthonormal system with inner product $\int_0^1 f(x)g(x)dx$.

P2.9 Suppose it is known that the impulse response of a system is of the form $h(t) = c_0 + c_1\exp(\alpha_1 t) + c_2\exp(\alpha_2 t)\sin(\omega t + \theta)$ with known parameters α_1, α_2, ω, and θ, and unknown coefficients c_0, c_1 and c_2.
Describe a method of determining the unknown coefficients.
Next, suppose that all the parameters of the impulse response are unknowns. Describe the methodology of determining all the parameters of the impulse response.

Bibliography

The list is indicative, not exhaustive. Similar references can be searched in the library.

Atkinson, K. A. 1988. *An Introduction to Numerical Analysis*. 2nd ed., New York: John Wiley.

Candy, J. V. 1986. *Signal Processing: The Model-Based Approach*. New York: McGraw-Hill.

Carnahan, B., H. A. Luther, and J. O. Wilkes. 1969. *Applied Numerical Methods*. New York: John Wiley.

Cheney, E. W. 1966. *Approximation Theory*, New York: McGraw-Hill.

Clark, J. J., M. R. Palmer, and P. D. Lawrence. 1985. 'A Transformation Method for the Reconstruction of Functions from Nonuniformly Spaced Samples.' *IEEE Trans Acoustic Speech Signal Process* ASSP-33: 1151–65.

Eveleigh, V. W. 1972. *Introduction to Control Systems Design*. New York: McGraw-Hill.

Farebrother, R. W. 1988. *Linear Least Squares Computations*. New York: Marcel Dekker.

Glashoff, K. S.-A. 1983. *Gustafson, Linear Optimization and Approximation*. New York: Springer-Verlag.

Gowar, J. 1984. *Optical Communication Systems*. London: Prentice-Hall.

Hildebrand, F. B. 1956. *Introduction to Numerical Analysis*. New York: McGraw-Hill.

Kay, S. M. 1988. *Modern Spectral Estimation: Theory and Application*. Englewood Cliffs, NJ: Prentice Hall.

Lorentz, G. G. 1966. *Approximation of Functions*. New York: Holt Rinehart Winston.

Pachner, J. 1984. *Handbook of Numerical Analysis Applications*. New York: McGraw-Hill.

Powell, M. J. D. 1981. *Approximation Theory and Methods*. Cambridge: Cambridge University Press.

Prony, R. 1975. 'Essai éxperimental et analytique ....' *J l'École Polytechn* 1 (22): 24–76.

Ralston, A., and P. Rabinowitz. 1978. *A First Course in Numerical Analysis*. 2nd ed., Singapore: McGraw-Hill.

Rice, J. R. 1964. *The Approximation of Functions: Linear Theory*. 1 vols. Reading, MA: Addison-Wesley.

———. 1969. *The Approximation of Functions: Non-linear and Multivariate Theory*. 2 vols. Reading, MA: Addison-Wesley.

Ricketts, L. W., J. E. Bridges, and J. Miletta. 1976. *EMP Radiation and Protective Techniques*. New York: John Wiley.

Rivlin, T. J. 2003. *An Introduction to the Approximation of Functions*. Mineola, NY: Dover.

Rutishauser, H. 1990. *Lectures on Numerical Mathematics*. Boston: Birkhauser.

Sircar, P. 1987. 'Accurate parameter estimation of damped sinusoidal signals sampled at non-uniform spacings.' Ph.D. dissertation, Syracuse, NY: Syracuse University.

Sircar, P., and V. Arvind. 1997. 'Numerical Analysis of Optical Fibers with Arbitrary Refractive Index Profile.' *J. Electromagn Waves Appl* 11: 1213–27.

Sircar, P., and N. Jethi. 2006. 'Methods for Analysis of Optical Fibers with Arbitrary Refractive Index Profile.' *Opt Eng* 45 (12): 5003–1–5003–7.

Sircar, P., and T. K. Sarkar. 1988. 'System Identification from Nonuniformly Spaced Signal Measurements.' *Signal Process* 14: 253–68.

Szego, G. 1967. *Orthogonal Polynomials*. 3rd ed., Providence, RI: American Mathematical Society.

Timan, A. F. 1994. *Theory of Approximation of Functions of a Real Variable*. Mineola, NY: Dover.

Walsh, J. L. 1965. *Interpolation and Approximation by Rational Functions in the Complex Domain*. 4th ed., Providence, RI: American Mathematical Society.

Watson, G. A. 1980. *Approximation Theory and Numerical Methods*. Chichester: John Wiley.

Weaver, H. J. 1983. *Applications of Discrete and Continuous Fourier Analysis*. New York: John Wiley.

Wilks, S. 1962. *Mathematical Statistics*. New York: John Wiley.

3 Generalized Inverse

3.1 Simultaneous Linear Equations

Let m linear equations in n unknowns be expressed in a matrix equation of the form

$$\mathbf{A}\mathbf{x} = \mathbf{b} \tag{3.1}$$

where $\mathbf{A}$ is the known system matrix of dimension $m \times n$, $\mathbf{b}$ is the known observation vector of dimension $m \times 1$, and $\mathbf{x}$ is the unknown solution vector of dimension $n \times 1$. Note that a matrix is a linear mapping from one finite-dimensional linear space into another finite-dimensional linear space, and a point in the space is represented by a complex vector which is an ordered set of complex numbers.

To investigate the availability of a solution vector of the matrix equation, we check the rank conditions. Let the ranks of the matrix $\mathbf{A}$ and the augmented matrix $(\mathbf{A}, \mathbf{b})$ be given respectively as $\text{rank}(\mathbf{A}) = k$, $\text{rank}(\mathbf{A}, \mathbf{b}) = l$.

We consider various cases which may arise in solving the above matrix equation:

Case 1 $k = l$

This condition indicates that some solutions of the linear set of equations will exist.

Moreover, if it is given

(I) $k = n \le m$,

then, it is possible to find a unique solution. Instead, if it is found

(II) $k < n$,

then, there exists an infinite number of solutions. We find a unique solution by minimizing the norm of the solution vector $\|\mathbf{x}\|_2 = \left(|x_1|^2 + |x_2|^2 + \cdots + |x_n|^2\right)^{1/2}$. This is the minimum-norm solution of the matrix equation.

Case 2 $k < l$,

This condition indicates that we have an inconsistent set of equations. Since the vector $\mathbf{b}$ does not belong to the column-space of the matrix $\mathbf{A}$, no linear combinations of the column vectors of $\mathbf{A}$ can be equal to $\mathbf{b}$. Therefore, no solution of the matrix equation will exist. However, if we change the notion of solution to the vector $\mathbf{x}$ which minimizes the norm of the residue vector,

$$\|\mathbf{r}\|_2 = \|\mathbf{A}\mathbf{x} - \mathbf{b}\|_2 \tag{3.2}$$

then, we obtain the least-squares solution(s).

Furthermore, if we have

(I) $k = n \leq m$,

then, it is possible to find a unique least-squares solution. Instead, if we find

(II) $k < n$,

then, there exists an infinite number of least-squares solutions. The minimum-norm least-squares solution is obtained by minimizing, in addition, the norm of the solution vector $\|\mathbf{x}\|_2$, and this is how we obtain a unique solution in this case.

3.2 Generalized Inverse of Matrix

In the matrix equation $\mathbf{Ax} = \mathbf{b}$, if $\mathbf{A}$ is a square non-singular matrix of dimension $n \times n$, then the solution is given by

$$\mathbf{x} = \mathbf{A}^{-1}\mathbf{b} \tag{3.3}$$

where $\mathbf{A}^{-1}$ is the inverse of $\mathbf{A}$, and $\mathbf{A}^{-1}\mathbf{A} = \mathbf{AA}^{-1} = \mathbf{I}_n$, $\mathbf{I}_n$ being the identity matrix of dimension n.

When the matrix $\mathbf{A}$ is rectangular full-rank of dimension $m \times n$, then the solution is given by

$$\mathbf{x} = \mathbf{A}^{\#}\mathbf{b} \tag{3.4}$$

where $\mathbf{A}^{\#}$ is the pseudo-inverse of $\mathbf{A}$. Given the conditions, $m > n$, and $\text{rank}(\mathbf{A}) = \min(m, n) = n$, we obtain the least-squares solution $\mathbf{x}$ by computing the pseudo-inverse

$$\mathrm{A}^{\#} = (\mathbf{A}^H\mathbf{A})^{-1}\mathbf{A}^H \tag{3.5}$$

where ^{H} stands for Hermitian or complex conjugate transpose. On the other hand, if $m < n$, and $\text{rank}(\mathbf{A}) = \min(m, n) = m$, then one finds the minimum-norm solution $\mathbf{x}$ by using the pseudo-inverse

$$\mathrm{A}^{\#} = \mathbf{A}^H(\mathbf{AA}^H)^{-1} \tag{3.6}$$

Let us now consider the cases where the matrix $\mathbf{A}$ is square singular, or it is rectangular rank-deficient. In these cases, we must find techniques to compute the generalized inverse $\mathbf{A}^+$ of the matrix, and the desired solution will be given by

$$\mathbf{x} = \mathbf{A}^+\mathbf{b} \tag{3.7}$$

The generalized inverse is defined by the following four Moore–Penrose properties:

Properties

(1) $\mathbf{AA}^+\mathbf{A} = \mathbf{A}$, (3.8)

$\mathbf{A}^+$ is a 1-inverse or inner inverse of $\mathbf{A}$; Moreover,

(2) $(\mathbf{AA}^+)^H = \mathbf{AA}^+$, (3.9)

$\mathbf{A}^+$ is a Hermitian 1-inverse of $\mathbf{A}$.

(3) $\mathbf{A}^+\mathbf{AA}^+ = \mathbf{A}^+$, (3.10)

$\mathbf{A}^+$ is a 2-inverse or outer inverse of $\mathbf{A}$; In addition,

(4) $(\mathbf{A}^+\mathbf{A})^H = \mathbf{A}^+\mathbf{A}$, (3.11)

$\mathbf{A}^+$ is a Hermitian 2-inverse of $\mathbf{A}$.

Theorem 3.1

If $\mathbf{A}^+$ is a Hermitian 1-inverse of $\mathbf{A}$, then $\mathbf{x} = \mathbf{A}^+\mathbf{b}$ is the least-squares solution. This implication is both necessary and sufficient.

Theorem 3.2

On the other hand, if $\mathbf{A}^+$ is a Hermitian 2-inverse of $\mathbf{A}$, then $\mathbf{x} = \mathbf{A}^+\mathbf{b}$ is the minimum-norm solution. This implication is again necessary and sufficient.

Note that the first theorem provides the solution in the least-squares sense whether it is a consistent set of equations or an inconsistence set of equations. For the consistent set of equations, the minimum residue is a zero vector. The second theorem provides the solution vector which has the minimum norm. Remember that the minimum-norm solution is sought when more than one solution is feasible.

When the above two theorems are applied simultaneously, we get the following theorem:

Theorem 3.3

If $\mathbf{A}^+$ is both a Hermitian 1-inverse and a Hermitian 2-inverse of $\mathbf{A}$, then $\mathbf{x} = \mathbf{A}^+\mathbf{b}$ is the minimum-norm least-squares solution. Once again, this implication is both necessary and sufficient.

When the properties (1)–(4) are satisfied, $\mathbf{A}^+$ is the Moore–Penrose generalized inverse of $\mathbf{A}$. According to the third theorem, we compute the minimum-norm least-squares solution in this case. It is easy to verify that the regular inverse $\mathbf{A}^{-1}$, or each of the pseudo-inverses $\mathbf{A}^{\#}$ used for computation of the least-squares or minimum-norm solution satisfies the Moore–Penrose properties. Hence, each of these matrices is the generalized inverse of $\mathbf{A}$ in the respective cases. Convincingly, if $\mathbf{A}^+$ is the generalized inverse of $\mathbf{A}$, then $\mathbf{A}$ is the generalized inverse of $\mathbf{A}^+$ by the reciprocity of the Moore–Penrose properties.

Theorem 3.4

There exists the Moore–Penrose generalized inverse $\mathbf{A}^+$ for every real (or complex) matrix $\mathbf{A}$, and the generalized inverse $\mathbf{A}^+$ is unique.

The last theorem guarantees the existence and uniqueness of the generalized inverse. Based on the nature of the matrix $\mathbf{A}$, it is always possible to find a method to compute the generalized inverse $\mathbf{A}^+$. Moreover, it suffices to have only one implementation of the generalized inverse because even if there are two or more methods to compute the generalized inverse, it is the unique generalized inverse which will be computed in every implementation.

3.3 Generalized Inverse by Orthogonal Bordering

To compute the Bjerhammar generalized inverse of a matrix $\mathbf{A}$ which may be rectangular full-rank, rectangular rank-deficient or square singular, the matrix $\mathbf{A}$ is first enclosed in a larger square matrix $\mathbf{B}$ which is non-singular. The regular inverse of $\mathbf{B}$ can then be computed, and the generalized inverse of **A** is obtained in the transposed partition of $\mathbf{B}^{-1}$. To ensure the non-singularity of **B**, the enclosing of **A** is completed by appending new columns and rows which are orthogonal to the columns and rows, respectively, of the primary matrix.

We consider the following three cases, where the first two refer to the standard least-squares and minimum-norm solutions, and the third case is the general one.

Case 1

When the matrix **A** is rectangular full-rank, and it represents an over-determined set of linear equations, i.e., the determinant $|\mathbf{A}^H\mathbf{A}| \neq 0$, we form the square matrix **B** by column bordering,

$$\mathbf{B} = [\mathbf{A} \mid \mathbf{C}] \tag{3.12}$$

such that **B** is non-singular, i.e., $|\mathbf{B}| \neq 0$. It is necessary that the matrix **C** is full-rank, and the column-spaces of **A** and **C** are orthogonal to each other, i.e., $|\mathbf{C}^H\mathbf{C}| \neq 0$ and $\mathbf{A}^H\mathbf{C} = \mathbf{0}$, respectively.

Utilizing the following identity of inversion,

$$\mathbf{B}^{-1} = (\mathbf{B}^H\mathbf{B})^{-1}\mathbf{B}^H, \tag{3.13}$$

the partitioned inverse $\mathbf{B}^{-1}$ can be expressed in the following way,

$$\mathbf{B}^{-1} = \left[\begin{array}{c} \mathbf{A}^+ \\ \hline \mathbf{C}^+ \end{array}\right] = \begin{bmatrix} \mathbf{A}^H\mathbf{A} & \mathbf{0} \\ \mathbf{0} & \mathbf{C}^H\mathbf{C} \end{bmatrix}^{-1} \begin{bmatrix} \mathbf{A}^H \\ \mathbf{C}^H \end{bmatrix} \tag{3.14}$$

Thus we obtain the generalized inverse $\mathbf{A}^+$ as given below,

$$\mathbf{A}^+ = (\mathbf{A}^H\mathbf{A})^{-1}\mathbf{A}^H, \tag{3.15}$$

which can be employed to compute the least-squares solution $\mathbf{x}_{LS}$ of the matrix equation $\mathbf{A}\mathbf{x} \approx \mathbf{b}$.

Case 2

In this case, the matrix **A** is rectangular full-rank, and it represents an under-determined set of linear equations, i.e., the determinant $|\mathbf{A}\mathbf{A}^H| \neq 0$. We form the square matrix **B** by row bordering,

$$\mathbf{B} = \left[\begin{array}{c} \mathbf{A} \\ \hline \mathbf{D} \end{array}\right] \tag{3.16}$$

such that **B** is non-singular, i.e., $|\mathbf{B}| \neq 0$. It is necessary now that the matrix **D** is full-rank, and the row-spaces of **A** and **D** are orthogonal to each other, i.e., $|\mathbf{D}\mathbf{D}^H| \neq 0$ and $\mathbf{A}\mathbf{D}^H = \mathbf{0}$, respectively.

We utilize the following identity of inversion here,

$$\mathbf{B}^{-1} = \mathbf{B}^H(\mathbf{B}\mathbf{B}^H)^{-1}, \tag{3.17}$$

and express the partitioned inverse $\mathbf{B}^{-1}$ as follows,

$$\mathbf{B}^{-1} = \left[\mathbf{A}^{+} \,|\, \mathbf{D}^{+}\right] = \left[\mathbf{A}^{H}\mathbf{D}^{H}\right]\begin{bmatrix} \mathbf{A}\mathbf{A}^{H} & \mathbf{0} \\ \mathbf{0} & \mathbf{D}\mathbf{D}^{H} \end{bmatrix}^{-1}. \tag{3.18}$$

In this way, we get the generalized inverse $\mathbf{A}^{+}$ as shown below,

$$\mathbf{A}^{+} = \mathbf{A}^{H}(\mathbf{A}\mathbf{A}^{H})^{-1}. \tag{3.19}$$

The generalized inverse $\mathbf{A}^{+}$ can be used to compute the minimum-norm solution $\mathbf{x}_{MN}$ of the matrix equation $\mathbf{Ax} = \mathbf{b}$. Note that an infinite number of solutions is feasible here since the matrix equation is under-determined, and we choose the minimum-norm solution.

Case 3

We now consider the general case where the matrix $\mathbf{A}$ is square singular or rectangular rank-deficient. We have the determinants $|\mathbf{A}^{H}\mathbf{A}| = 0$ and $|\mathbf{A}\mathbf{A}^{H}| = 0$ simultaneously. The square matrix $\mathbf{B}$ is formed in this case by both column and row bordering,

$$\mathbf{B} = \begin{bmatrix} \mathbf{A} & \mathbf{C} \\ \mathbf{D} & \mathbf{0} \end{bmatrix}, \tag{3.20}$$

such that $\mathbf{B}$ is non-singular, i.e., $|\mathbf{B}| \neq 0$.

In order to ensure that $\mathbf{B}$ has full column-rank, it is necessary that the matrix $\mathbf{C}$ is full-rank and the column-spaces of $\mathbf{A}$ and $\mathbf{C}$ are orthogonal to each other. Similarly, $\mathbf{B}$ has full row-rank when the matrix $\mathbf{D}$ is full-rank and the row-spaces of $\mathbf{A}$ and $\mathbf{D}$ are orthogonal to each other.

By utilizing the matrix identity $\mathbf{B}^{-1} = (\mathbf{B}^{H}\mathbf{B})^{-1}\mathbf{B}^{H}$, the partitioned inverse $\mathbf{B}^{-1}$ is written as follows,

$$\mathbf{B}^{-1} = \begin{bmatrix} \mathbf{A}^{+} & \mathbf{D}^{+} \\ \mathbf{C}^{+} & \mathbf{0} \end{bmatrix} = \begin{bmatrix} \mathbf{A}^{H}\mathbf{A} + \mathbf{D}^{H}\mathbf{D} & \mathbf{0} \\ \mathbf{0} & \mathbf{C}^{H}\mathbf{C} \end{bmatrix}^{-1} \begin{bmatrix} \mathbf{A}^{H} & \mathbf{D}^{H} \\ \mathbf{C}^{H} & \mathbf{0} \end{bmatrix}, \tag{3.21}$$

which provides the first expression of the Bjerhammar generalized inverse $\mathbf{A}^{+}$ as given below,

$$\mathbf{A}^{+} = (\mathbf{A}^{H}\mathbf{A} + \mathbf{D}^{H}\mathbf{D})^{-1}\mathbf{A}^{H}, \tag{3.22}$$

where the determinant $|\mathbf{A}^{H}\mathbf{A} + \mathbf{D}^{H}\mathbf{D}| \neq 0$.

On the other hand, when we use the matrix identity $\mathbf{B}^{-1} = \mathbf{B}^{H}(\mathbf{B}\mathbf{B}^{H})^{-1}$, and express the partitioned inverse $\mathbf{B}^{-1}$ as follows,

$$\mathbf{B}^{-1} = \begin{bmatrix} \mathbf{A}^{+} & \mathbf{D}^{+} \\ \mathbf{C}^{+} & \mathbf{0} \end{bmatrix} = \begin{bmatrix} \mathbf{A}^{H} & \mathbf{D}^{H} \\ \mathbf{C}^{H} & \mathbf{0} \end{bmatrix} \begin{bmatrix} \mathbf{A}\mathbf{A}^{H} + \mathbf{C}\mathbf{C}^{H} & \mathbf{0} \\ \mathbf{0} & \mathbf{D}\mathbf{D}^{H} \end{bmatrix}^{-1}, \tag{3.23}$$

we get the second expression of the Bjerhammar generalized inverse $\mathbf{A}^+$ as shown below,

$$\mathbf{A}^+ = \mathbf{A}^H(\mathbf{A}\mathbf{A}^H + \mathbf{C}\mathbf{C}^H)^{-1}, \tag{3.24}$$

where the determinant $\left|\mathbf{A}\mathbf{A}^H + \mathbf{C}\mathbf{C}^H\right| \neq 0$.

Interestingly it can be argued that since by construction the matrix **B** is non-singular, if only one of the dummy matrices **C** or **D** is changed, then the two expressions of the Bjerhammar generalized inverse must still provide identical results. This fact proves that the orthogonal bordering technique gives one unique inverse for any **A** independent of the choice of bordering matrices. Moreover, since the computation of the generalized inverse is carried out by means of the construction of a non-singular matrix **B**, the existence of the generalized inverse for every matrix **A** is verified.

Indeed by utilizing the two orthogonal conditions as needed in this case, e.g., $\mathbf{A}^H\mathbf{C} = \mathbf{0}$, $\mathbf{A}\mathbf{D}^H = \mathbf{0}$, it is not difficult to prove that both the expressions of the Bjerhammar generalized inverse satisfy the four Moore–Penrose properties, and hence each of them is identical to the unique generalized inverse of the matrix **A**.

3.4 Generalized Inverse by Matrix Factorization

For computing the generalized inverse of the matrix **A** of dimension $m \times n$ and rank k, the matrix is first factorized in the form given below,

$$\mathbf{A} = \mathbf{B}\mathbf{C} \tag{3.25}$$

where **B** is a matrix of dimension $m \times k$, **C** is of dimension $k \times n$, and **B**, **C** are each of rank k. Without any loss of generality, it is assumed that the first k columns of **A** are linearly independent.

The generalized inverse $\mathbf{A}^+$ of the matrix **A** is then computed by utilizing the pseudo-inverses of **B** and **C**,

$$\mathbf{A}^+ = \mathbf{C}^+\mathbf{B}^+ \tag{3.26}$$

where $\mathbf{C}^+ = \mathbf{C}^H(\mathbf{C}\mathbf{C}^H)^{-1}$ and $\mathbf{B}^+ = (\mathbf{B}^H\mathbf{B})^{-1}\mathbf{B}^H$.

Note that the pseudo-inverses $\mathbf{C}^+$ and $\mathbf{B}^+$ are the matrices employed for computing the minimum-norm and least-squares solutions respectively, and the matrix equation $\mathbf{Ax} = \mathbf{b}$ is now effectively being solved in the following two steps,

$$\mathbf{Bu} = \mathbf{b} \tag{3.27}$$

and

$$\mathbf{Cx} = \mathbf{u} \tag{3.28}$$

where the vector **u** is the intermediate solution.

We present here two versions of the above procedure of full-rank factorization.

Process 1

The matrix **A** is factorized as follows,

$$\mathbf{A} = \mathbf{LU} \tag{3.29}$$

where **L** is a lower trapezoidal matrix with units on the principal diagonal, and **U** is an upper trapezoidal matrix.

The generalized inverse $\mathbf{A}^+$ is computed by employing the pseudo-inverses $\mathbf{U}^+$ and $\mathbf{L}^+$,

$$\mathbf{A} = \mathbf{U}^+\mathbf{L}^+ = \mathbf{U}^H(\mathbf{U}\mathbf{U}^H)^{-1}(\mathbf{L}^H\mathbf{L})^{-1}\mathbf{L}^H \tag{3.30}$$

The **LU**-decomposition of the matrix as shown above is carried out by various methods based on Gauss eliminations.

Process 2

Alternatively, the matrix **A** is factorized as given below,

$$\mathbf{A} = \mathbf{QR} \tag{3.31}$$

where **Q** is a matrix with orthonormal columns satisfying $\mathbf{Q}^H\mathbf{Q} = \mathbf{I}_k$, and **R** is a upper trapezoidal matrix.

In this case, the generalized inverse $\mathbf{A}^+$ is determined by the following expression,

$$\mathbf{A}^+ = \mathbf{R}^H\left(\mathbf{R}\mathbf{R}^H\right)^{-1}\mathbf{Q}^H. \tag{3.32}$$

The **QR**-factorization of the matrix as needed here is obtained by methods based on Householder transformation or Gram-Schmidt orthogonalization.

3.4.1 Singular Value Decomposition

In an approach related to the modal analysis of matrices, the matrix **A** is decomposed as follows,

$$\mathbf{A} = \mathbf{U}\Sigma\mathbf{V}^H \tag{3.33}$$

where **U** is a matrix of dimension $m \times k$, **V** is a matrix of dimension $n \times k$, both are having orthonormal columns, and Σ is a diagonal matrix of dimension k with positive diagonal elements known as the singular values.

Utilizing the above decomposition of the matrix, the generalized inverse $\mathbf{A}^+$ is computed in this case as shown below,

$$\mathbf{A}^+ = \mathbf{V}\Sigma^{-1}\mathbf{U}^H \tag{3.34}$$

3.5 Generalized Inverse by Matrix Partitioning

The technique of computing the generalized inverse by matrix partitioning is closely related to the presented technique of computing the same by full-rank factorization of the system matrix. We consider here, case by case, the full-rank over-determined and underdetermined systems, and then the general one where the system matrix is rank-deficient.

Case 1 The determinant $\left|\mathbf{A}^H\mathbf{A}\right| \neq 0$
Without any loss of generality, the matrix $\mathbf{A}$ of dimension $m \times n$ is partitioned in the form,

$$\mathbf{A} = \begin{bmatrix} \mathbf{A}_1 \\ \mathbf{A}_2 \end{bmatrix} = \begin{bmatrix} \mathbf{I}_n \\ \mathbf{W} \end{bmatrix} \mathbf{A}_1 \tag{3.35}$$

where $\mathbf{W} = \mathbf{A}_2\mathbf{A}_1^{-1}$, and it is assumed that the first n rows of $\mathbf{A}$ are linearly independent, i.e., $\mathbf{A}_1$ is non-singular. The generalized inverse $\mathbf{A}^+$ is given by

$$\mathbf{A}^+ = \mathbf{A}_1^{-1}\left(\mathbf{I}_n + \mathbf{W}^H\mathbf{W}\right)^{-1}\begin{bmatrix} \mathbf{I}_n & \mathbf{W}^H \end{bmatrix} \tag{3.36}$$

where the determinant $\left|\mathbf{I}_n + \mathbf{W}^H\mathbf{W}\right| \neq 0$.

Now using the matrix equation, $\mathbf{Ax} = \mathbf{b}$, and the partition of the excitation vector, $\mathbf{b}^T = \begin{bmatrix} \mathbf{b}_1^T & \mathbf{b}_2^T \end{bmatrix}$ where ^{T} stands for transpose, it is easy to show that the least-squares solution $\mathbf{x}$ is computed in this case by solving the matrix equation

$$\mathbf{A}_1\mathbf{x} = \mathbf{b}_1 + \left(\mathbf{I}_n + \mathbf{W}^H\mathbf{W}\right)^{-1}\mathbf{W}^H\left(\mathbf{b}_2 - \mathbf{W}\mathbf{b}_1\right) \tag{3.37}$$

It is apparent that if we have a consistent set of linear equations, and $\mathbf{b}_2 = \mathbf{W}\mathbf{b}_1$, then we get the solution by solving the square system $\mathbf{A}_1\mathbf{x} = \mathbf{b}_1$. However, if the equations are not consistent, then there will be a correction term in the excitation as shown in the complete matrix equation presented above.

Case 2 The determinant $\left|\mathbf{A}\mathbf{A}^H\right| \neq 0$
In this case without any loss of generality, the matrix $\mathbf{A}$ of dimension $m \times n$ is partitioned in the form,

$$\mathbf{A} = \begin{bmatrix} \mathbf{A}_1 & \mathbf{A}_2 \end{bmatrix} = \mathbf{A}_1\begin{bmatrix} \mathbf{I}_m & \mathbf{Z} \end{bmatrix} \tag{3.38}$$

where $\mathbf{Z} = \mathbf{A}_1^{-1}\mathbf{A}_2$, and it is assumed that the first m columns of $\mathbf{A}$ are linearly independent, i.e., $\mathbf{A}_1^{-1}$ exists. The generalized inverse $\mathbf{A}^+$ is given by

$$\mathbf{A}^+ = \begin{bmatrix} \mathbf{I}_m \\ \mathbf{Z}^H \end{bmatrix}\left(\mathbf{I}_m + \mathbf{Z}\mathbf{Z}^H\right)^{-1}\mathbf{A}_1^{-1} \tag{3.39}$$

where the determinant $\left|\mathbf{I}_m + \mathbf{Z}\mathbf{Z}^H\right| \neq 0$.

Introducing the partition of the solution vector $\mathbf{x}^T = \begin{bmatrix} \mathbf{x}_1^T & \mathbf{x}_2^T \end{bmatrix}$ in the expression $\mathbf{x} = \mathbf{A}^+\mathbf{b}$, we find the relation between the two components of the solution as

$$\mathbf{x}_2 = \mathbf{Z}^H\mathbf{x}_1 \tag{3.40}$$

Indeed, it is not difficult to show here that the squared-norm of the solution vector, $\mathbf{x}_1^H\mathbf{x}_1 + \mathbf{x}_2^H\mathbf{x}_2$, is minimized for the class of vectors satisfying $\mathbf{x}_1 + \mathbf{Z}\mathbf{x}_2 = \mathbf{A}_1^{-1}\mathbf{b} = \mathbf{c}$ only when the above relation between the two components holds. The inner product and norm are defined in the linear vector space (linear space whose points are vectors) as

$\langle \mathbf{x}, \mathbf{y} \rangle = \mathbf{x}^H\mathbf{y}$ and $\|\mathbf{x}\| = \langle \mathbf{x}, \mathbf{x} \rangle^{1/2}$, respectively.

Case 3 The general case when the determinants $\left|\mathbf{A}^H\mathbf{A}\right| = 0$ and $\left|\mathbf{A}\mathbf{A}^H\right| = 0$

We consider the matrix $\mathbf{A}$ of dimension $m \times n$, and rank($\mathbf{A}$) = k. Let $\mathbf{A}_{11}$ be a non-singular $k \times k$ submatrix of $\mathbf{A}$, and $\mathbf{A}_{11}$ is brought to the top left corner of $\mathbf{A}$ by row and column permutations,

$$\mathbf{PAQ} = \begin{bmatrix} \mathbf{A}_{11} & \mathbf{A}_{12} \\ \mathbf{A}_{21} & \mathbf{A}_{22} \end{bmatrix} \tag{3.41}$$

where $\mathbf{P}$ and $\mathbf{Q}$ are permutation matrices. When the partitioned matrix $\mathbf{A}$ is factorized, it takes the following form,

$$\mathbf{A} = \mathbf{P}^T \begin{bmatrix} \mathbf{I}_k \\ \mathbf{W} \end{bmatrix} \mathbf{A}_{11} \begin{bmatrix} \mathbf{I}_k & \mathbf{Z} \end{bmatrix} \mathbf{Q}^T, \tag{3.42}$$

where $\mathbf{W} = \mathbf{A}_{21}\mathbf{A}_{11}^{-1}$, $\mathbf{Z} = \mathbf{A}_{11}^{-1}\mathbf{A}_{12}$, and $\mathbf{A}_{22}$ is substituted as $\mathbf{A}_{22} = \mathbf{A}_{21}\mathbf{A}_{11}^{-1}\mathbf{A}_{12}$. The relation for $\mathbf{A}_{22}$ can be derived from the conditions of the above determinants.

The Noble generalized inverse $\mathbf{A}^+$ of the matrix $\mathbf{A}$ is then computed as

$$\mathbf{A}^+ = \mathbf{Q} \begin{bmatrix} \mathbf{I}_k \\ \mathbf{Z}^H \end{bmatrix} \left(\mathbf{I}_k + \mathbf{Z}\mathbf{Z}^H\right)^{-1} \mathbf{A}_{11}^{-1} \left(\mathbf{I}_k + \mathbf{W}^H\mathbf{W}\right)^{-1} \begin{bmatrix} \mathbf{I}_k & \mathbf{W}^H \end{bmatrix} \mathbf{P}. \tag{3.43}$$

It is not difficult to verify that the Noble generalized inverse satisfies the four Moore–Penrose properties, and hence it is identical to the unique generalized inverse of the matrix $\mathbf{A}$. Furthermore, since the computation of the generalized inverse is carried out in the above procedure by constructing a full-rank factorization, $\mathbf{A} = \mathbf{BC}$, where $\mathbf{B} = \mathbf{P}^T \begin{bmatrix} \mathbf{I}_k \\ \mathbf{W} \end{bmatrix} \mathbf{A}_{11}$ and $\mathbf{C} = \begin{bmatrix} \mathbf{I}_k & \mathbf{Z} \end{bmatrix} \mathbf{Q}^T$, the existence of the generalized inverse for every real (or complex) matrix $\mathbf{A}$ is verified.

3.6 Perturbation Analysis of Solution Vector

Let $\hat{\mathbf{x}}$ be the solution vector that is computed while solving the matrix equation $\mathbf{Ax} = \mathbf{b}$ whose true solution is $\mathbf{x}$. This can be caused due to computational errors in the adopted procedure of solution. Alternatively, $\hat{\mathbf{x}}$ can be considered as the solution vector of the perturbed system, whose system matrix or/and excitation vector have been changed due to measurement errors in the input data. In this case, $\mathbf{x}$ is the desired solution of the system of linear equations being considered.

The condition number, $\mathrm{cond}(\mathbf{A})$, of the system matrix $\mathbf{A}$ plays a vital role in perturbation analysis for solution, and the number in 2-norm is defined as follows,

$$\mathrm{cond}(\mathbf{A}) = \|\mathbf{A}\|_2 \cdot \left\|\mathbf{A}^+\right\|_2 \tag{3.44}$$

where the matrix 2-norm is defined in terms of the vector 2-norm as shown below,

$$\|\mathbf{A}\|_2 = \sup_{\mathbf{x}\neq 0} \frac{\|\mathbf{Ax}\|_2}{\|\mathbf{x}\|_2} = \max_{\|\mathbf{x}\|_2=1} \|\mathbf{Ax}\|_2 \text{, and } \|\mathbf{x}\|_2 = \left(|x_1|^2 + |x_2|^2 + \cdots + |x_n|^2\right)^{1/2}.$$

Similarly, the condition number can be defined in 1-norm or ∞-norm. Note that an estimate of $\|\mathbf{A}^+\|$ is needed for computing $\text{cond}(\mathbf{A})$ when the generalized inverse $\mathbf{A}^+$ may not be available beforehand.

First, consider the perturbed system $(\mathbf{A}+\mathbf{E})\hat{\mathbf{x}} = \mathbf{b}$, where $\mathbf{E}$ is the error matrix and $\hat{\mathbf{x}}$ is the perturbed solution. The solution of the unperturbed system $\mathbf{Ax} = \mathbf{b}$ is the desired solution $\mathbf{x}$.

Theorem 3.5

Given that $\text{cond}(\mathbf{A})\frac{\|\mathbf{E}\|}{\|\mathbf{A}\|} < 1$, then

$$\frac{\|\mathbf{x}-\hat{\mathbf{x}}\|}{\|\mathbf{x}\|} \leq \frac{\text{cond}(\mathbf{A})\frac{\|\mathbf{E}\|}{\|\mathbf{A}\|}}{1-\text{cond}(\mathbf{A})\frac{\|\mathbf{E}\|}{\|\mathbf{A}\|}} \tag{3.45}$$

where $\text{cond}(\mathbf{A})$ is the condition number of $\mathbf{A}$ with respect to inversion.

Corollary 3.6

Given that $\text{cond}(\mathbf{A})\frac{\|\mathbf{E}\|}{\|\mathbf{A}\|} \ll 1$, then

$$\frac{\|\mathbf{x}-\hat{\mathbf{x}}\|}{\|\mathbf{x}\|} \leq \text{cond}(\mathbf{A})\frac{\|\mathbf{E}\|}{\|\mathbf{A}\|} \tag{3.46}$$

The above theorem together with its corollary provides an upper bound for the norm of the deviation of solution vector relative to the norm of the desired solution.

Next, we consider the perturbed system $\mathbf{A}\hat{\mathbf{x}} = \mathbf{b}+\mathbf{e}$, where $\mathbf{e}$ is the error vector and $\hat{\mathbf{x}}$ is the solution of the perturbed system. The solution of the system $\mathbf{Ax} = \mathbf{b}$ is the desired solution.

Theorem 3.7

$$\frac{\|\mathbf{x}-\hat{\mathbf{x}}\|}{\|\mathbf{x}\|} \leq \text{cond}(\mathbf{A})\frac{\|\mathbf{e}\|}{\|\mathbf{b}\|} \tag{3.47}$$

Once again, the above theorem provides an upper bound for the relative deviation of solution vector.

In essence, the last two theorems state that the maximum relative error in the computed solution vector is determined by the product of the relative error in the system matrix or in the excitation vector, and the condition number of the system of linear equations. In this sense, the condition number measures the sensitivity of the solution of a matrix equation to errors in the input data. A large condition number indicates that the results from matrix inversion and computation of solution may be erroneous. When $\text{cond}(\mathbf{A})$ is a large number, the system of linear equations is said to be ill-conditioned.

Theorem 3.8

If $\mathbf{A}$ is a matrix of dimension $m \times n$, and $\text{rank}(\mathbf{A}) = n \leq m$, then

$$\text{cond}\left(\mathbf{A}^H\mathbf{A}\right) = \{\text{cond}(\mathbf{A})\}^2 \tag{3.48}$$

A direct consequence of the above theorem is that if the full-rank least-squares problem $\mathbf{Ax} = \mathbf{b}$ is ill-conditioned, then the computing problem $\left(\mathbf{A}^H\mathbf{A}\right)\mathbf{x} = \mathbf{A}^H\mathbf{b}$ worsens the condition unnecessarily.

Indeed, the generalized inverse $\mathbf{A}^+$ of the matrix $\mathbf{A}$ should be computed by matrix partitioning in the above case. Note that although the system matrix is full-rank and over-determined, it is not advisable to form the normal equation as shown above due to consideration of required accuracy in the computed solution.

Estimate of Condition Number

An estimate of the condition number of system matrix is needed to assess whether an accurate solution of a set of linear equations is feasible. Usually, one computes an estimate of $\text{cond}(\mathbf{A}) = \|\mathbf{A}\| \cdot \|\mathbf{A}^+\|$ before or independent of the computation of the inverse-matrix $\mathbf{A}^+$. Hence, an estimate of $\|\mathbf{A}^+\|$ is required for the estimation of $\text{cond}(\mathbf{A})$.

If for a suitable excitation $\mathbf{u}$ the solution $\mathbf{w}$ is obtained for the system $\mathbf{Aw} = \mathbf{u}$, then a fair estimate of $\|\mathbf{A}^+\|$ will be

$$\|\mathbf{A}^+\| = \frac{\|\mathbf{w}\|}{\|\mathbf{u}\|} \tag{3.49}$$

In order to solve the above system one may first get an **LU**-decomposition of the matrix $\mathbf{A}$, and then solve the couple of equations $\mathbf{Lv} = \mathbf{u}$, $\mathbf{Uw} = \mathbf{v}$ by the methods of forward and backward substitutions. In the serial computations shown above, an appropriate vector $\mathbf{u}$ can be chosen adaptively in such a way which enhances the growth of the intermediate solution $\mathbf{v}$. This step improves the achievable accuracy in the final solution $\mathbf{w}$.

Next, a second pass is made with $\mathbf{A}^H$ for the estimation of $\|\mathbf{A}^+\|$. If now $\mathbf{w}$ is the excitation and $\mathbf{x}$ is the solution of the system $\mathbf{A}^H\mathbf{x} = \mathbf{w}$, then the second estimate of $\|\mathbf{A}^+\|$ will be

$$\|\mathbf{A}^+\| = \frac{\|\mathbf{x}\|}{\|\mathbf{w}\|}. \tag{3.50}$$

The average of the two estimates of $\|\mathbf{A}^+\|$ can be taken for computation of $\text{cond}(\mathbf{A})$.

3.7 Stability of Least Squares Solution

The stability condition of least-squares procedure implies that a finite incremental change in the input data results in a finite incremental change in the output solution. The stability check indicates whether a small error at one stage of computation can magnify and propagate to the later stage, and thus produce an unacceptably large deviation in the final output. An unstable solution is obtained when an adopted procedure becomes unstable for a given problem.

When the least-squares solution is obtained by the **LU**-decomposition of the system matrix **A**, the stability check is carried out by comparing the product $\hat{\mathbf{L}}\hat{\mathbf{U}}$ of the computed factors with **A**. In this case the rest of the method for calculating the solution is transparent.

For a variety of other methods for computing the least-squares solution, the intermediate results are difficult to interpret, and the stability check is not straight forward. In all these cases for a given computed solution $\hat{\mathbf{x}}$, the residual vector $\mathbf{r} = \mathbf{b} - \mathbf{A}\hat{\mathbf{x}}$ can be used to check stability.

Theorem 3.9 (Stewart)

Let $\dfrac{\|\mathbf{r}\|}{\|\mathbf{A}\|\cdot\|\hat{\mathbf{x}}\|} = \varepsilon$, then there is a matrix **E** with $\dfrac{\|\mathbf{E}\|}{\|\mathbf{A}\|} = \varepsilon$ such that

$$(\mathbf{A}+\mathbf{E})\hat{\mathbf{x}} = \mathbf{b} \tag{3.51}$$

Conversely, if $\hat{\mathbf{x}}$ satisfies the above equation, then

$$\frac{\|\mathbf{r}\|}{\|\mathbf{A}\|\cdot\|\hat{\mathbf{x}}\|} \le \frac{\|\mathbf{E}\|}{\|\mathbf{A}\|} \tag{3.52}$$

The theorem states that if for a given computed solution the number ε is found to be large, then something has gone wrong in the procedure. An appropriate measure is the ratio $\rho = \dfrac{\varepsilon}{\varepsilon_m}$, where ε_m is the machine constant due to rounding error. A computed solution is unacceptable where ρ is large (say, 100 or more).

Next, we consider the stability check for the perturbed system. Let **x** be the solution of minimizing $\|\mathbf{b}-\mathbf{A}\mathbf{x}\|$, and let $\mathbf{r} = \mathbf{b} - \mathbf{A}\mathbf{x}$ be the residual of the unperturbed system. Let $\hat{\mathbf{x}}$ be the computed solution, and let $\hat{\mathbf{r}} = \mathbf{b} - \mathbf{A}\hat{\mathbf{x}}$ be the residual of the perturbed system.

Theorem 3.10 (Stewart)

The vector $\hat{\mathbf{x}}$ exactly minimizes

$$\|\mathbf{b}-(\mathbf{A}+\mathbf{E}_i)\hat{\mathbf{x}}\|;\quad i=1,2 \tag{3.53}$$

where $\mathbf{E}_1 = -\dfrac{\hat{\mathbf{r}}\hat{\mathbf{r}}^H\mathbf{A}}{\|\hat{\mathbf{r}}\|^2}$, and $\mathbf{E}_2 = \dfrac{(\hat{\mathbf{r}}-\mathbf{r})\hat{\mathbf{x}}^H}{\|\hat{\mathbf{x}}\|^2}$.

The norms of the matrices are given by

$$\|\mathbf{E}_1\| = \frac{\|\mathbf{A}^H\hat{\mathbf{r}}\|}{\|\hat{\mathbf{r}}\|} \tag{3.54}$$

and,

$$\|\mathbf{E}_2\| \simeq \frac{\left(\|\hat{\mathbf{r}}\|^2 - \|\mathbf{r}\|^2\right)^{1/2}}{\|\mathbf{x}\|} = \frac{\|\hat{\mathbf{r}} - \mathbf{r}\|}{\|\mathbf{x}\|} \tag{3.55}$$

The norm of the first error matrix is small when the residual $\hat{\mathbf{r}}$ is almost orthogonal to the column space of **A**. This condition is easy to check. The norm of the second matrix is small when $\|\hat{\mathbf{r}}\|$ is almost equal to $\|\mathbf{r}\|$. This condition is difficult to check because $\|\mathbf{r}\|$ may not be known. An estimate of $\|\mathbf{r}\|$ can be found from the machine constant as implied by the previous theorem of Stewart.

Both of the above criteria can be used for software testing of the stability of least-squares solution. Unfortunately, the criteria are sufficient, but not necessary, which means that many cases of unstable solution cannot be identified by the stability check alone.

3.8 Linear Operator Equation and Method of Orthogonal Projection

We now generalize our consideration for computing inverse to a linear operator equation of the form

$$Lu = g \tag{3.56}$$

where L is a linear operator, u is the unknown to be solved, and g is the known excitation.

Previously we have considered linear operator equation involving matrix where the matrix operator is the mapping from one linear vector space to a second linear vector space, and each vector is represented by an ordered set of complex numbers. Among other possibilities, the operator in consideration may be a linear differential or integral operator. In these cases, the linear spaces concerned are the function spaces where each point of a function space represents a piecewise continuous complex-valued function of a real variable.

A linear integro-differential operator equation can be solved by the method of orthogonal projection. We seek an approximate solution u_n in a known n-dimensional function space as shown below,

$$u \simeq u_n = \sum_{j=1}^{n} \alpha_j q_j \tag{3.57}$$

where α_j are unknown complex numbers, q_j are known coordinating functions, and n is set by the desired accuracy of the solution. Any set of n linearly independent functions can be chosen as the coordinating functions in the n-dimensional function space. Note that the exact solution u is a point in an infinite dimensional function space.

The residual r_n is given by

$$r_n = g - Lu_n = g - \sum_{j=1}^{n} \alpha_j \left(Lq_j \right) \tag{3.58}$$

and the coefficients $\alpha_1, \ldots, \alpha_n$ are determined from the condition that the above residual is orthogonal to a set of n linearly independent weighting functions $w_1, \ldots, w_n$. One may choose $w_i = q_i$, but other choices are equally convenient. The orthogonality conditions are satisfied when the following inner products are set to zero,

$$\langle w_i, r_n \rangle = 0 \quad \text{for } i = 1, \ldots, n \tag{3.59}$$

where the inner product and norm in the function space are defined as $\langle y, z \rangle = \int_a^b y(t) z^*(t) dt$, and $\|y\| = \langle y, y \rangle^{1/2}$ respectively, $[a, b]$ being the interval of the real variable t over which the functions are defined and * stands for complex conjugation.

The matrix equation to be solved in this case is expressed as

$$\left[\langle w_i, Lq_j \rangle \right] \begin{bmatrix} \alpha_1 \\ \vdots \\ \alpha_n \end{bmatrix} = \begin{bmatrix} \langle w_1, g \rangle \\ \vdots \\ \langle w_n, g \rangle \end{bmatrix} \tag{3.60}$$

where $\langle w_i, Lq_j \rangle$ denotes an element of the matrix at i row and j column. Note that the integro-differential equation is transformed into a matrix equation with elements of the matrix as inner products. The method described above is more often referred to as the weighting-function method of Galerkin. In order to determine the coefficients $\alpha_1, \ldots, \alpha_n$ uniquely, it is necessary that the determinant of the system matrix is non-zero.

While solving a boundary-value problem, the solution must satisfy the differential operator equation together with a set of boundary conditions. In this case, the coordinating functions q_j are usually chosen to satisfy the boundary conditions, and then the method of orthogonal projection is applied to compute the coefficient α_j for an optimal approximate solution.

3.8.1 Integral Equation and Method of Galerkin

Linear continuous processes are represented by linear differential or integral equations. An important distinction between the two approaches is that in the differential equation formulation of a physical problem the boundary conditions are imposed separately, whereas the integral equation formulation contains the boundary conditions implicitly. A linear differential equation with specified boundary conditions can be transformed into an equivalent linear integral equation.

A linear integral equation involving the integral operator L is defined as follows,

$$L\left[u(t) \right] = \int_a^b h(t, \tau) u(t) dt \tag{3.61}$$

where the kernel $h(t,\tau)$ is a measurable piecewise continuous complex-valued function defined for $t \in [a,b]$ and $\tau \in [c,d]$. The operator satisfies the linearity condition

$$L[\lambda_1 u_1(t) + \lambda_2 u_2(t)] = \lambda_1 L[u_1(t)] + \lambda_2 L[u_2(t)] \quad (3.62)$$

where λ_1 and λ_2 are constants.

The Fredholm equation of the first kind has the form

$$L[u(t)] = g(\tau), \quad c \le \tau \le d \quad (3.63)$$

The Laplace integral, the Fourier sine and cosine integrals, and several other transform integrals are examples of the Fredholm equation of the first kind.

The Fredholm equation of the second kind has the form

$$L[u(t)] + g(\tau) = u(\tau), \quad a \le \tau \le b \quad (3.64)$$

while the corresponding homogeneous equation is

$$L[u(t)] = u(\tau), \quad a \le \tau \le b \quad (3.65)$$

The Volterra equations of the first kind and the second kind can be obtained from the corresponding Fredholm equations by setting $h(t,\tau) = 0$ for $a \le \tau < t \le b$.

Previously, we have calculated the solution of the operator equation of the form $Lu = g$, let us now investigate how an operator equation of the form

$$u = \lambda Lu + g, \qquad \lambda \text{ constant}, \quad (3.66)$$

can be solved. We seek an approximate solution

$$u_n = g + \sum_{j=1}^{n} \alpha_j q_j \quad (3.67)$$

where q_j are the chosen expansion functions which are linearly independent, and α_j are unknown complex numbers.

Defining the residual $r[u] = u - \lambda Lu - g$ which is identically zero, we find the residual for the approximate solution as,

$$r[u_n] = \sum_{j=1}^{n} \{\alpha_j q_j - \lambda \alpha_j L q_j\} - \lambda Lg = r_n \quad (3.68)$$

Now the coefficients α_i are determined from the condition that the above residual is orthogonal to all chosen expansion function $q_1, \ldots, q_n$. The orthogonality conditions provide the relations, $\langle q_i, r_n \rangle = 0$ for $i = 1, \ldots, n$, which in turn leads to the following matrix equation,

$$\left[\langle q_i, q_j \rangle - \lambda \langle q_i, Lq_j \rangle\right] \begin{bmatrix} \alpha_1 \\ \vdots \\ \alpha_n \end{bmatrix} = \lambda \begin{bmatrix} \langle q_1, Lg \rangle \\ \vdots \\ \langle q_n, Lg \rangle \end{bmatrix} \quad (3.69)$$

If the determinant of the above system matrix is non-zero, then it uniquely determines the coefficients $\alpha_1, \ldots, \alpha_n$. The method described above is known as the method of Galerkin, which finds an optimal approximate solution by utilizing the principle of orthogonal projection.

3.9 Conjugate Direction Methods

In order to solve the linear operator equation $Lu = g$, first we select a set of vectors (functions) p_i which are L-conjugate,

$$\left\langle Lp_i, p_j \right\rangle = 0 \text{ for } i \neq j \tag{3.70}$$

Theorem 3.11

If the directions p_i are mutually conjugate, then they are linearly independent.
Using the above theorem and assuming that the approximate solution u_n of the operator equation belongs to the finite dimensional linear space spanned by the directions p_i; $i = 1, \ldots, n$, we write the expression,

$$u \simeq u_n = \sum_{i=1}^{n} \alpha_i p_i \tag{3.71}$$

or

$$Lu \simeq Lu_n = \sum_{i=1}^{n} \alpha_i \, Lp_i \simeq g \tag{3.72}$$

From the above equation using the property of L-conjugate directions one finds, $\alpha_i \langle Lp_i, p_i \rangle = \langle g, p_i \rangle$, which provides the coefficients as

$$\alpha_i = \frac{\langle g, p_i \rangle}{\langle Lp_i, p_i \rangle} \tag{3.73}$$

and thus the solution is expressed as

$$u \simeq \sum_{i=1}^{n} \frac{\langle g, p_i \rangle}{\langle Lp_i, p_i \rangle} p_i \tag{3.74}$$

(A) Gauss Elimination

In the direct method, the L-conjugate vectors (functions) p_i are generated starting from the standard basis vectors (functions) f_i of the n-dimensional linear space as follows,

$$\begin{aligned} p_1 &= f_1 \\ p_2 &= f_2 - \frac{\langle Lf_2, p_1 \rangle}{\langle Lp_1, p_1 \rangle} p_1 \\ p_3 &= f_3 - \frac{\langle Lf_3, p_1 \rangle}{\langle Lp_1, p_1 \rangle} p_1 - \frac{\langle Lf_3, p_2 \rangle}{\langle Lp_2, p_2 \rangle} p_2 \\ &\vdots \end{aligned} \tag{3.75}$$

The above procedure is a variant of the Gram-Schmidt decomposition technique. Although the entire set of L-conjugate directions p_i, and the hence the approximate solution u_n are obtained in a finite number of steps, the quantization noise can be a serious problem in the computation when n is large. This is so because the computation related to the decomposition procedure is done in a serial fashion, and the error introduced at one stage can accumulate in the proceeding stages.

(B) Conjugate Gradient Method

In the iterative method, the L-conjugate directions p_i are obtained from the residual r_i at the ith stage,

$$r_i = g - Lu_i \tag{3.76}$$

where u_i is the partial solution,

$$u_i = u_0 + \sum_{j=0}^{i-1} \alpha_j p_j \tag{3.77}$$

u_0 being the initial guess.

At each iteration ($i = 1, 2, \ldots$), the cost function

$$\phi(u_i) = \|r_i\|^2 \tag{3.78}$$

is minimized with respect to the constant α_{i-1}, when the search directions are generated recursively by the relation,

$$p_i = v_i + \beta_{i-1} p_{i-1} \tag{3.79}$$

with $v_i = L^{\#} r_i$, where $L^{\#}$ denotes the adjoint operator defined by $\langle Lx, y\rangle = \langle x, L^{\#} y\rangle$ for all x, y and the constant β_{i-1} is chosen such that

$$\langle Lp_i, Lp_{i-1}\rangle = 0 \tag{3.80}$$

The constants α_{i-1} and β_{i-1} are calculated as follows,

$$\alpha_{i-1} = \frac{\|v_{i-1}\|^2}{\|Lp_{i-1}\|^2} \tag{3.81}$$

$$\beta_{i-1} = \frac{\|v_i\|^2}{\|v_{i-1}\|^2} \tag{3.82}$$

while the initial guess u_0 is employed to compute the initial values,

$$r_0 = g - Lu_0 \tag{3.83}$$

and

$$p_0 = v_0 = L^{\#} r_0 \tag{3.84}$$

Note that the generating directions v_i are proportional to the gradient of the function $\phi(u_i)$. Indeed, the sequence u_i converges to the exact solution in a finite number of steps as indicated by the difference,

$$\phi(u_i)-\phi(u_{i+1})=\alpha_i^2 \| Lp_i\|^2 \tag{3.85}$$

which is always positive.

3.10 Ill-posed Operator Equations and Regularization Theory

A linear operator equation $Lu = g$, where L is a known continuous linear operator from one separable Hilbert space (complete inner product space with countable basis vectors/ functions) $\mathscr{H}_1$ into another separable Hilbert space $\mathscr{H}_2$, is well-posed (unique solution, well-conditioned) if it satisfies the following three Hadamard criteria:

1. For every observation g in $\mathscr{H}_2$, there exists a solution u in the domain space $\mathscr{D}(L) \subset \mathscr{H}_1$;
2. The solution u is unique in $\mathscr{D}(L)$;
3. The dependence of u upon g is continuous.

Criterion (1) implies that the range space $\mathscr{R}(L) = \mathscr{H}_2$ and the mapping is onto, whereas (2) indicates that the mapping is one-to-one. Thus when combined, criteria (1) and (2) assure that the mapping related to the operator equation is invertible, and we can always compute a unique solution of the operator equation, although we may not necessarily compute the inverse operator explicitly. Moreover when criterion (3) is satisfied, we know that the inverse mapping is continuous or the inverse operator is continuous.

When the mapping and inverse mapping both are continuous, the inverse images of neighbourhoods of g are neighbourhoods of u and vice versa. This feature is the key requirement of well-conditioning of an operator equation. Furthermore, we know that the operator and thus the inverse operator both are linear, and for linear operators the continuity and boundedness are equivalent properties. Therefore, we can compute the norm of the bounded (integro-differential) operator L as defined by

$$\|L\| = \sup_{\substack{u\in\mathfrak{D}(L) \\ u\neq 0}} \frac{\|Lu\|}{\|u\|} \tag{3.86}$$

where $\|u\|=\langle u,u\rangle^{1/2}$, and then using $\|L\|$ and an estimate for the norm of the inverse operator which may not be available, the well-conditioning of the operator equation can be checked by forming the product of the two norms exactly the way it is done for matrix operators in the finite dimensional spaces.

An ill-posed problem arises when one or more of the Hadamard criteria are violated. An ill-posed operator equation is converted into a well-posed operator equation whose solution is acceptable by the method of regularization. The various approaches to regularization as summarized by Nashed involve one or more of the following ideas:

- A change of the concept of a solution (or reformulation of the problem considering the uncertainty in observation);

- A restriction of the data (or preprocessing to reduce the system interference);
- A change of the spaces and/or topologies (or recasting the problem relative to new spaces in which the operator has a closed range);
- A modification of the operator itself (approximation by a compact linear operator whose high frequency modes are filtered out);
- The concept of regularization operators (admissible solution from compact subset of the solution space);
- Probabilistic methods or well-posed stochastic extensions of ill-posed problems (statistical methods when noise statistics is known).

We discuss the first five approaches in the following sections and refer to the other approach in the subsequent chapters.

3.10.1 Regularization Theory of Miller

We consider an additive model for noise by which the observation is corrupted,

$$Lu = g + w \tag{3.87}$$

Since both of the solution u and noise w are unknowns, and the equation $Lu = g$ may not have any solution, we reformulate the problem as follows:

Find u such that

$$\|Lu - g\|_{\mathscr{R}(L)} \leq \varepsilon \tag{3.88}$$

where ε is a positive constant depending on the level of noise, and the norm is defined in $\mathscr{R}(L)$.

We then concentrate on the requirement that the inverse mapping should also be continuous together with the forward mapping whose continuity is already assumed. In order to assure the continuity of the inverse mapping, we restrict the class of admissible solutions to a compact subset $\mathscr{S}$ of $\mathscr{D}(L)$, see Figure 3.1 [Bertero et al. 1980].

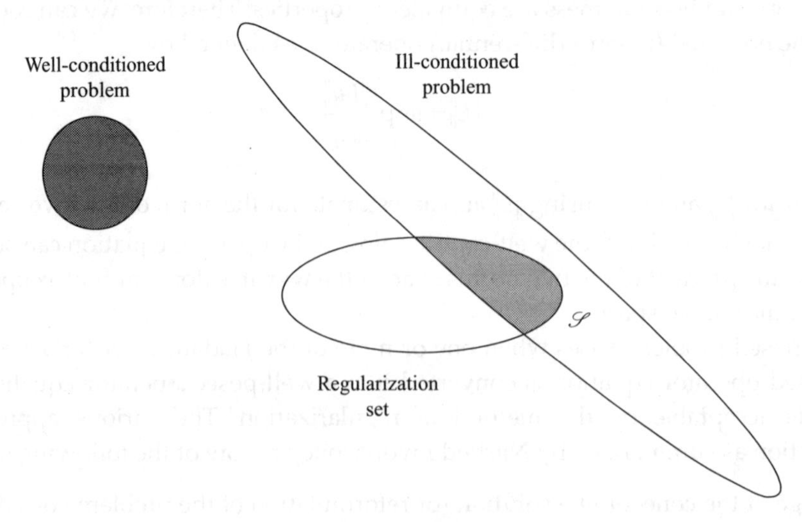

FIGURE 3.1 Idea of regularization in achieving a compact solution space $\mathscr{S}$

The basic idea of the regularization theory of Miller is to achieve the required compactness of the solution space by a constraint operator C as follows,

$$\|Cu\|_{\mathscr{D}(L)} \leq E \tag{3.89}$$

where E is some positive value deciding the size of the regularization set which in turn fixes the size of the solution space $\mathscr{S}$, the linear continuous operator C is to be assigned on $\mathscr{D}(L)$ into itself such that the inverse operator C^{-1} is bounded, and the prescribed norm is defined in $\mathscr{D}(L)$.

In order to find a solution u which will satisfy both the inequality constraints given above, Miller proposed the function φ and it is minimized with respect to u,

$$\varphi(u;\alpha)=\|Lu-g\|^2+\alpha\|Cu\|^2 \tag{3.90}$$

where $\alpha=\left(\varepsilon/E\right)^2$ is a positive parameter, and the norms are defined consistently in respective topological spaces.

The minimization problem of φ can be shown to be equivalent to solving the extended operator equation as given below,

$$\left(L^{\#}L+\alpha C^{\#}C\right)u=L^{\#}g \tag{3.91}$$

where $L^{\#}$, $C^{\#}$ are the adjoint operators, and the operator $B(\alpha)=L^{\#}L+\alpha C^{\#}C$ is defined on the domain of $C^{\#}C$.

Furthermore, the operator B is self-adjoint $\left(B^{\#}=B\right)$, positive definite (for any v in its domain, $\langle Bv,v\rangle>0$), and it has a continuous inverse B^{-1} (from the analogous property assumed for C). Therefore, we can compute the unique solution u of the extended operator equation $Bu=f$ where $f=L^{\#}g$, which equivalently minimizes the function φ. The desired solution u is given by

$$u=B^{-1}f \tag{3.92}$$

Alternatively, an iterative method can be used here as follows,

$$u_{i+1}=u_i+\gamma\left(f-Bu_i\right),\quad i=1,2,\ldots,\quad \text{with } u_0=0, \tag{3.93}$$

where γ is the step size which can be constant or it can depend on the iteration index. When f is in the range space of B, the above algorithm of successive approximation converges to the minimum-norm solution for $\gamma>0$ and $\gamma<2/\|B\|$, according to Bialy's theorem. An extension of the theorem finds the same bound for γ when the observation is noise-corrupted. In the later case the algorithm converges to the minimum-norm least squares solution.

An essential part of the approach of Miller–Tikhonov regularization operators is to choose a permissible couple $\{\varepsilon,E\}$ for which the solution space $\mathscr{S}$ is not empty. For a constant parameter $\alpha\geq 0$, if u_α is the solution of minimizing φ, then we have

$$\|Lu_\alpha-g\|^2=\varepsilon_\alpha^2\leq\varepsilon^2 \tag{3.94}$$

and,

$$\|Cu_\alpha\|^2=E_\alpha^2\leq E^2. \tag{3.95}$$

The ε-value is given by the accuracy of data. The E-value, however, is to be chosen, which selects the regularization set. Note that the approach of regularization operators converts the operator equation $Lu = g$ (an operator equation of first kind) into the extended operator equation $Bu = f$ (an operator equation of second kind when $C = I$, the identity operator). The solution of the extended operator equation converges to the solution of the original operator equation when $\alpha \to 0$ or $E \to \infty$, which in effect removes the constraint of regularization. The iterative method is well suited for this purpose, where the α-parameter is reduced ($\alpha \to 0$) in steps as the recursive solution u_α tends to the true solution u.

3.10.2 Landweber Iteration

An iterative method of regularization has been suggested by rewriting the operator equation $Lu = g$ in the form $u = \left(I - aL^{\#}L\right)u + aL^{\#}g$ for some $a > 0$ and computing iteratively for $k = 1, 2, \ldots$

$$u_k = \left(I - aL^{\#}L\right)u_{k-1} + aL^{\#}g \tag{3.96}$$

with the initial value $u_0 = 0$. This iteration scheme is known as the Landweber iteration, and the method can be interpreted as the steepest descent method applied to the quadratic function $\frac{1}{2}\|Lu - g\|^2$.

Using the linear recursion formula (3.96) for u_k, it is easily seen that u_k has the form

$$u_k = R_k g \tag{3.97}$$

where the operator R_k is defined by

$$R_k = a\sum_{j=0}^{k-1}\left(I - aL^{\#}L\right)^j L^{\#} \quad \textit{for } k = 1, 2, \ldots \tag{3.98}$$

It can be shown that for $0 < a < 1/\|L\|^2$, the operator R_k define a regularization strategy, $u = \left(\alpha I + L^{\#}L\right)^{-1} L^{\#}g$, with the discrete regularization parameter $\alpha = 1/k$, and $\|R_k\| \le \sqrt{ak}$. In addition, let $r > 1$ and g_δ be perturbations of g such that $\|g - g_\delta\| \le \delta$ and $\|g_\delta\| \ge r\delta$ for all $\delta \in [0, \delta_0]$. Let the sequence $\{u_{k,\delta};\ k = 1, 2, \ldots\}$ be determined by the iteration

$$u_{k,\delta} = u_{k-1,\delta} + aL^{\#}\left(g_\delta - Lu_{k-1,\delta}\right) \tag{3.99}$$

Then, the following assertions hold: (Kirsch)

1. $\lim_{k\to\infty}\left\|Lu_{k,\delta} - g_\delta\right\| = 0$ for every $\delta > 0$; thus, the stopping rule is well-defined, that is, let $k = k_\delta$ be the smallest integer with $\left\|Lu_{k,\delta} - g_\delta\right\| \le r\delta$.
2. $\sup\left\|R_{k,\delta}g_\delta - u\right\| \to 0$ for $\|Lu - g_\delta\| \le \delta$ as $\delta \to 0$, that is, the choice of k_δ is admissible. Thus, the sequence $u_{k,\delta}$ converges to u.

Therefore, the iteration method of Landweber leads to an admissible and optimal regularization strategy.

3.10.3 Regularization Method of Tikhonov

We consider the following Fredholm integral equation of the first kind,

$$(Lu)(\tau) = L[u(t)] = \int_a^b k(t,\tau)u(t)dt = g(\tau), \quad c \le \tau \le d \tag{3.100}$$

with the kernel function $k(t,\tau)$ being piecewise continuous over the region $a \le t \le b$ and $c \le \tau \le d$. If it is assumed that there exists a unique solution $u(t)$ corresponding to the observation $g(\tau)$, we may add to that solution a function $u^{(n)}(t) = \lambda \sin(nt)$, where λ is an arbitrary constant. It is known from the Riemann–Lebesgue theorem that

$$\lim_{n\to\infty} \int_a^b k(t,\tau)\sin(nt)dt = 0 \tag{3.101}$$

Therefore, we see that widely different functions u can be the solution of the integral equation with slightly perturbed g, and thus we need to regularize the procedure of solution.

The requirement of smoothness for the solution of the integral equation can be achieved by means of a regularizing function $\Omega(u)$ as given below,

$$\Omega(u) = \int_a^b \left\{ p(t)|u'(t)|^2 + q(t)|u(t)|^2 \right\} dt \tag{3.102}$$

where $u'(t)$ is the derivative of $u(t)$, and the weight functions $p(t)$ and $q(t)$ are strictly positive over the region $t \in [a,b]$. It can be proved by application of the Ascoli–Arzela theorem that the set $\Omega(u) \le E^2$ is a compact subset of the function space in which the solution u is sought. We impose the regularizing constraint simultaneously with the following requirement for an acceptable solution,

$$\int_c^d |Lu(\tau) - g(\tau)|^2 d\tau \le \varepsilon^2 \tag{3.103}$$

when the observation is noise-corrupted.

3.11 Signal Processing Applications III: Non-uniform Sampling and Signal Parameter Estimation

3.11.1 Transient Signal Analysis II

The problem of parameter estimation of a sum of complex exponentials $g(t) = \sum_{i=1}^{M} b_{ci} \exp(s_i t)$, $t \ge 0$ is closely related to the problem of system identification. Indeed, the impulse response of a linear system can be characterized by a sum of weighted complex exponentials where the complex frequencies s_i are the distinct poles of the system and the complex amplitudes b_{ci} are the residues at the poles.

The characterization of the impulse response of a linear system by a sum of weighted complex exponentials, and then estimating the complex frequencies and complex amplitudes of the signal with high degree of accuracy has its special importance in a wide variety of applications. The study of the transient behavior of a system is one such application. Modelling the impulse response of a system by the sum of complex exponentials, and then estimating the parameters of the complex exponentials is equivalent to extracting the distinct poles and residues at the poles of a linear system.

The sum of complex exponential signals provides a more general model than the sum of complex sinusoidal signals. The complex sinusoidal signal model with harmonically or non-harmonically related frequencies has been used in various applications. However, the complex sinusoidal signal model is applicable where the signal is stationary, whereas the complex exponential signal model can fit a class of non-stationary signals whose characteristics are changing over time.

In the following examples, we investigate the problem of parameter estimation of complex exponentials when the modelled signal is sampled at non-uniformly distributed abscissas. Note that the uniform sampling is a special case of non-uniform sampling. Therefore, our approach applies equally well when the sampling grid is uniformly spaced.

Our main concern in the parameter estimation problem is the accuracy that can be achieved when the signal values are corrupted with noise. The noise mixed with the signal values may be measurement noise, or may be in the form of approximation error. We like to investigate how the Prony's method or the Derivative and the Integral methods developed for parameter estimation of the sum of complex exponential signals perform when the signal values are corrupted with noise. It turns out that the accuracy of parameter estimation depends mainly on the conditioning of the system matrix that is formed for each method and on the order of linear prediction modelling.

3.11.2 Simulation Study: Example 1 (Numerical Stability)

We consider the transient signal $g(t)$ corrupted with additive noise $w(t)$, $y(t)=g(t)+w(t)=\exp(-0.3t)\sin 2t+\exp(-0.5t)\sin 5t+w(t)$ corresponding to the complex conjugate pole-pairs $s_{1,2}=-0.3\pm j2.0$ and $s_{3,4}=-0.5\pm j5.0$, $M=4$, sampled at 60 non-uniformly distributed points over the span [0–4.8] units of time. The distribution of sampled points is completely arbitrary except that no interval of sampling is allowed to exceed 0.14 units. It is necessary that the sampling theorem be satisfied locally while employing non-uniform sampling.

Using the signal samples $\{y(t_n); n=1,2,\ldots,N\}$ over the non-uniformly distributed points t_n, we find a closed-form expression for the signal as

$$y(t)=\sum_{k=1}^{K} c_k p_{k-1}(t), \tag{3.104}$$

where $p_{k-1}(t)$ is a polynomial of degree $(k-1)$ in t, and c_k is its coefficient. We choose the polynomials p_k which are orthogonal over an arbitrary set of sampled points and can be generated by recurrence relation as given in Chapter 2. The order of approximation k is decided by the minimum error-variance criterion as discussed in the same chapter.

Once the closed-form expression for the signal is available, we can compute the signal values y_n at equally spaced points $t = nT$, T being the equal spacing, the signal derivatives y_n^k which are the kth order derivative evaluated at $t = t_n$, and the zero-initial conditioned integrals $y_0^{-k}(t_n)$ which are the kth order integral evaluated over the time span $[t_0, t_n]$, where t_0 is the initial time.

We compute 40 points of y_n and y_n^k in the middle of the sampling span, and use these values for Prony's method and the Derivative method, respectively. We compute 40 points of $y_0^{-k}(t_n)$ at the beginning of the sampling span and use these values for the Integral method. When noise is present, the approximation error in the derivatives may become large at the ends of sampling span, whereas the error in the integral values may accumulate over the sampling span.

We mix zero-mean white Gaussian noise sequence $w(t_n)$ with the signal values $g(t_n)$ setting the signal-to-noise ratio (SNR), defined as

$$\text{SNR} = 10\log_{10}\left[\sum_n g^2(t_n) \Big/ \sum_n w^2(t_n)\right] \text{ (dB)} \tag{3.105}$$

at 10, 20, 30, and 40 dB respectively, and estimate the s_i-parameters employing Prony's method, the Derivative method, and the Integral method in turn. We compute the average distance in the s-plane between the actual and estimated s_i-values as a measure of the accuracy of estimation,

$$\text{Average distance} = \frac{1}{M}\sum_{i=1}^{M}\left|s_i - \hat{s}_i\right| \tag{3.106}$$

where s_i is the actual value and $\hat{s}_i$ is the estimated value. Statistically the accuracy of estimation is demonstrated in terms of the confidence interval given by the (mean ± standard deviation) of the average distance, computed using independent noise sequences for 100 trials. The corresponding plots for Prony's method, the Derivative method and the Integral method are shown in Figures 3.2–3.4.

We observe from these plots that Prony's method and the Derivative method perform comparably in estimation of s_i-parameters, whereas the Integral method performs far better compared to the other two methods at low SNR level. However, it is also observed that the Integral method shows some instability at high SNR level, and the mean and the standard deviation of average distance do not converge to zero even when the additive noise becomes insignificant. Similarly, Prony's method does not show zero error of estimation when the additive noise is insignificant.

We compute the condition number of the matrix whose inverse is to be calculated in determining the linear prediction coefficients in each method. The condition number is found to be ~10^6 for Prony's method, ~10^3 for the Derivative method, and ~10^5 for the Integral method. The derivative values cannot be computed as accurately as the signal values in presence of noise using polynomial approximation over non-uniformly sampled data. Nevertheless, the Derivative method performs similar to Prony's method at low SNR level because the matrix involved in the Derivative method is far better conditioned than the matrix formed in Prony's method. The zero-initial conditioned integrals can be evaluated very accurately even at low SNR level when the additive noise is zero-mean.

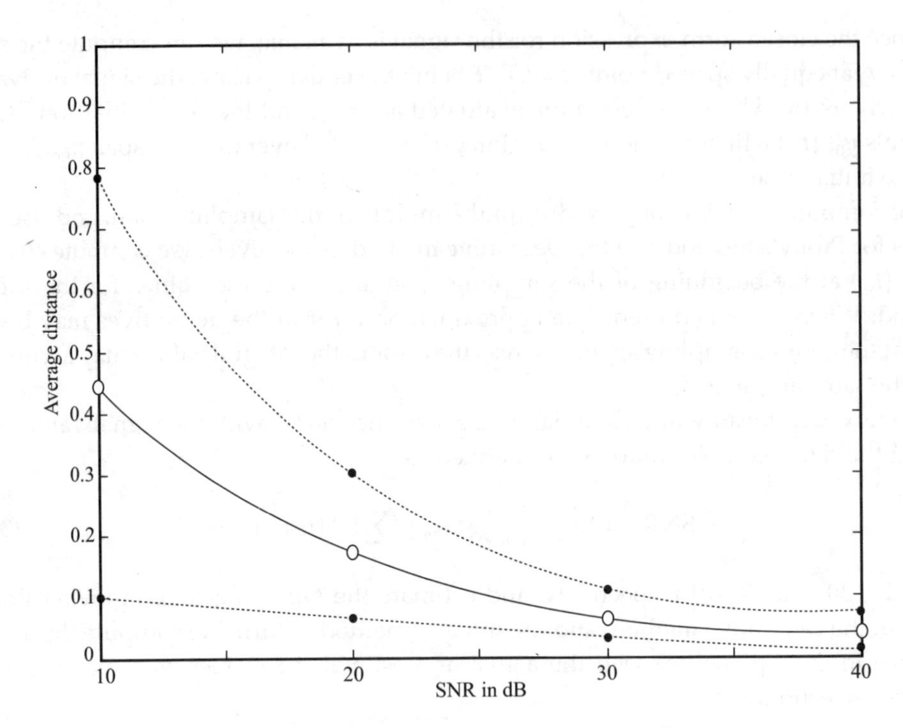

FIGURE 3.2 Prony's method: SNR vs. Average distance

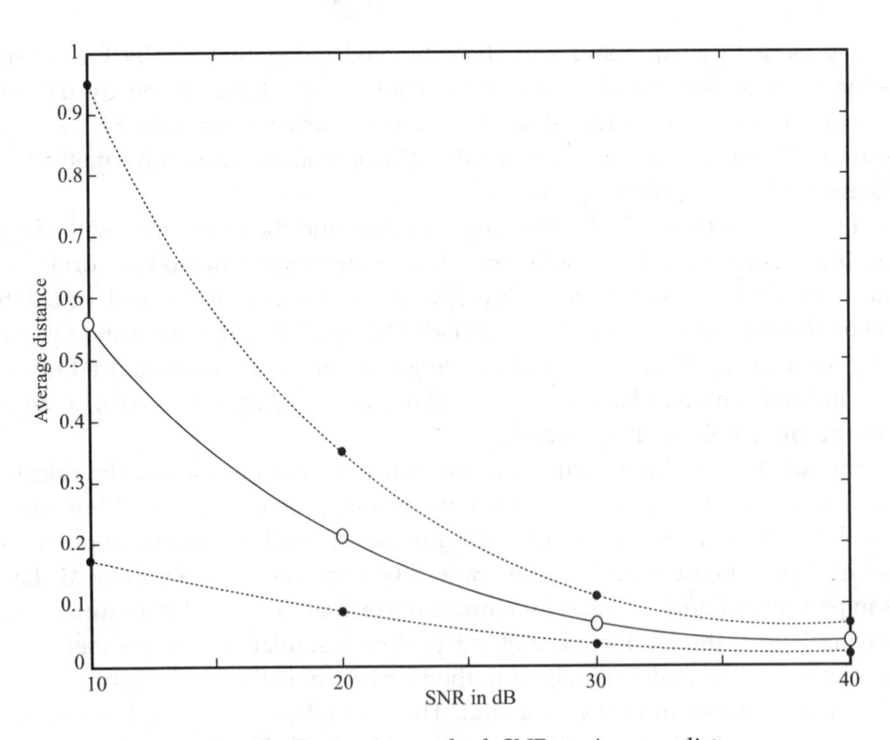

FIGURE 3.3 The Derivative method: SNR vs. Average distance

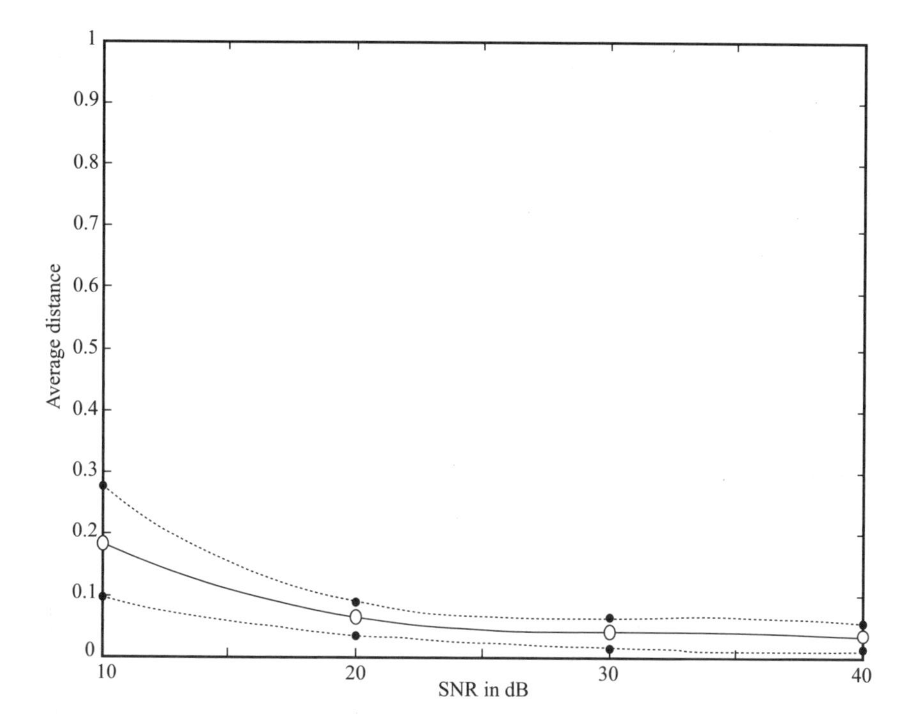

FIGURE 3.4 The Integral method: SNR vs. Average distance (Damped sinusoids)

Since the accuracy of estimation depends on the condition number of the matrix to be inverted as well as the accuracy of the data to be fed in the estimation process, the Integral method performs better than the other two methods. However, the comparative edge of the Integral method does not remain at high SNR level, when the data fed for other two methods also become reasonably accurate, and the accuracy of estimation mainly depends on the conditioning of the matrix to be inverted. The Derivative method performs best at high SNR level because it deals with the matrix which is best conditioned for inverse operation.

We now consider another transient signal with additive noise, $y(t) = \exp(-0.3t)\cos 2t + \exp(-0.5t)\cos 5t + w(t)$, which corresponds to the same complex conjugate pole-pairs as the previous signal. Note that the previous signal is a sum of damped sinusoids and the present signal is a sum of damped cosinusoids. We follow the same steps in simulation as done with the previous signal, and get the confidence interval of estimation of the average distance between the actual and estimated poles. The results are very similar for Prony's method and the Derivative method, and the corresponding plots are not shown here because of this similarity. Interestingly, we notice significant improvement in accuracy of estimation for the Integral method, and the plot of confidence interval is shown in Figure 3.5. It is particularly noticeable from this plot that the instability of the Integral method at high SNR level what we have observed in the simulation with the damped sinusoids, does not appear again in the case of the damped cosinusoids.

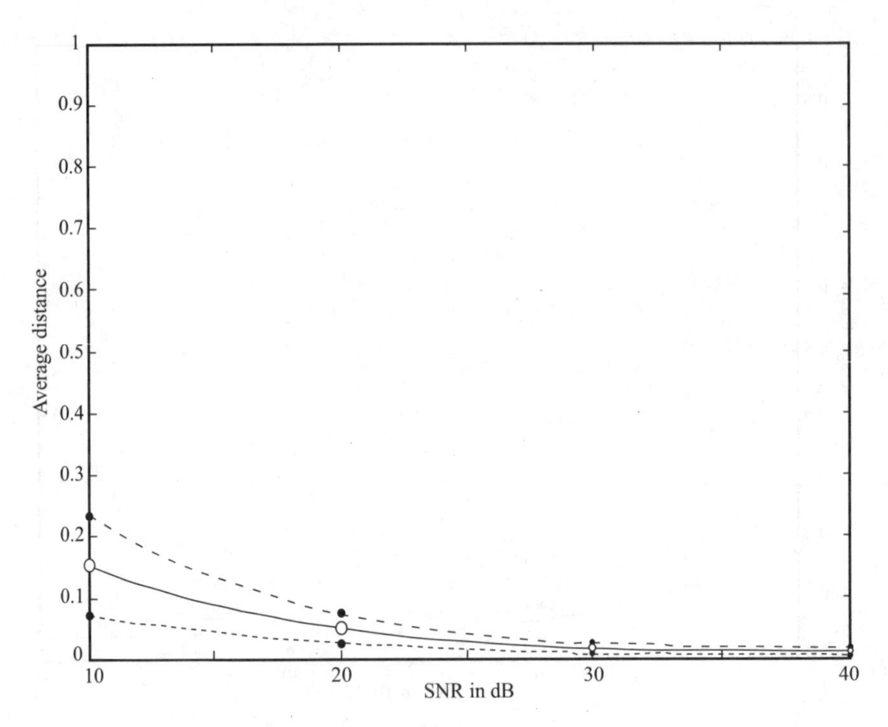

FIGURE 3.5 The Integral method: SNR vs. Average distance (Damped cosinusoids)

We compute the condition number of the matrix to be inverted in the Integral method, and the condition number is found to be ~10^4 in this case. An improvement in the condition number by a factor of 10 compared to the previous signal case plays the trick. Now the performance of the Integral method in estimating the s_i-parameters becomes perfect. At this point, we explore a bit further and find that the integrals computed for the damped sinusoids almost monotonically increase over the time span of processing, which perhaps contribute to the severe ill-conditioning of the problem. In contrast, the integrals computed for the damped cosinusoids are relatively evenly distributed between small positive and negative values, which control the ill-conditioning of the problem to some extent, and we get the desired results in this case.

The outcome of distinctly different results applying the same method with minor change in the problem demonstrates some intricate issues related to the Monte Carlo simulation. Note that the only difference between the two signals that we have considered for the simulation is the change of phase in each of their corresponding components. In fact, we can get somewhat similar result with the damped sinusoidal signal just by shifting the time origin or the sampling span such that the computed integrals distribute evenly between small positive and negative values. In essence, the Monte Carlo simulation shows the conditioning of a specific problem, and the evaluation of performance demonstrated by simulation is valid for the specific problem. The conditioning of a specific problem changes not only with the method being employed, but also with the signal being processed and many other numerical conditions. Therefore, the comparison of performance based on the Monte Carlo simulation cannot be inferred as one method performing better than the other method(s) for all problems. The generalization from a specific problem to all possible problems is simply not valid.

3.11.3 Simulation Study: Example 2 (Extended-Order Modelling)

While using Prony's method for parameter estimation of complex exponential signals in presence of noise, we need to use extended-order modelling for desired accuracy. In this example, we demonstrate how the accuracy of estimation changes when the noise level is varied or when the model-order setting is altered.

Let the test data be the damped sinusoidal signal $g(t)$ expressed as $g(t) = \exp(-0.3t)\sin 2t + \exp(-0.5t)\sin 5t$, corresponding to the complex conjugate pole-pairs $s_{1,2} = -0.3 \pm j2.0$ and $s_{3,4} = -0.5 \pm j5.0$, $M = 4$, sampled at $N = 60$ uniformly spaced points $\{nT;\, n = 0, 1, \ldots, N-1\}$, and $T = 0.08$.

The samples of zero-mean white Gaussian noise $w(t)$ are mixed with the signal-samples to fix the SNR level at 10, 20, 30 and 40 dB in turn. We estimate the signal poles $z_i = \exp(s_i T)$ using Prony's method with the extended-model order $L = 18$ in each case. The signal poles are distinguished from the noise poles by computation of the residues at the poles. Using independent noise sequences we estimate the signal poles in 40 realizations and plot the estimated poles z_i (crosses) for each SNR level as shown in Figures 3.6–3.9. The true signal poles z_i (circles) are shown in the same figures to give an indication of the accuracy of

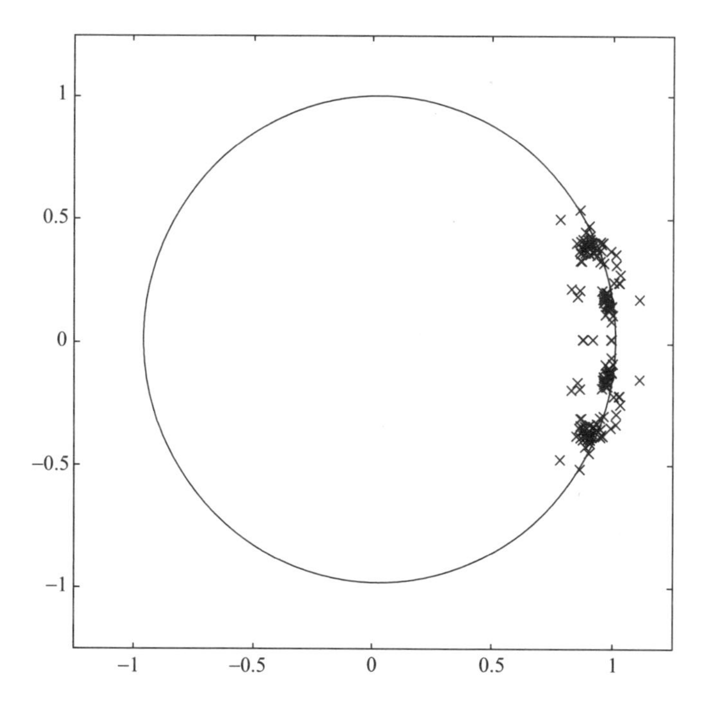

FIGURE 3.6 Signal poles: Prony's method, model order = 18; SNR = 10 dB

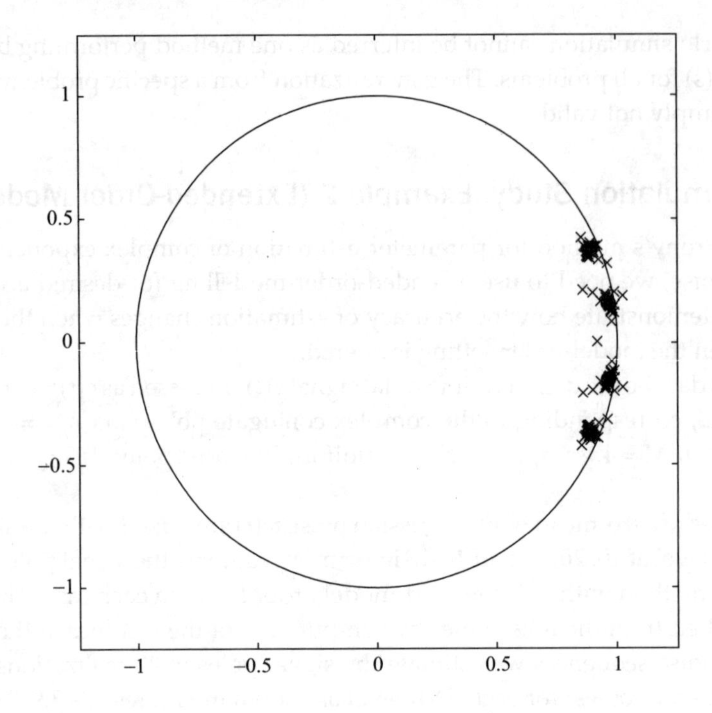

FIGURE 3.7 Signal poles: Prony's method, model order = 18; SNR = 20 dB

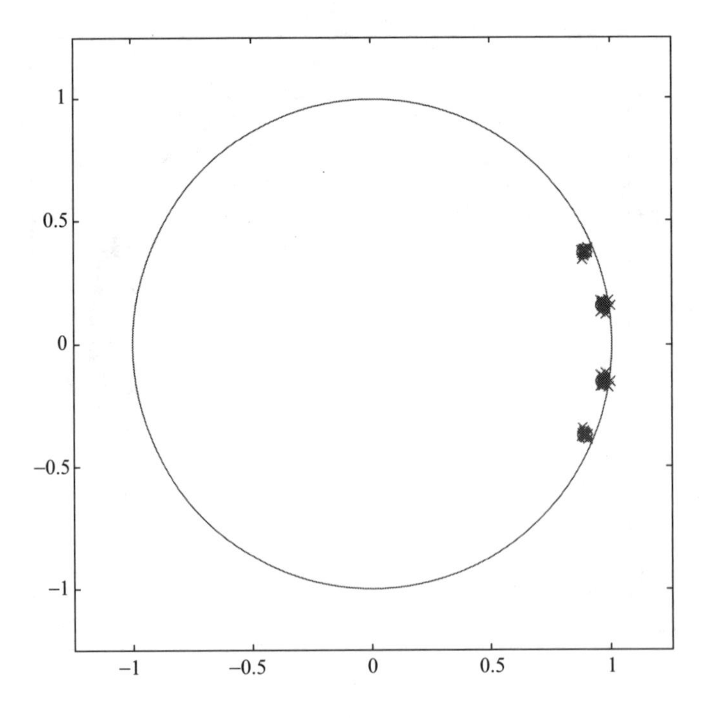

FIGURE 3.8 Signal poles: Prony's method, model order = 18; SNR = 30 dB

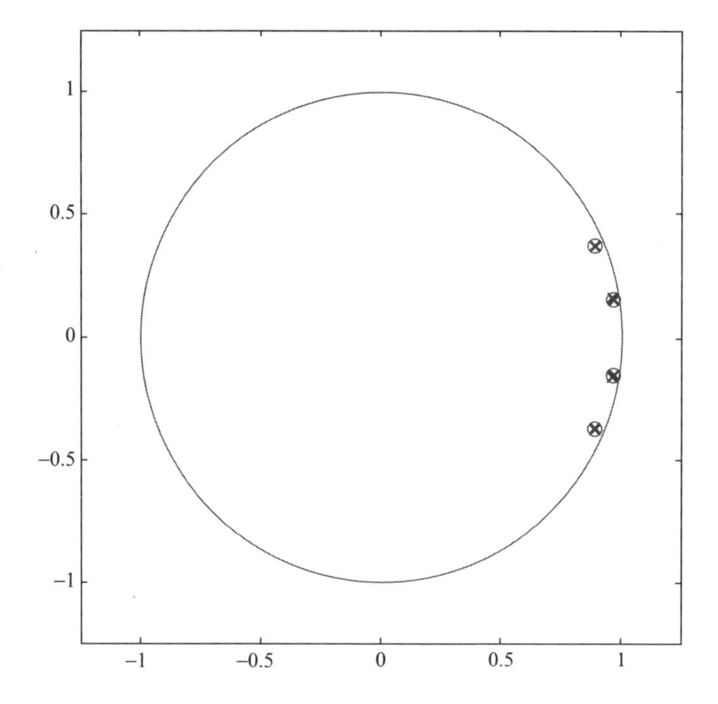

FIGURE 3.9 Signal poles: Prony's method, model order = 18; SNR = 40 dB

estimation. Note that the deviation of the centroid of pole-estimates from the true position of a pole shows the presence of a bias of estimate, whereas the scattering of pole-estimates around the centroid is indicative of a variance of estimate. Apparently with the extended model order set at $L = 18$, Prony's method provides high accuracy of estimation for the number exponentials $M = 4$ at SNR around 30 dB and above.

In the next phase of simulation, we demonstrate how the accuracy of estimation varies with the extended-model order. We estimate the signal poles applying Prony's method setting the extended-model order at 16, 18, 20 and 22 in turn, while the SNR level is fixed at 20 dB. The estimated signal poles z_i (crosses) for 40 realizations at each extended-model order are plotted in Figures 3.10–3.13 along with the true signal poles z_i (circles). It is clear that for getting high accuracy of estimation with the SNR level equal to 20 dB, the extended-model order should be close to 22 or more when the number of exponentials is 4.

We thus demonstrate that for accurate estimation of the parameters of complex exponential signals, Prony's method should be used with extended-order modelling when data are noise-corrupted. We may argue that the inclusion of a large number of additional complex exponential components can model the additive noise closely, and hence, the signal parameters can be estimated accurately in this case using the signal components. In other words, the extended-order modelling reduces the model-mismatch problem when the signal samples are mixed with noise. In broad sense, the technique of extended-order modelling can be referred to as a method of regularization by a change of solution space.

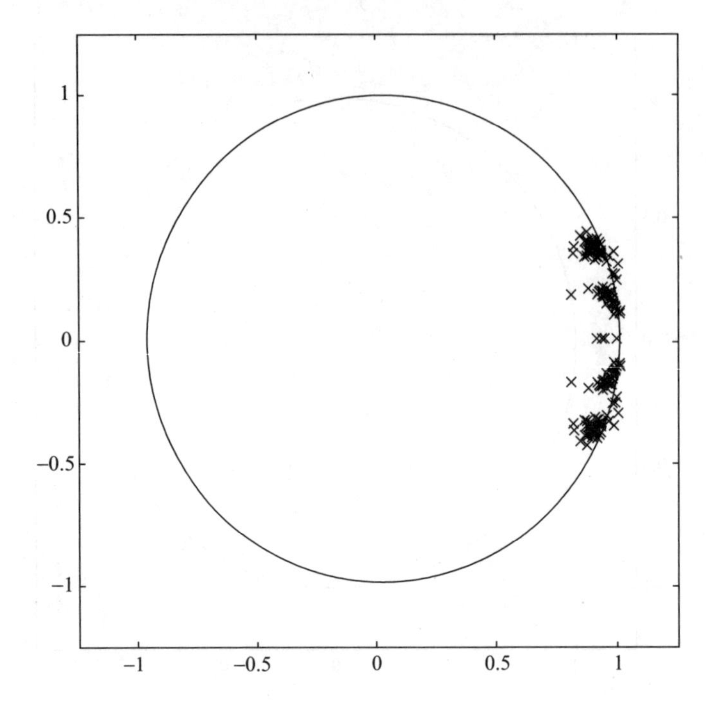

FIGURE 3.10 Signal poles: SNR = 20 dB; Prony's method, model order = 16

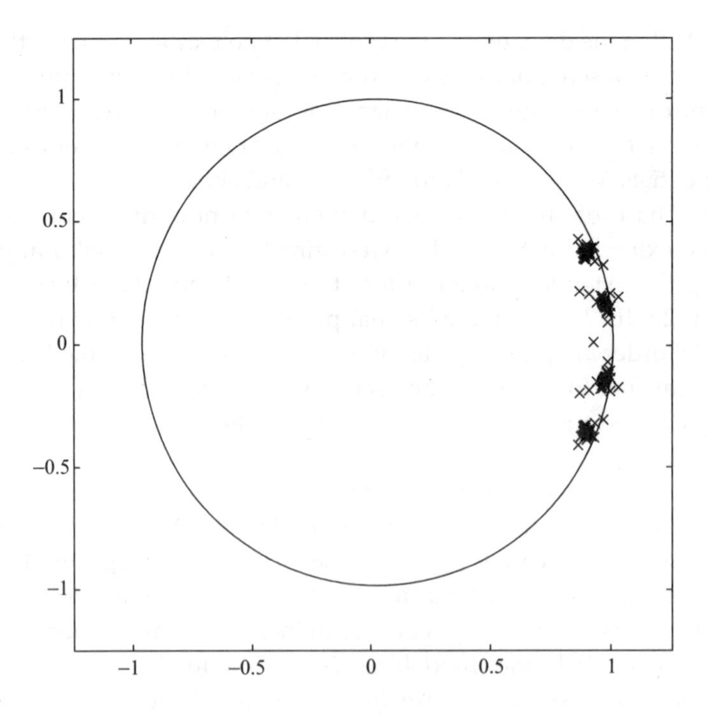

FIGURE 3.11 Signal poles: SNR = 20 dB; Prony's method, model order = 18

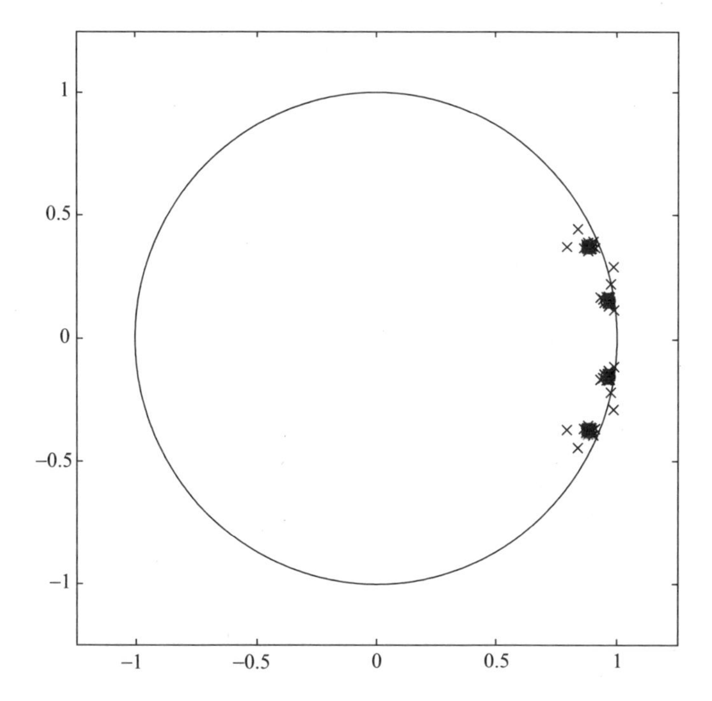

Figure 3.12 Signal poles: SNR = 20 dB; Prony's method, model order = 20

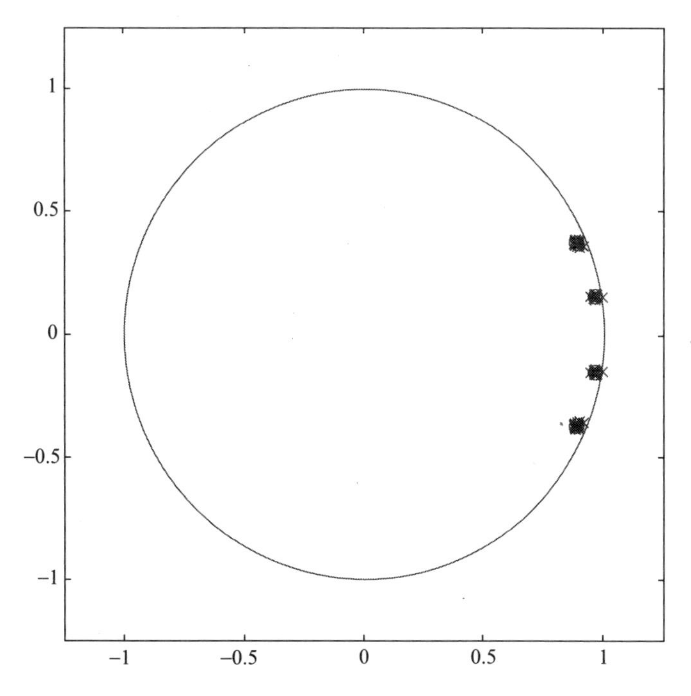

Figure 3.13 Signal poles: SNR = 20 dB; Prony's method, model order = 22

Problems

Some sample problems are given below. Additional problems can be obtained from standard texts on matrix analysis, generalized inverses, and operator theory.

P3.1 Let $\mathbf{A}$ be a matrix of order $m \times m$ given by

$$\mathbf{A} = \begin{bmatrix} 1 & \rho & \rho & \cdots & \rho \\ \rho & 1 & \rho & \cdots & \rho \\ \rho & \rho & 1 & \cdots & \rho \\ \vdots & \vdots & \vdots & \ddots & \vdots \\ \rho & \rho & \rho & \cdots & 1 \end{bmatrix}$$

show that $\mathbf{A}$ is positive definite if and only if $-\frac{1}{m-1} < \rho < 1$.

P3.2 (Hadamard inequality) Show that if $\mathbf{A} = [a_{ij}] \in \mathbb{C}^{n\times n}$ is positive definite, then $\det(\mathbf{A}) \le \prod_{i=1}^{n} a_{ii}$. Moreover, show that equality holds if and only if $\mathbf{A}$ is diagonal.

P3.3 Using the Hadamard inequality, show that for any matrix $\mathbf{B} = [b_{ij}] \in \mathbb{C}^{n\times n}$, $|\det(\mathbf{B})| \le \prod_{i=1}^{n}\left(\sum_{j=1}^{n}|b_{ij}|^2\right)^{1/2}$ and $|\det(\mathbf{B})| \le \prod_{j=1}^{n}\left(\sum_{i=1}^{n}|b_{ij}|^2\right)^{1/2}$. Moreover, show that equality holds if and only if the rows (columns, respectively) of $\mathbf{B}$ are orthogonal.

P3.4 The generalized inverse of the matrix $\mathbf{A}$ is to be computed. Carrying out orthogonal bordering of $\mathbf{A}$, we obtain a non-singular matrix $\mathbf{U}$ as shown below:

$$\mathbf{A} = \begin{bmatrix} 1 & 2 & 1 \\ 2 & 4 & 2 \\ 3 & 6 & 3 \end{bmatrix}, \quad \mathbf{U} = \begin{bmatrix} 1 & 2 & 1 & -1 & -3 \\ 2 & 4 & 2 & 5 & -6 \\ 3 & 6 & 3 & -3 & 5 \\ 5 & -2 & -1 & 0 & 0 \\ -1 & 1 & -1 & 0 & 0 \end{bmatrix}$$

Determine the generalized matrix of $\mathbf{A}$. Is it unique?

P3.5 Prove that the equation $\mathbf{Ax} = \mathbf{b}$, $\mathbf{A} \in \mathbb{C}^{m\times n}$, $\mathbf{b} \in \mathbb{C}^m$ is solvable if and only if $\mathbf{b}$ is orthogonal to all solutions of the homogeneous equation $\mathbf{A}^H\mathbf{y} = \mathbf{0}$.

P3.6 Let $\mathbf{A}$ be a matrix of dimension $m \times n$, and $\mathbf{a}$, $\mathbf{b}$ are column vectors of dimension $m \times 1$ and $n \times 1$, respectively. Let $\mathbf{G}$ be any 1-inverse of $\mathbf{A}$. If it is given that $\mathbf{a} \in \text{span}(\mathbf{A})$ and $\mathbf{b} \in \text{span}(\mathbf{A}^T)$, then show that

$$\mathbf{G}_1 = \mathbf{G} - \frac{(\mathbf{Ga})(\mathbf{b}^T\mathbf{G})}{(1 + \mathbf{b}^T\mathbf{Ga})}$$

is a 1-inverse of $(\mathbf{A} + \mathbf{ab}^T)$, provided $(1 + \mathbf{b}^T\mathbf{Ga}) \ne 0$.

P3.7 Let $\mathbf{A} \in \mathbb{C}^{m\times n}$ have the decomposition given by

$$\mathbf{A} = \mathbf{P}\begin{bmatrix} \mathbf{D} & \mathbf{0} \\ \mathbf{0} & \mathbf{0} \end{bmatrix}\mathbf{Q}$$

where $\mathbf{P}$ and $\mathbf{Q}$ are non-singular matrices of dimensions $m \times m$ and $n \times n$, respectively, and $\mathbf{D}$ is a diagonal matrix of dimension $l \times l$ with rank l, then show that

(i) For any three matrices $\mathbf{E}_1$, $\mathbf{E}_2$ and $\mathbf{E}_3$ of appropriate dimensions,

$$\mathbf{G}_1 = \mathbf{Q}^{-1}\begin{bmatrix} \mathbf{D}^{-1} & \mathbf{E}_1 \\ \mathbf{E}_2 & \mathbf{E}_3 \end{bmatrix}\mathbf{P}^{-1}$$

is a 1-inverse of $\mathbf{A}$.

(ii) For any two matrices $\mathbf{E}_1$ and $\mathbf{E}_2$ of appropriate dimensions,

$$\mathbf{G}_2 = \mathbf{Q}^{-1}\begin{bmatrix} \mathbf{D}^{-1} & \mathbf{E}_1 \\ \mathbf{E}_2 & \mathbf{E}_2\mathbf{D}\mathbf{E}_1 \end{bmatrix}\mathbf{P}^{-1}$$

is a 2-inverse of $\mathbf{A}$. What are the ranks of $\mathbf{A}$, $\mathbf{G}_1$ and $\mathbf{G}_2$?

Bibliography

The list is indicative, not exhaustive. Similar references can be searched in the library.

Ben-Israel, A., and T. N. E. Greville. 2003. *Generalized Inverses: Theory and Applications*. 2nd ed., New York: Springer-Verlag.

Bertero, M., C. De Mol, and G. A. Viano. 1980. 'The Stability of Inverse Problems.' In *Inverse Scattering Problems*, edited by H. P. Baltes, 161–214. Berlin: Springer-Verlag.

Bjerhammar, A. 1973. *Theory of Errors and Generalized Matrix Inverses.* Amsterdam: Elsevier Scientific.

Cadzow, J. A., and M.-M. Wu. 1987. 'Analysis of Transient Data in Noise.' *IEE Proceedings F, Commun., Radar & Signal Process* 134 (1): 69–78.

Campbell, S. L., and C. D. Meyer. 1991. *Generalized Inverses of Linear Transformations*. Mineola, NY: Dover.

Golub, G. H., and C. F. Van Loan. 1996. *Matrix Computations*. 3rd ed., Baltimore: The Johns Hopkins University Press.

Hestenes, M. R., and E. Stiefel. 1952. 'Methods of Conjugate Gradient for Solving Linear Systems.' *J. Res. Nat. Bureau Standards*, 49 (6): 409–436.

Katsaggelos, A. K., J. Biemond, R. M. Mersereau, and R. W. Schafer. 1985. 'A General Formulation of Constrained Iterative Restoration Algorithms.' In Proc. IEEE Int. Conf. Acoustics, Speech, Signal Process. ICASSP.' 85 (10): 700–703.

Kirsch, A. 1996. *An Introduction to the Mathematical Theory of Inverse Problems*. New York: Springer-Verlag.

Kumaresan, R., D. W. Tufts, and L. L. Scharf. 1984. 'A Prony Method for Noisy Data: Choosing the Signal Components and Selecting the Order in Exponential Signal Models.' *Proceedings IEEE* 72 (2): 230–233.

Lanczos, C. 1964. *Applied Analysis*. Englewood Cliffs, NJ: Prentice-Hall.

Moiseiwitsch, B. L. 1977. *Integral Equations*. London: Longman.

Nashed, M. Z. 1976. 'Aspects of Generalized Inverses in Analysis and Regularization.' In *Generalized Inverses and Applications*, edited by M. Z. Nashed, 193–244. Academic Press: New York.

Noble, B. 1976. 'Methods for Computing the Moore–Penrose Generalized Inverse, and Related Matters.' In *Generalized Inverses and Applications*, edited by M. Z. Nashed, 245–301. Academic Press: New York.

Polyanin, A. D., and A. V. Manzhirov. 1998. *Handbook of Integral Equations*, Boca Raton, FL: CRC Press.

Ralston, A., and P. Rabinowitz. 1978. *A First Course in Numerical Analysis*. 2nd ed., Singapore: McGraw-Hill.

Rao, C. R., and S. K. Mitra. 1971. *Generalized Inverse of Matrices and Its Applications*. New York: John Wiley.

Sarkar, T. K., and E. Arvas. 1985. 'On a Class of Finite Step Iterative Methods (Conjugate Directions) for the Solution of an Operator Equation Arising in Electromagnetic.' *IEEE Trans. Antennas & Propagation*, AP-33 (10): 1058–1065.

Sircar, P., and T. K. Sarkar. 1987a. 'Further Investigation on System Identification from Nonuniformly Sampled Data.', In Proc. 30th Midwest Symp. Circuits and Systems, pp. 398–401, Syracuse Univ., NY, 17–18 August.

———. 1987b. 'Some aspects of spectral estimation from non-uniformly spaced signal measurements.' In. Proc. Int. Symp. Electronic Devices, Circuits & Systems, Indian Institute of Technology, Kharagpur, India, 16–18 December.

———. 1988. 'System Identification from Nonuniformly Spaced Signal Measurements.' *Signal Processing*, 14, 253–268.

Sobol, I. 1974. *The Monte Carlo Method*. Chicago: The University of Chicago Press.

Stewart, G. W. 1977. 'Research, Development, and LINPACK.' In Mathematical Software III, pp. 1–14, Academic Press, New York, 1977.

Tikhonov, A. N. 1963. 'Solution of Incorrectly Formulated Problems and the Regularization Method', edited by J. R. Rice. Soviet Math. Doklady, 5, 1035–1038.

Walsh, J. 1971. 'Direct and Indirect Methods.' In *Large Sparse Sets of Linear Equations*, edited by J. K. Reid, New York: Academic Press.

4 Modal Decomposition

4.1 Schmidt Pair and Singular Value Decomposition

Let **A** be a complex matrix of dimension $m \times n$. The vectors $\{\mathbf{u}, \mathbf{v}\}$, where **u** is a complex vector of dimension $m \times 1$ and **v** is complex vector of dimension $n \times 1$, will be called a Schmidt pair for **A** with singular value σ $(\sigma \geq 0)$ if

$$\mathbf{Av} = \sigma \mathbf{u} \tag{4.1}$$

$$\mathbf{A}^H \mathbf{u} = \sigma \mathbf{v} \tag{4.2}$$

where $\mathbf{A}^H$ is the Hermitian or complex conjugate transpose of **A**.

From (4.1) and (4.2), we readily obtain

$$\mathbf{AA}^H \mathbf{u} = \sigma^2 \mathbf{u} \tag{4.3}$$

$$\mathbf{A}^H \mathbf{Av} = \sigma^2 \mathbf{v} \tag{4.4}$$

Therefore, **u** is an eigenvector of $\mathbf{AA}^H$ with eigenvalue σ^2, and **v** is an eigenvector of $\mathbf{A}^H\mathbf{A}$ with the same eigenvalue. Note that the matrices $\mathbf{AA}^H$ and $\mathbf{A}^H\mathbf{A}$ will have the same number of non-zero eigenvalues, and the number is equal to rank (**A**).

Note that $\mathbf{AA}^H$ and $\mathbf{A}^H\mathbf{A}$ are Hermitian positive semi-definite matrices having the following important properties:

Lemma 4.1

If $\mathbf{S} \in \mathbb{C}^{n \times n}$ is Hermitian and positive semi-definite (definite), then all the eigenvalues of **S** are real and non-negative (positive).

Lemma 4.2

If $\mathbf{S} \in \mathbb{C}^{n \times n}$ is Hermitian, then all its eigenvalues are real, and the eigenvectors associated with two different eigenvalues are orthogonal.

Lemma 4.3

If $\mathbf{S} \in \mathbb{C}^{n \times n}$ is Hermitian, then **S** has a full set of n linearly independent eigenvectors, regardless of the multiplicity of the eigenvalues.

Lemma 4.4

If $\mathbf{S} \in \mathbb{C}^{n\times n}$ is Hermitian, then there exists a unitary matrix $\mathbf{U} \in \mathbb{C}^{n\times n}$ and a diagonal matrix $\Lambda \in \mathbb{R}^{n\times n}$ such that $\mathbf{S} = \mathbf{U}\Lambda\mathbf{U}^H$.

Therefore, the eigenvectors of $\mathbf{A}\mathbf{A}^H$ and $\mathbf{A}^H\mathbf{A}$ associated with two different eigenvalues are complex orthogonal; i.e., for $\sigma_i^2 \neq \sigma_j^2$, the complex inner products of the eigenvectors are vanishing:

$$\langle \mathbf{u}_i, \mathbf{u}_j \rangle = \mathbf{u}_i^H \mathbf{u}_j = 0 \tag{4.5}$$

$$\langle \mathbf{v}_i, \mathbf{v}_j \rangle = \mathbf{v}_i^H \mathbf{v}_j = 0 \tag{4.6}$$

where $\mathbf{u}_i$ and $\mathbf{v}_i$ are the eigenvectors of $\mathbf{A}\mathbf{A}^H$ and $\mathbf{A}^H\mathbf{A}$, respectively, with the same eigenvalue $\sigma_i^2 \geq 0$, and $\mathbf{u}_j$, $\mathbf{v}_j$ are the eigenvectors of the matrices with eigenvalue $\sigma_j^2 \geq 0$. Furthermore, each of the matrices $\mathbf{A}\mathbf{A}^H$ and $\mathbf{A}^H\mathbf{A}$ has a full set of linearly independent eigenvectors, even when some of the eigenvalues are repeated.

Suppose now that $\text{rank}(\mathbf{A}) = l \leq q = \min(m,n)$. Let $\{\mathbf{u}_i; i = 1, 2, \ldots, l\}$ be the complex orthonormal ordered set of eigenvectors of $\mathbf{A}\mathbf{A}^H$ corresponding to the eigenvalues $\sigma_1^2 \geq \sigma_2^2 \geq \cdots \geq \sigma_l^2 > 0$, and let $\{\mathbf{v}_i; i = 1, 2, \ldots, l\}$ be the ordered set of eigenvectors of $\mathbf{A}^H\mathbf{A}$ for the same eigenvalues. The orthonormal sets $\{\mathbf{u}_i\}$ and $\{\mathbf{v}_i\}$ can be completed to form the bases of the m- and n-dimensional complex vector spaces $\mathbb{C}^m$ and $\mathbb{C}^n$, respectively.

The unitary matrices $\mathbf{U} \in \mathbb{C}^{m\times m}$ and $\mathbf{V} \in \mathbb{C}^{n\times n}$ are formed by stacking orthonormal vectors as shown below,

$$\mathbf{U} = [\mathbf{u}_1, \ldots, \mathbf{u}_l, \mathbf{u}_{l+1}, \ldots, \mathbf{u}_m] \tag{4.7}$$

$$\mathbf{V} = [\mathbf{v}_1, \ldots, \mathbf{v}_l, \mathbf{v}_{l+1}, \ldots, \mathbf{v}_n] \tag{4.8}$$

The matrices, in turn, can be used to transform $\mathbf{A}$ into a diagonal matrix $\Sigma \in \mathbb{R}^{m\times n}$ as revealed in the following theorem:

Theorem 4.5

Let $\mathbf{A}$ be a complex matrix of dimension $m \times n$, and $\text{rank}(\mathbf{A}) = l \leq q = \min(m,n)$, then there exist unitary matrices

$\mathbf{U} = [\mathbf{u}_1, \ldots, \mathbf{u}_m] \in \mathbb{C}^{m\times m}$ and $\mathbf{V} = [\mathbf{v}_1, \ldots, \mathbf{v}_n] \in \mathbb{C}^{n\times n}$ such that

$$\mathbf{U}^H\mathbf{A}\mathbf{V} = \Sigma \tag{4.9}$$

where $\Sigma \in \mathbb{R}^{m\times n}$ is given by $\Sigma = \begin{cases} \text{diag}(\sigma_1, \ldots, \sigma_l) & \text{for } l = q \\ \text{diag}(\sigma_1, \ldots, \sigma_l, 0, \ldots) & \text{for } l < q \end{cases}$

and, $\sigma_1 \geq \sigma_2 \geq \cdots \geq \sigma_l > 0$. Note that Σ is a pseudo-diagonal matrix which may not necessarily be a square matrix, but it has non-zero entries only on the main diagonal.

In solving many problems, diagonalization of a matrix, i.e., transforming a matrix into a diagonal matrix becomes a desirable step. For an arbitrary matrix $\mathbf{A}$, the Hermitian matrices $\mathbf{A}\mathbf{A}^H$ and $\mathbf{A}^H\mathbf{A}$ are diagonalizable by Lemma 4.4. However, Theorem 4.5 provides a means to diagonalize an arbitrary matrix without the necessity of forming a Hermitian matrix.

Rewriting (4.9) in the form of matrix decomposition, we get,

$$\mathbf{A} = \mathbf{U}\Sigma\mathbf{V}^H \tag{4.10}$$

which is the singular value decomposition (SVD) of $\mathbf{A}$. Note that (4.10) can be expressed as an expansion of $\mathbf{A}$ in terms of matrices of unit rank,

$$\mathbf{A} = \sum_{i=1}^{l} \sigma_i \mathbf{u}_i \mathbf{v}_i^H \tag{4.11}$$

where $\mathbf{u}_i$ is the left singular vector and $\mathbf{v}_i$ is the right singular vector, corresponding to the singular value $\sigma_i > 0$.

4.2 Rank Property and Subspaces

Various properties related to the structure of a matrix are revealed in the SVD of the matrix. Consider the matrix $\mathbf{A} \in \mathbb{C}^{m\times n}$ with the SVD given by (4.10) or (4.11). Then, $\text{rank}(\mathbf{A}) = l$, where l is the number of non-zero singular values of $\mathbf{A}$. Moreover, the ordered bases of various spaces of the matrices $\mathbf{A}$ and $\mathbf{A}^H$ are found in terms of the singular vectors as follows:

Proposition 4.6

Let $\mathbf{A} = \mathbf{U}\Sigma\mathbf{V}^H$ be the SVD of $\mathbf{A} \in \mathbb{C}^{m\times n}$. Then

$$\mathscr{R}\left(\mathbf{A}^H\right) = \text{range space of } \mathbf{A}^H = \text{span}\{\mathbf{v}_1, \mathbf{v}_2, \ldots, \mathbf{v}_l\} \tag{4.12}$$

$$\mathscr{N}(\mathbf{A}) = \text{null space of } \mathbf{A} = \mathbb{C}^n - \mathscr{R}\left(\mathbf{A}^H\right) = \text{span}\{\mathbf{v}_{l+1}, \ldots, \mathbf{v}_n\} \tag{4.13}$$

$$\mathscr{R}(\mathbf{A}) = \text{range space of } \mathbf{A} = \text{span}\{\mathbf{u}_1, \mathbf{u}_2, \ldots, \mathbf{u}_l\} \tag{4.14}$$

$$\mathscr{N}\left(\mathbf{A}^H\right) = \text{null space of } \mathbf{A}^H = \mathbb{C}^m - \mathscr{R}(\mathbf{A}) = \text{span}\{\mathbf{u}_{l+1}, \ldots, \mathbf{u}_m\} \tag{4.15}$$

and

$$\text{rank}(\mathbf{A}) = l = \dim(\mathscr{R}(\mathbf{A})) = \dim\left(\mathscr{R}\left(\mathbf{A}^H\right)\right) \tag{4.16}$$

Figure 4.1 shows the range and null spaces of the matrices $\mathbf{A}$ and $\mathbf{A}^H$, where $\mathbf{0}$ denotes the null vector of appropriate dimension, and the domain spaces of $\mathbf{A}$ and $\mathbf{A}^H$ are $\mathbb{C}^n$ and $\mathbb{C}^m$, respectively.

An important property of the SVD of a matrix is that the norm of the matrix is expressed in terms of the singular values as follows:

Proposition 4.7

Let $\mathbf{A} = \mathbf{U}\Sigma\mathbf{V}^H$ be the SVD of $\mathbf{A} = \left[a_{ij}\right] \in \mathbb{C}^{m\times n}$. Then,

$$\|\mathbf{A}\|_2 = \text{Matrix 2-norm} = \sup_{\mathbf{x}\neq\mathbf{0}} \frac{\|\mathbf{Ax}\|_2}{\|\mathbf{x}\|_2} = \max_{\|\mathbf{x}\|_2=1} \|\mathbf{AX}\|_2 = \sigma_1 \tag{4.17}$$

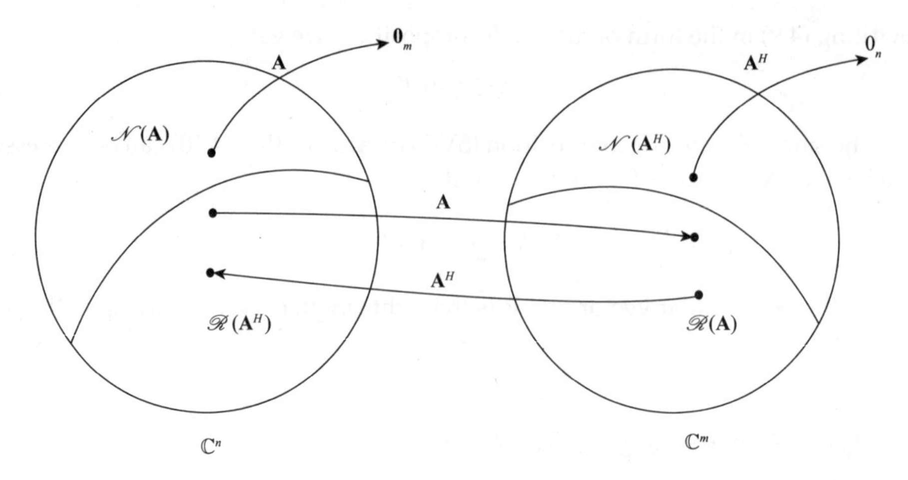

FIGURE 4.1 Range and null spaces of $\mathbf{A}$ and $\mathbf{A}^H$

$$\|\mathbf{A}\|_F = \text{Frobenius norm} = \left(\sum_{i=1}^{m}\sum_{j=1}^{n}|a_{ij}|^2\right)^{1/2} = \left(\sum_{i=1}^{l}\sigma_i^2\right)^{1/2} \tag{4.18}$$

The above results follow directly from the property that the matrix norm is invariant with respect to unitary transformations, i.e.,

$$\|\mathbf{A}\| = \|\Sigma\| \tag{4.19}$$

for **U** and **V** being unitary matrices. Since the matrix norm provides a measure of distance on the space of matrices, the singular values being related to the matrix norm, can be used to determine the *effective* rank of a matrix or to quantify the *closeness* of singularity of a matrix.

Theorem 4.8

Let the SVD of $\mathbf{A} = [a_{ij}] \in \mathbb{C}^{m\times n}$ is given $\mathbf{U}\Sigma\mathbf{V}^H$.

If $k < l = \text{rank}(\mathbf{A})$ and

$$\mathbf{A}_k = \sum_{i=1}^{k}\sigma_i\mathbf{u}_i\mathbf{v}_i^H, \tag{4.20}$$

then

$$\min_{\text{rank}(\mathbf{B})=k}\|\mathbf{A}-\mathbf{B}\|_2 = \sigma_{k+1}, \tag{4.21}$$

and

$$\mathbf{B} = \mathbf{A}_k. \tag{4.22}$$

An application of the above theorem is useful when it is known that the elements a_{ij} of the matrix **A** is accurate to within $\pm\varepsilon$, then it will make sense to work with the approximate matrix **B**, instead of **A**, such that the effective rank k is chosen by using a threshold on the singular values:

$$\sigma_1 \geq \cdots \geq \sigma_k > \varepsilon \geq \sigma_{k+1} \geq \cdots \geq \sigma_l > 0 \tag{4.23}$$

Theorem 4.9

Let the SVD of $\mathbf{A} \in \mathbb{C}^{m \times n}$ is given $\mathbf{U\Sigma V}^H$.

If $k < l = \text{rank}(\mathbf{A})$ and

$$\mathbf{A}_k = \sum_{i=1}^{k} \sigma_i \mathbf{u}_i \mathbf{v}_i^H, \tag{4.24}$$

then

$$\min_{\text{rank}(\mathbf{B})=k} \|\mathbf{A} - \mathbf{B}\|_F = \left(\sum_{i=k+1}^{l} \sigma_i^2 \right)^{1/2}, \tag{4.25}$$

and

$$\mathbf{B} = \mathbf{A}_k \tag{4.26}$$

which is unique when $\sigma_{k+1} < \sigma_k$.

An application of the above theorem is useful in identifying the components of a matrix, which are significant, and the components which are insignificant. In other words, we can measure the closeness of a matrix from rank-deficiency (singularity for a square matrix). Specify the singular value normalized error criterion as the ratio

$$\rho_k = \frac{\sum_{i=k+1}^{l} \sigma_i^2}{\sum_{i=1}^{l} \sigma_i^2} \tag{4.27}$$

to be bounded on the upper side, and use this error criterion to separate the *principal components* (for $i = 1, \ldots, k$) of a matrix from its *non-principal components* (for $i = k + 1, \ldots, l$). The constrained-rank k is determined as the least value for which the normalized error ρ_k is less than an error bound δ. The error bound δ is determined based on knowledge of data imprecision and computational error.

4.3 Matrix Inverse and Condition Number

Consider a linear system $\mathbf{Ax} = \mathbf{b}$, where $\mathbf{A} \in \mathbb{C}^{n \times n}$ is non-singular and $\mathbf{b} \in \mathbb{C}^n$. We want to investigate how perturbations in $\mathbf{A}$ and $\mathbf{b}$ will deviate the solution $\mathbf{x} \in \mathbb{C}^n$. Applying the variational calculus approach, we get the deviation of $\mathbf{x} = \mathbf{A}^{-1}\mathbf{b}$ as follows,

$$\begin{aligned} d\mathbf{x} &= d\left(\mathbf{A}^{-1}\mathbf{b}\right) = -\mathbf{A}^{-1}(d\mathbf{A})\mathbf{A}^{-1}\mathbf{b} + \mathbf{A}^{-1}d\mathbf{b} \\ &= \mathbf{A}^{-1}\left[-(d\mathbf{A})\mathbf{x} + d\mathbf{b}\right] \end{aligned} \tag{4.28}$$

and hence, using the Cauchy–Schwartz inequality, we obtain

$$\|d\mathbf{x}\| \le \|\mathbf{A}^{-1}\| \left[\|d\mathbf{A}\| \cdot \|\mathbf{x}\| + \|d\mathbf{b}\|\right] \tag{4.29}$$

Now, from $\mathbf{Ax} = \mathbf{b}$, we have

$$\|\mathbf{b}\| \le \|\mathbf{A}\| \cdot \|\mathbf{x}\| \tag{4.30}$$

Combining (4.29) and (4.30), we derive the following upper bound for the relative deviation in $\mathbf{x}$,

$$\frac{\|d\mathbf{x}\|}{\|\mathbf{x}\|} \leq \|\mathbf{A}\| \cdot \left\|\mathbf{A}^{-1}\right\| \left[\frac{\|d\mathbf{A}\|}{\|\mathbf{A}\|} + \frac{\|d\mathbf{b}\|}{\|\mathbf{b}\|}\right] \tag{4.31}$$

which holds true for any vector norm and the consistent matrix norm.

Define the condition number, cond($\mathbf{A}$), as

$$\operatorname{cond}(\mathbf{A}) = \|\mathbf{A}\| \cdot \left\|\mathbf{A}^{-1}\right\| \tag{4.32}$$

Let $\{\sigma_i, \mathbf{u}_i, \mathbf{v}_i\}$ be the SVD of $\mathbf{A}$, i.e., $\mathbf{A} = \mathbf{U}\Sigma\mathbf{V}^H$. Since $\mathbf{A}$ is non-singular, $\mathbf{A}^{-1} = \mathbf{V}\Sigma^{-1}\mathbf{U}^H$, which implies that $\{1/\sigma_i, \mathbf{v}_i, \mathbf{u}_i\}$ will be the SVD of $\mathbf{A}^{-1}$. Hence, the norms of the matrices are $\|\mathbf{A}\|_2 = \sigma_1$ and $\left\|\mathbf{A}^{-1}\right\|_2 = 1/\sigma_n$, which provide the condition number in 2-norm as

$$\operatorname{cond}(\mathbf{A}) = \frac{\sigma_1}{\sigma_n} \tag{4.33}$$

We now summarize the above results in the following theorem:

Theorem 4.10

Let the linear system $\mathbf{Ax} = \mathbf{b}$ be solved with error in $\mathbf{A}$ and $\mathbf{b}$. Then, the relative deviation in the solution $\mathbf{x}$ in 2-norm will be bounded as

$$\frac{\|d\mathbf{x}\|}{\|\mathbf{x}\|} \leq \operatorname{cond}(\mathbf{A}) \left[\frac{\|d\mathbf{A}\|}{\|\mathbf{A}\|} + \frac{\|d\mathbf{b}\|}{\|\mathbf{b}\|}\right] \tag{4.34}$$

where $\|d\mathbf{A}\|/\|\mathbf{A}\|$ is the relative error in $\mathbf{A}$, $\|d\mathbf{b}\|/\|\mathbf{b}\|$ is the relative error in $\mathbf{b}$, and cond($\mathbf{A}$) is the ratio of the largest to smallest singular values of the matrix $\mathbf{A}$.

The implication of the above theorem is that the relative error in $\mathbf{A}$ or in $\mathbf{b}$ can be magnified by the condition number to provide the worst case relative deviation in $\mathbf{x}$. Note that cond ($\mathbf{A}$) = ∞ for $\mathbf{A}$ being singular, and cond($\mathbf{A}$) = 1 for $\mathbf{A}$ being a unitary matrix. A high value of the condition number indicates an *ill-conditioned* linear system, whereas a *well-conditioned* system is characterized by a low value. It is difficult to solve an ill-conditioned system accurately.

Next, consider the general case where $\mathbf{A} \in \mathbb{C}^{m\times n}$ and $\operatorname{rank}(\mathbf{A}) = l \leq q = \min(m, n)$. Let $\mathbf{U}\Sigma\mathbf{V}^H$ be the SVD of $\mathbf{A}$ with $\Sigma \in \mathbb{R}^{m\times n}$ as given by (4.9). The Moore–Penrose inverse $\mathbf{A}^+$ of $\mathbf{A}$ is expressed as

$$\mathbf{A}^+ = \mathbf{V}\Sigma^+\mathbf{U}^H \tag{4.35}$$

with $\Sigma^+ \in \mathbb{R}^{m\times n}$ as given by $\Sigma^+ = \begin{cases} \operatorname{diag}(1/\sigma_1, \ldots, 1/\sigma_l) & \text{for } l = q \\ \operatorname{diag}(1/\sigma_1, \ldots, 1/\sigma_l, 0, \ldots) & \text{for } l < q \end{cases}$

Accordingly, the solution of the linear system is computed as $\mathbf{x} = \mathbf{A}^+\mathbf{b}$.

Suppose now that $\operatorname{rank}(\mathbf{A}) = l = q = \min(m, n)$, however, the matrix $\mathbf{A}$ is *nearly* rank-deficient. We identify the principal components of the matrix $\mathbf{A}$ by using the singular value normalized error criterion (4.27), and form the matrix $\mathbf{B}$ with principal components.

The rank-constrained matrix $\mathbf{B} = \mathbf{U}\Sigma_c\mathbf{V}^H$ with $\Sigma_c = \text{diag}(\sigma_1, \ldots, \sigma_k, 0, \ldots)$ thus formed is then employed to compute the Moore–Penrose inverse $\mathbf{B}^+$ expressed as

$$\mathbf{B}^+ = \mathbf{V}\Sigma_c^+\mathbf{U}^H \tag{4.36}$$

where $\Sigma_c^+ \in \mathbb{R}^{m\times n}$ and it is given by $\Sigma_c^+ = \text{diag}(1/\sigma_1, \ldots, 1/\sigma_k, 0, \ldots)$. The solution $\mathbf{x} = \mathbf{B}^+\mathbf{b}$ is called the *principal component solution*.

Theorem 4.11

Let the linear system $\mathbf{Ax} = \mathbf{b}$ be solved where $\mathbf{A} \in \mathbb{C}^{m\times n}$ is nearly rank-deficient. Let $\mathbf{U}\Sigma\mathbf{V}^H$ be the SVD of $\mathbf{A}$. The principal components of $\mathbf{A}$ are identified by the least value of k for which the singular value normalized error $\rho_k = \sum_{i=k+1}^{l}\sigma_i^2 / \sum_{i=1}^{l}\sigma_i^2$ is less than a specified error bound δ. The matrix $\mathbf{B}$ formed by the principal components of $\mathbf{A}$ is used to compute the principal component solution $\mathbf{x}$ as

$$\mathbf{x} = \mathbf{B}^+\mathbf{b}, \tag{4.37}$$

where the Moore–Penrose inverse $\mathbf{B}^+$ is given by $\mathbf{B}^+ = \mathbf{V}\Sigma_c^+\mathbf{U}^H$ with $\Sigma_c^+ = \text{diag}(1/\sigma_1, \ldots, 1/\sigma_k, 0, \ldots)$.

4.4 Generalized Singular Value Decomposition

Simultaneous diagonalization of two matrices is an important step in solving many problems. Consider a complex matrix $\mathbf{A}$ of dimension $m \times n$ with $m \le n$, and another complex matrix $\mathbf{B}$ of dimension $m \times p$. We form the Hermitian positive semi-definite matrices $\mathbf{AA}^H$ and $\mathbf{BB}^H$, and focus on the problem of simultaneous diagonalization of the Hermitian matrices, both of dimension $m \times m$.

Lemma 4.12

If $\mathbf{S}, \mathbf{R} \in \mathbb{C}^{m\times m}$ are two Hermitian matrices and there is a real weighted linear combination of $\mathbf{S}$ and $\mathbf{R}$, which is positive definite, then there exists a non-singular $\mathbf{C} \in \mathbb{C}^{m\times m}$ such that both $\mathbf{CSC}^H$ and $\mathbf{CRC}^H$ are diagonal.

Corollary 4.13

If $\mathbf{S}, \mathbf{R} \in \mathbb{C}^{m\times m}$ are two Hermitian matrices and $\mathbf{R}$ is positive definite, then there exists a non-singular $\mathbf{C} \in \mathbb{C}^{m\times m}$ such that $\mathbf{CSC}^H$ is diagonal and $\mathbf{CRC}^H = \mathbf{I}$, $\mathbf{I}$ being the identity matrix.

We now explore the relationship between the problem of simultaneous diagonalization of two matrices and the generalized eigenvalue problem as described below. Given a Hermitian matrix $\mathbf{S} \in \mathbb{C}^{m\times m}$ and a Hermitian positive definite $\mathbf{R} \in \mathbb{C}^{m\times m}$, we get the Hermitian-definite generalized eigenequation,

$$\mathbf{Sc} = \lambda\mathbf{Rc} \quad \Rightarrow \quad (\mathbf{S} - \lambda\mathbf{R})\mathbf{c} = \mathbf{0} \tag{4.38}$$

which is to be solved for a non-zero vector $\mathbf{c}$ and a scalar λ. The scalar λ is the eigenvalue to be determined by setting the matrix pencil $(\mathbf{S} - \lambda\mathbf{R})$ singular, i.e.,

$$\lambda(\mathbf{S}, \mathbf{R}) = \{\lambda \mid \det(\mathbf{S} - \lambda\mathbf{R}) = 0\} \tag{4.39}$$

Lemma 4.14

If the matrix pencil $(\mathbf{S}-\lambda\mathbf{R}) \in \mathbb{C}^{m\times m}$ is Hermitian-definite, i.e., $\mathbf{S}$ is Hermitian and $\mathbf{R}$ is Hermitian positive definite, then there exists a non-singular $\mathbf{C}=[\mathbf{c}_1,\ldots,\mathbf{c}_m]$ such that

$$\mathbf{CSC}^H = \text{diag}(\alpha_1,\ldots,\alpha_m), \quad \alpha_i \text{ real} \tag{4.40}$$

and

$$\mathbf{CRC}^H = \text{diag}(\beta_1,\ldots,\beta_m), \quad \beta_i \text{ real}, \ \beta_i > 0 \tag{4.41}$$

Moreover,

$$\mathbf{Sc}_i = \lambda_i \mathbf{Rc}_i \quad \text{for } i=1,\ldots,m \tag{4.42}$$

where $\lambda_i = \alpha_i/\beta_i$ is real.

The simultaneous diagonalization problem of the Hermitian-definite matrix pencil is equivalent to the eigenvalue problem:

$$\mathbf{R}^{-1}\mathbf{Sc} = \lambda\mathbf{c} \tag{4.43}$$

where $\mathbf{S},\mathbf{R} \in \mathbb{C}^{m\times m}$ are Hermitian matrices, and in addition, $\mathbf{R}$ is positive definite, ensuring that $\mathbf{R}^{-1}\mathbf{S}$ is also Hermitian.

Corollary 4.15

If the matrix pencil $(\mathbf{S}-\lambda\mathbf{R}) \in \mathbb{C}^{m\times m}$ is Hermitian-semidefinite, i.e., $\mathbf{S}$ is Hermitian and $\mathbf{R}$ is Hermitian positive semidefinite, then there exists a non-singular $\mathbf{C}=[\mathbf{c}_1,\ldots,\mathbf{c}_m]$ such that

$$\mathbf{CSC}^H = \text{diag}(\alpha_1,\ldots,\alpha_n), \quad \alpha_i \text{ real} \tag{4.44}$$

and

$$\mathbf{CRC}^H = \text{diag}(\beta_1,\ldots,\beta_l,0,\ldots), \quad \beta_i \text{ real}, \ \beta_i > 0 \tag{4.45}$$

where $\text{rank}(\mathbf{R}) = l$.

Moreover,

$$\mathbf{Sc}_i = \lambda_i \mathbf{Rc}_i \quad \text{for } i=1,\ldots,l \tag{4.46}$$

where $\lambda_i = \alpha_i/\beta_i$ is real.

Furthermore, the full set of linearly independent vectors $\{\mathbf{c}_1,\ldots,\mathbf{c}_m\}$ are $\mathbf{R}$-orthogonal, i.e.,

$$\mathbf{c}_i^H\mathbf{Rc}_j = 0 \quad \text{for } i \neq j \tag{4.47}$$

The simultaneous diagonalization problem of the Hermitian-semidefinite matrix pencil is equivalent to the eigenvalue problem:

$$\mathbf{R}^{+}\mathbf{Sc} = \lambda\mathbf{c} \tag{4.48}$$

where $\mathbf{S},\mathbf{R} \in \mathbb{C}^{m\times m}$ are Hermitian matrices, $\mathbf{R}$ is positive semidefinite and $\mathbf{R}^+$ is the Moore–Penrose inverse of $\mathbf{R}$.

It is clear now that the matrices $\mathbf{AA}^H$ and $\mathbf{BB}^H$ can be simultaneously diagonalized, as intended at the beginning of this section, by applying Corollary 4.15. In this case, the generalized eigenvalues $\lambda(\mathbf{AA}^H, \mathbf{BB}^H)$ are real and non-negative. Note that it is often preferable to find a means for simultaneous diagonalization of complex matrices $\mathbf{A}$ and $\mathbf{B}$

without forming the Hermitian matrices $\mathbf{AA}^H$ and $\mathbf{BB}^H$. This is achieved by the following theorem.

Theorem 4.16

Let $\mathbf{A}$ be a complex matrix of dimension $m \times n$ with $m \leq n$ and $\mathbf{B}$ be a complex matrix of dimension $m \times p$, $q = \min(m,p)$, then there exist unitary matrices $\mathbf{U} \in \mathbb{C}^{n\times n}$ and $\mathbf{V} \in \mathbb{C}^{p\times p}$, and an invertible matrix $\mathbf{C} \in \mathbb{C}^{m\times m}$ such that

$$\mathbf{CAU}^H = \mathbf{D}_\mathrm{A} = \mathrm{diag}(a_1, \ldots, a_m), \quad a_i \text{ real}, \ a_i \geq 0 \tag{4.49}$$

and

$$\mathbf{CBV}^H = \mathbf{D}_\mathrm{B} = \left\{ \begin{matrix} \mathrm{diag}(b_1, \ldots, b_l) & \text{for } l = q \\ \mathrm{diag}(b_1, \ldots, b_l, 0, \ldots) & \text{for } l < q \end{matrix} \right\} \ b_i \text{ real}, \ b_i > 0 \tag{4.50}$$

where $\mathrm{rank}(\mathbf{B}) = l$.

Moreover, the set of values $\sigma(\mathbf{A}, \mathbf{B}) = \{a_1/b_1, \ldots, a_l/b_l\}$ are related to the set of generalized eigenvalues λ of the matrix-pair $\mathbf{AA}^H$ and $\mathbf{BB}^H$ as

$$\sigma^2(\mathbf{A}, \mathbf{B}) = \lambda\left(\mathbf{AA}^H, \mathbf{BB}^H\right). \tag{4.51}$$

and, the column of $\mathbf{C} = [\mathbf{c}_1, \ldots, \mathbf{c}_m]$ satisfy

$$\mathbf{AA}^H \mathbf{c}_i = \sigma_i^2 \mathbf{BB}^H \mathbf{c}_i \quad \text{for } i = 1, \ldots, l \tag{4.52}$$

where $\sigma_i^2 = a_i^2 / b_i^2 \geq 0$, and

$$\mathbf{c}_i^H \mathbf{BB}^H \mathbf{c}_j = 0 \quad \text{for } i \neq j \tag{4.53}$$

The simultaneous decomposition of the matrices $\mathbf{A}$ and $\mathbf{B}$ as

$$\mathbf{A} = \mathbf{C}^{-1} \mathbf{D}_\mathrm{A} \mathbf{U} \tag{4.54}$$

and

$$\mathbf{B} = \mathbf{C}^{-1} \mathbf{D}_\mathrm{B} \mathbf{V} \tag{4.55}$$

is referred to as the generalized singular value decomposition (GSVD) of the matrix-pair $(\mathbf{A}, \mathbf{B})$. The elements of the set $\sigma(\mathbf{A}, \mathbf{B})$ are the generalized singular values, and the vectors $\{\mathbf{c}_1, \ldots, \mathbf{c}_n\}$ are the generalized singular vectors of the pair $(\mathbf{A}, \mathbf{B})$. The GSVD is a generalization of the SVD in the sense that if $\mathbf{B} = \mathbf{I}_m$, $\mathbf{I}_m$ being the identity matrix of dimension $m \times m$, then $\sigma(\mathbf{A}, \mathbf{B}) = \sigma(\mathbf{A})$.

4.5 Energy Distribution in Matrix Space

Consider a sequence of complex m-vectors $\{\mathbf{a}_j; j = 1, \ldots, n\}$ and the associated matrix $\mathbf{A} = [\mathbf{a}_1, \ldots, \mathbf{a}_n] \in \mathbb{C}^{m\times n}$. The total energy of the vector sequence is defined by the Frobenius norm of $\mathbf{A} = [a_{ij}]$ as

$$E[\mathbf{A}] = \|\mathbf{A}\|_F^2 = \sum_{i=1}^{m} \sum_{j=1}^{n} |a_{ij}|^2 \tag{4.56}$$

4.5.1 Oriented Energy

Let $\mathbf{A}$ be a complex matrix of dimension $m \times n$ and $\{\mathbf{a}_j; j = 1, \ldots, n\}$ be its column-vectors. For any unit vector $\mathbf{d} \in \mathbb{C}^m$, the energy E_{d} of the vector-set $\{\mathbf{a}_j\}$, $\mathbf{a}_j \in \mathbb{C}^m$, measured in the direction $\mathbf{d}$, is defined as

$$E_{\text{d}}[\mathbf{A}] = \sum_{j=1}^{n} \left|\mathbf{d}^H \mathbf{a}_j\right|^2 \tag{4.57}$$

Theorem 4.17

Let $\mathbf{A}$ be a complex matrix of dimension $m \times n$ with $m \le n$, and $\mathbf{A} = \sum_{i=1}^{l} \sigma_i \mathbf{u}_i \mathbf{v}_i^H$ be its SVD. Then each direction, along which the oriented energy is maximal, is generated by a left singular vector, and the maximal energy is equal to the corresponding singular value squared:

$$E_{\text{u}_i}[\mathbf{A}] = \sigma_i^2 \tag{4.58}$$

Corollary 4.18

Let $\mathbf{A}$ be a complex matrix of dimension $m \times n$ with $m \le n$, and $\mathbf{A} = \sum_{i=1}^{l} \sigma_i \mathbf{u}_i \mathbf{v}_i^H$ be its SVD. For all complex m-vectors $\mathbf{d}$ with $\|\mathbf{d}\|_2 = 1$, if $\mathbf{d} = \sum_{i=1}^{l} \gamma_i \mathbf{u}_i$, then

$$E_{\text{d}}[\mathbf{A}] = \sum_{i=1}^{l} \gamma_i \sigma_i^2 \tag{4.59}$$

Corollary 4.19

Consider a sequence of complex m-vectors $\{\mathbf{a}_j; j = 1, \ldots, n\}$ and the associated matrix $\mathbf{A} = [\mathbf{a}_1, \ldots, \mathbf{a}_n] \in \mathbb{C}^{m \times n}$, $m \le n$ with SVD $\mathbf{A} = \sum_{i=1}^{l} \sigma_i \mathbf{u}_i \mathbf{v}_i^H$. Then the sequence of m-vectors $\{\sigma_1 \mathbf{u}_1, \sigma_2 \mathbf{u}_2, \ldots, \sigma_l \mathbf{u}_l\}$ has the same oriented energy distribution as that of $\{\mathbf{a}_j\}$.

According to Theorem 4.17, the SVD of a matrix gives indication about the distribution of energy in the range space of the matrix. In other words, the SVD provides the spatial energy spectral analysis of the matrix, and reveals that a spectral peak occurs along each direction of a left singular vector with energy equal to the corresponding singular value squared. Thus, the SVD separates the matrix space into one-dimensional subspaces whose contributions can be separately computed and superimposed as shown by (4.59). Moreover, various properties of the vector sequence $\{\mathbf{a}_j\}$, which may not necessarily be a finite sequence, will be preserved in the finite set of the left singular vectors scaled by the corresponding singular values, as given by Corollary 4.19. Thus, we obtain the decomposition of a matrix into fundamental modes through the SVD. Note that alternatively we can form the Hermitian matrix $\mathbf{AA}^H$ and maximize the quadratic term $\mathbf{u}^H\mathbf{AA}^H\mathbf{u}$ subject to the constraint $\mathbf{u}^H\mathbf{u} = 1$, which leads to the eigenequation $\mathbf{AA}^H\mathbf{u} = \sigma^2\mathbf{u}$ to get the fundamental modes of the matrix. However, the advantage of the SVD is that the decomposition provides the fundamental modes without the need for computing the Hermitian matrix $\mathbf{AA}^H$.

The oriented energy distribution of a sequence of m-vectors $\{\mathbf{a}_j\}$ is called *isotropic* when the singular values of the corresponding matrix $\mathbf{A}$ are all equal. The range space of such matrices is spatially white, which means that the energy distribution is same along any direction of the matrix space. Note that the notion of *matrix space* refers to a multidimensional complex (or real) linear space with an assigned oriented energy distribution.

4.5.2 Oriented Signal-to-Signal Ratio

Let $\mathbf{A}$ and $\mathbf{B}$ be two complex matrices of dimension $m \times n$, and let $\{\mathbf{a}_j\}$ and $\{\mathbf{b}_j\}$ be their column-vectors, respectively. For any unit vector $\mathbf{d} \in \mathbb{C}^m$, the signal-to-signal ratio R_d of the vector-sets $\{\mathbf{a}_j\}$, $\mathbf{a}_j \in \mathbb{C}^m$ and $\{\mathbf{b}_j\}$, $\mathbf{b}_j \in \mathbb{C}^m$, measured in the direction $\mathbf{d}$, is defined as

$$R_\mathrm{d}[\mathbf{A},\mathbf{B}] = \frac{E_\mathrm{d}[\mathbf{A}]}{E_\mathrm{d}[\mathbf{B}]} \tag{4.60}$$

The signal-to-signal ratio R_d of two vector sequences, measured in a direction $\mathbf{d}$ is the ratio of oriented energies E_d's of the vector sequences in that direction.

Theorem 4.20

Let $\mathbf{A}$ and $\mathbf{B}$ be two complex matrices of dimension $m \times n$ with $m \leq n$, and

$$\mathbf{A} = \mathbf{C}^{-1}\mathbf{D}_\mathrm{A}\mathbf{U}, \quad \mathbf{D}_\mathrm{A} = \mathrm{diag}\{a_i\}, \ a_i \geq 0,$$
$$\mathbf{B} = \mathbf{C}^{-1}\mathbf{D}_\mathrm{B}\mathbf{V}, \quad \mathbf{D}_\mathrm{B} = \mathrm{diag}\{b_i\}, \ b_i \geq 0$$

be their GSVD, where the generalized singular values are ordered such that $(a_1/b_1) \geq (a_2/b_2) \geq \cdots \geq (a_l/b_l) \geq 0$, and rank $(\mathbf{B}) = l$. Then each direction along which the signal-to-signal ratio is maximal, is generated by a row of the matrix $\mathbf{C}$, and the maximal signal-to-signal ratio is equal to the corresponding generalized singular value squared:

$$R_{\mathrm{e}_i}[\mathbf{A},\mathbf{B}] = (a_i/b_i)^2 \tag{4.61}$$

where $\mathbf{e}_i = \mathbf{c}_i/\|\mathbf{c}_i\|$, $\mathbf{c}_i^T$ being the ith row of $\mathbf{C}$.

Corollary 4.21

Let $\mathbf{A}$ and $\mathbf{B}$ be two complex matrices of dimension $m \times n$ with $m \leq n$, and

$$\mathbf{A} = \mathbf{C}^{-1}\mathbf{D}_\mathrm{A}\mathbf{U}, \quad \mathbf{D}_\mathrm{A} = \mathrm{diag}\{a_i\}, \ a_i \geq 0,$$
$$\mathbf{B} = \mathbf{C}^{-1}\mathbf{D}_\mathrm{B}\mathbf{V}, \quad \mathbf{D}_\mathrm{B} = \mathrm{diag}\{b_i\}, \ b_i \geq 0$$

be their GSVD, where the generalized singular values are ordered such that $(a_1/b_1) \geq (a_2/b_2) \geq \cdots \geq (a_l/b_l) \geq 0$, and rank($\mathbf{B}$) = l. For all complex m-vectors $\mathbf{d}$ with $\|\mathbf{d}\|_2 = 1$, if $\mathbf{d} = \sum_{i=1}^{l} \gamma_i \mathbf{c}_i$ where $\mathbf{c}_i^T$ is the ith row of $\mathbf{C}$, then

$$R_\mathrm{d}[\mathbf{A},\mathbf{B}] = \frac{\sum_{i=1}^{l} \gamma_i^2 (a_i/b_i)^2}{\sum_{i=1}^{l} \gamma_i^2} \tag{4.62}$$

The oriented signal-to-signal ratio correlates the energy distributions in the range spaces of two matrices by taking the ratio of oriented energies along any direction. Thus, the measure of signal-to-signal ratio of two matrices gives a comparative analysis of spatial energy distributions in the matrix spaces. According to Theorem 4.20, the GSVD provides the directions of maximal signal-to-signal ratio together with the values of maximal ratios as the generalized singular value squared.

Moreover, the directions of maximal signal-to-signal ratio constitute a set of linearly independent vectors, and this fact ensures that the differences of spatial energy distributions in the two matrix spaces are captured for the complete space. Corollary 4.21 reveals that the signal-to-signal ratio along any arbitrary direction in the l-dimensional space can be obtained using the GSVD of the two matrices. In other words, the GSVD provides all information for comparison of the spatial energy spectra of two matrix spaces. An application of the above concept will be useful in separating the desired vector sequence from a mixer sequence of desired and undesired vectors, provided the energy distribution of the undesired vector sequence is known *a priori*. Another application of the GSVD can be implemented for estimation of signal or system parameters using the information of signal-to-signal ratio of two correlated system matrices.

Note that the comparison of oriented energy distributions of two vector sequences $\{\mathbf{a}_j\}$ and $\{\mathbf{b}_j\}$ can be done by finding a linear transformation which will convert the sequence $\{\mathbf{b}_j\}$ into an isotropic sequence, by applying the same transformation on the sequence $\{\mathbf{a}_j\}$, and then by determining the energy distribution of the resulting sequence. The advantage of the GSVD is that it combines the above three processes into one process efficiently.

4.5.3 Minimax and Maximin Characterization Properties

Let $\mathbf{A}$ be a matrix of dimension $m \times n$. The column space of $\mathbf{A}$ is the space of m-vectors with an assigned energy distribution, and the row space of $\mathbf{A}$ is the space of n-vectors with similar energy distribution. The energy distribution reveals the modes of the matrix. If the data matrix is generated from a function sequence, then the modes of the matrix may be related to the constituents of the function. It is this specific reason why we are often interested in modal decomposition of a matrix.

In order to study the energy distribution in the matrix space, we can form the Hermitian matrices $\mathbf{A}^H\mathbf{A}$ and $\mathbf{A}\mathbf{A}^H$, and compute the eigenvalues which are the measures of energy along the eigenvectors. The minimax and maximin characterization properties are applied to extract the eigenvalues of $\mathbf{A}^H\mathbf{A}$ and $\mathbf{A}\mathbf{A}^H$, as stated in the following theorems.

Theorem 4.22 (Rayleigh–Ritz)

Let $\mathbf{S} \in \mathbb{C}^{n\times n}$ be a Hermitian matrix, and let the eigenvalues of $\mathbf{S}$ be ordered as

$$\lambda_{\min} = \lambda_1 \le \lambda_2 \le \cdots \le \lambda_{n-1} \le \lambda_n = \lambda_{\max}.$$

Then, $\lambda_1 \mathbf{x}^H\mathbf{x} \le \mathbf{x}^H\mathbf{S}\mathbf{x} \le \lambda_n \mathbf{x}^H\mathbf{x}$ for all $\mathbf{x} \in \mathbb{C}^n$, and

$$\lambda_{\max} = \max_{\mathbf{x}\neq\mathbf{0}} \frac{\mathbf{x}^H\mathbf{S}\mathbf{x}}{\mathbf{x}^H\mathbf{x}} = \max_{\mathbf{x}^H\mathbf{x}=1} \mathbf{x}^H\mathbf{S}\mathbf{x} \tag{4.63}$$

$$\lambda_{\min} = \min_{\mathbf{x} \neq \mathbf{0}} \frac{\mathbf{x}^H \mathbf{S} \mathbf{x}}{\mathbf{x}^H \mathbf{x}} = \min_{\mathbf{x}^H \mathbf{x} = 1} \mathbf{x}^H \mathbf{S} \mathbf{x} \tag{4.64}$$

Theorem 4.23 (Courant–Fischer)

Let $\mathbf{S} \in \mathbb{C}^{n \times n}$ be a Hermitian matrix with the eigenvalues $\lambda_1 \leq \lambda_2 \leq \cdots \leq \lambda_{n-1} \leq \lambda_n$.

Then, given $1 \leq k \leq n$,

$$\lambda_k = \min_{\mathbf{s}_1, \mathbf{s}_2, \ldots, \mathbf{s}_{n-k} \in \mathbb{C}^n} \; \max_{\substack{\mathbf{x} \neq \mathbf{0}, \, \mathbf{x} \in \mathbb{C}^n \\ \mathbf{x} \perp \mathbf{s}_1, \mathbf{s}_2, \ldots, \mathbf{s}_{n-k}}} \frac{\mathbf{x}^H \mathbf{S} \mathbf{x}}{\mathbf{x}^H \mathbf{x}} \tag{4.65}$$

$$\lambda_k = \max_{\mathbf{s}_1, \mathbf{s}_2, \ldots, \mathbf{s}_{k-1} \in \mathbb{C}^n} \; \min_{\substack{\mathbf{x} \neq \mathbf{0}, \, \mathbf{x} \in \mathbb{C}^n \\ \mathbf{x} \perp \mathbf{s}_1, \mathbf{s}_2, \ldots, \mathbf{s}_{k-1}}} \frac{\mathbf{x}^H \mathbf{S} \mathbf{x}}{\mathbf{x}^H \mathbf{x}} \tag{4.66}$$

Corollary 4.24

Let $\mathbf{S} \in \mathbb{C}^{n \times n}$ be a Hermitian matrix with the eigenvalues $\lambda_1 \leq \lambda_2 \leq \cdots \leq \lambda_{n-1} \leq \lambda_n$.

Then, given $1 \leq k \leq n$,

$$\lambda_k = \min_{\mathcal{S}_k \subset \mathbb{C}^n} \max_{\substack{\mathbf{x} \neq \mathbf{0} \\ \mathbf{x} \in \mathcal{S}_k}} \frac{\mathbf{x}^H \mathbf{S} \mathbf{x}}{\mathbf{x}^H \mathbf{x}} \tag{4.67}$$

$$\lambda_k = \max_{\mathcal{S}_{n-k+1} \subset \mathbb{C}^n} \min_{\substack{\mathbf{x} \neq \mathbf{0} \\ \mathbf{x} \in \mathcal{S}_{n-k+1}}} \frac{\mathbf{x}^H \mathbf{S} \mathbf{x}}{\mathbf{x}^H \mathbf{x}} \tag{4.68}$$

where $\mathcal{S}_j$ denotes a subspace of dimension j and the outer optimization is over all subspaces of the indicated dimension.

The energy distribution in the matrix space can be estimated by extracting the squared singular values of $\mathbf{A}$ instead. In this way, we avoid forming the Hermitian matrices $\mathbf{A}^H\mathbf{A}$ and $\mathbf{A}\mathbf{A}^H$, and evade some issues related to ill-conditioning. In the following theorem, the minimax and maximin characterization properties are applied to extract the singular values of $\mathbf{A}$.

Theorem 4.25

Let $\mathbf{A}$ be a complex matrix of dimension $m \times n$ and let $\sigma_1 \geq \sigma_2 \geq \cdots \geq \sigma_q$, $q = \min(m, n)$, be the ordered singular values of $\mathbf{A}$.

Then, given $1 \leq k \leq q$,

$$\sigma_k = \min_{\mathcal{S}_k \subset \mathbb{C}^n} \max_{\substack{\mathbf{x} \neq \mathbf{0} \\ \mathbf{x} \in \mathcal{S}_k}} \frac{\|\mathbf{A}\mathbf{x}\|_2}{\|\mathbf{x}\|_2} \tag{4.69}$$

$$\sigma_k = \max_{\mathcal{S}_{n-k+1} \subset \mathbb{C}^n} \min_{\substack{\mathbf{x} \neq \mathbf{0} \\ \mathbf{x} \in \mathcal{S}_{n-k+1}}} \frac{\|\mathbf{A}\mathbf{x}\|_2}{\|\mathbf{x}\|_2} \tag{4.70}$$

The measure of signal-to-signal ratio of two matrices gives a comparative analysis of spatial energy distributions in the matrix spaces. We are interested to find the directions of maximal signal-to-signal ratio together with the values of maximal ratios as the generalized singular value squared. The minimax and maximin characterization properties are applied in the theorem below to extract the generalized singular values. We define the following two measures for this application.

Definition
The maximal-minimal signal-to-signal ratio of two m-vector sequences $\{\mathbf{a}_k\}$ and $\{\mathbf{b}_k\}$, $k=1, \ldots, n$ over all possible r-dimensional subspaces $(0 < r \le m < n)$, denoted MmR $[\mathbf{A}, \mathbf{B}, r]$ is defined as

$$\mathrm{MmR}[\mathbf{A}, \mathbf{B}, r] = \max_{\mathscr{S}_r \subset \mathbb{C}^m} \min_{\mathbf{e} \in \mathscr{S}_r} \mathrm{R}_\mathrm{e}[\mathbf{A}, \mathbf{B}] \tag{4.71}$$

The minimal–maximal signal-to-signal ratio of two m-vector sequences $\{\mathbf{a}_k\}$ and $\{\mathbf{b}_k\}$, $k=1, \ldots, n$ over all possible r-dimensional subspaces $(0 < r \le m < n)$, denoted $\mathrm{mMR}[\mathbf{A},\mathbf{B},r]$ is defined as

$$\mathrm{mMR}[\mathbf{A}, \mathbf{B}, r] = \min_{\mathscr{S}_r \subset \mathbb{C}^m} \max_{\mathbf{e} \in \mathscr{S}_r} \mathrm{R}_\mathrm{e}[\mathbf{A}, \mathbf{B}] \tag{4.72}$$

Theorem 4.26

Consider two m-vector sequences $\{\mathbf{a}_k\}$ and $\{\mathbf{b}_k\}$, $k = 1, \ldots, n$ and associated $m \times n$ matrices $\mathbf{A}$ and $\mathbf{B}$. Consider the GSVD of $\mathbf{A}$ and $\mathbf{B}$ as $\mathbf{A} = \mathbf{C}^{-1}\mathbf{D}_\mathrm{A}\mathbf{U}$ and $\mathbf{B} = \mathbf{C}^{-1}\mathbf{D}_\mathrm{B}\mathbf{V}$.

Let (a_i/b_i) be the generalized singular values of $\mathbf{A}$ and $\mathbf{B}$, arranged in non-increasing order. Denote by $\mathbf{c}_i^T$, the ith row vector of $\mathbf{C}$.

Then, given $0 < r \le m < n$,

$\mathrm{MmR}[\mathbf{A}, \mathbf{B}, r] = a_r/b_r$, and the corresponding subspace is generated by the first r row vectors of $\mathbf{C}$.

Similarly, given $0 < r \le m < n$,

$\mathrm{mMR}[\mathbf{A}, \mathbf{B}, r] = a_{m-r+1}/b_{m-r+1}$, and the corresponding subspace is generated by the last r row vectors of $\mathbf{C}$.

In the terminology of signal processing, we say that by using the GSVD, the spatial spectral information for the signal vector sequence can be extracted by processing the signal-plus-noise vector sequence together with the noise-alone vector sequence.

4.6 Least Squares Solutions

Consider a linear system $\mathbf{Ax} = \mathbf{b}$ where $\mathbf{A} \in \mathbb{C}^{m\times n}$ is the system matrix, $\mathbf{x} \in \mathbb{C}^n$ is the excitation vector, and $\mathbf{b} \in \mathbb{C}^m$ is the observation vector. Suppose $\mathbf{A}$ and $\mathbf{b}$ are known, and we want to determine $\mathbf{x}$. When $m > n$, we often cannot find an exact solution for $\mathbf{x}$, because $\mathbf{b}$ does not belong to $\mathscr{R}(\mathbf{A})$, the range space of $\mathbf{A}$, where $\mathscr{R}(\mathbf{A}) \subset \mathbb{C}^m$. Note that the dimension of $\mathscr{R}(\mathbf{A})$ is n or less depending on whether the matrix $\mathbf{A}$ is full rank or rank-deficient. Usually we find a solution in this case by minimizing the p-norm of the residual vector $\|\mathbf{Ax} - \mathbf{b}\|_p$. The most popular is the 2-norm minimization for the reason that it is easily tractable, and this leads to the least squares (LS) problem

$$\min_{\mathbf{x} \in \mathbb{C}^n} \|\mathbf{Ax} - \mathbf{b}\|_2 \tag{4.73}$$

The solution of (4.73) is unique when rank(**A**) = n. However, there will be an infinite number of solutions when rank(**A**) < n, because if **x** is a solution of (4.73) and $\mathbf{z} \in \mathcal{N}(\mathbf{A})$, the null space of **A**, then **x** + **z** is also a solution of the LS problem. Incidentally, the set of all solutions of (4.73) constitute a convex set, and a unique solution is obtained by searching the solution having minimum 2-norm $\|\mathbf{x}\|_2$. We denote this unique solution as $\mathbf{x}_{\text{LS}}$.

The SVD of **A** provides a complete orthogonal decomposition of the matrix, and we get a closed-form expression for the minimum-norm least squares solution $\mathbf{x}_{\text{LS}}$ and the norm of the minimum residual $\rho_{\text{LS}} = \|\mathbf{A}\mathbf{x}_{\text{LS}} - \mathbf{b}\|_2$ as given in the following theorem.

Theorem 4.27

Consider the linear system **Ax** = **b** where the matrix $\mathbf{A} \in \mathbb{C}^{m \times n}$ with rank(**A**) = l, and the vector $\mathbf{b} \in \mathbb{C}^m$ are known. Let $\mathbf{U}^H \mathbf{A} \mathbf{V} = \Sigma$ be the SVD of **A** where $\mathbf{U} = [\mathbf{u}_1, \ldots, \mathbf{u}_m]$, $\mathbf{V} = [\mathbf{v}_1, \ldots, \mathbf{v}_n]$, and $\Sigma = \text{diag}(\sigma_1, \ldots, \sigma_l, 0, \ldots)$. Then, the solution

$$\mathbf{x}_{\text{LS}} = \sum_{i=1}^{l} \frac{\mathbf{u}_i^H \mathbf{b}}{\sigma_i} \mathbf{v}_i \tag{4.74}$$

minimizes $\|\mathbf{A}\mathbf{x} - \mathbf{b}\|_2$, and it has the minimum 2-norm $\|\mathbf{x}\|_2$. Moreover, the norm of residual ρ_{LS} is given by

$$\rho_{\text{LS}} = \|\mathbf{A}\mathbf{x}_{\text{LS}} - \mathbf{b}\|_2 = \left(\sum_{i=l+1}^{m} \left|\mathbf{u}_i^H \mathbf{b}\right|^2 \right)^{1/2} \tag{4.75}$$

We define the matrix $\mathbf{A}^+ \in \mathbb{C}^{n \times m}$ as $\mathbf{A}^+ = \mathbf{V}\Sigma^+\mathbf{U}^H$ where $\Sigma^+ = \text{diag}(1/\sigma_1, \ldots, 1/\sigma_l, 0, \ldots) \in \mathbb{C}^{n \times m}$, l = rank(**A**), then $\mathbf{x}_{\text{LS}} = \mathbf{A}^+\mathbf{b}$ and $\rho_{\text{LS}} = \left\|(\mathbf{A}\mathbf{A}^+ - \mathbf{I}_m)\mathbf{b}\right\|_2$. The matrix $\mathbf{A}^+$ is the Moore–Penrose inverse of **A**. It is the unique solution to the problem

$$\min_{\mathbf{B} \in \mathbb{C}^{n \times m}} \|\mathbf{A}\mathbf{B} - \mathbf{I}_m\|_F \tag{4.76}$$

where $\| \|_F$ denotes the Frobenius norm of a matrix. The matrix $\mathbf{A}^+$ satisfies the Moore–Penrose conditions as described by the following lemma.

Lemma 4.28

Let $\mathbf{A} \in \mathbb{C}^{m \times n}$ have the SVD $\mathbf{A} = \mathbf{U}\Sigma\mathbf{V}^H$ where $\Sigma = \text{diag}(\sigma_1, \ldots, \sigma_l, 0, \ldots) \in \mathbb{C}^{m \times n}$, l = rank(**A**). Define $\mathbf{A}^+ = \mathbf{V}\Sigma^+\mathbf{U}^H$ where $\Sigma^+ = \text{diag}(1/\sigma_1, \ldots, 1/\sigma_l, 0, \ldots) \in \mathbb{C}^{n \times m}$. Then,

(i) $\mathbf{A}\mathbf{A}^+$ and $\mathbf{A}^+\mathbf{A}$ are Hermitian,
(ii) $\mathbf{A}\mathbf{A}^+\mathbf{A} = \mathbf{A}$, and
(iii) $\mathbf{A}^+\mathbf{A}\mathbf{A}^+ = \mathbf{A}^+$.

If **A** is square and non-singular, then $\mathbf{A}^+ = \mathbf{A}^{-1}$. Note that for the conditions of Lemma 4.28 to be satisfied, it is necessary that $\mathbf{A}\mathbf{A}^+$ and $\mathbf{A}^+\mathbf{A}$ are orthogonal projections onto $\mathcal{R}(\mathbf{A})$ and $\mathcal{R}(\mathbf{A}^H)$, the range spaces of **A** and $\mathbf{A}^H$, respectively. Indeed, $\mathbf{A}\mathbf{A}^+ = \mathbf{U}_1\mathbf{U}_1^H$ and $\mathbf{A}^+\mathbf{A} = \mathbf{V}_1\mathbf{V}_1^H$ where $\mathbf{U}_1 = [\mathbf{u}_1, \ldots, \mathbf{u}_l]$ and $\mathbf{V}_1 = [\mathbf{v}_1, \ldots, \mathbf{v}_l]$, l = rank(**A**).

4.7 Constrained Least Squares Solutions

The constrained least squares (CLS) problem refers to minimizing over a proper subset of $\mathbb{C}^n$. For example, we may like to find the solution $\mathbf{x}$ such that $\|\mathbf{x}\|_2 = 1$. Another example of the CLS problem arises when we wish to solve $\mathbf{x}$ such that $\mathbf{x}$ belongs to a bounded subspace of $\mathbb{C}^n$. While solving an ill-conditioned linear system with error in $\mathbf{A}$ or $\mathbf{b}$, we use such constraints in the LS problem, and the method is known as *regularization*. In this section, we present the GSVD based techniques to solve the regularized LS problems as proposed by Golub and Van Loan.

4.7.1 Quadratic Inequality Constraint

The LS problem with quadratic inequality constraint, referred as the LSQI problem, is formulated as

$$\text{minimize } \|\mathbf{Ax}-\mathbf{b}\|_2 \quad \text{subject to } \|\mathbf{Bx}\|_2 \le \delta, \quad \delta \ge 0 \tag{4.77}$$

where $\mathbf{A} \in \mathbb{C}^{m\times n}$, $\mathbf{b} \in \mathbb{C}^m$, and $\mathbf{B} \in \mathbb{C}^{n\times n}$ is non-singular. More generally, we have the problem as stated below

$$\text{minimize } \|\mathbf{Ax}-\mathbf{b}\|_2 \quad \text{subject to } \|\mathbf{Bx}-\mathbf{d}\|_2 = \delta, \quad \delta \ge 0 \tag{4.78}$$

where $\mathbf{A} \in \mathbb{C}^{m\times n}$ with $m \ge n$, $\mathbf{b} \in \mathbb{C}^m$, $\mathbf{B} \in \mathbb{C}^{p\times n}$, and $\mathbf{d} \in \mathbb{C}^p$. We assume that the matrices $\mathbf{A}$ and $\mathbf{B}$ are full rank for clarity.

Let the GSVD of the matrix pair ($\mathbf{A}$, $\mathbf{B}$) be given by the relations:

$$\begin{aligned} \mathbf{U}^H\mathbf{AC} = \mathbf{D}_\text{A} = \operatorname{diag}(\alpha_1,\ldots,\alpha_n) \in \mathbb{R}^{m\times n}, \quad \mathbf{U}^H\mathbf{U} = \mathbf{I}_m \\ \mathbf{V}^H\mathbf{BC} = \mathbf{D}_\text{B} = \operatorname{diag}(\beta_1,\ldots,\beta_q) \in \mathbb{R}^{p\times n}, \quad \mathbf{V}^H\mathbf{V} = \mathbf{I}_p \end{aligned} \tag{4.79}$$

where $q = \min(p,n)$. Then, employing the GSVD, the CLS problem (4.78) is rewritten as

$$\text{minimize } \|\mathbf{D}_\text{A}\mathbf{y}-\mathbf{s}\|_2 \quad \text{subject to } \|\mathbf{D}_\text{B}\mathbf{y}-\mathbf{r}\|_2 = \delta \tag{4.80}$$

where $\mathbf{s} = \mathbf{U}^H\mathbf{b}$, $\mathbf{r} = \mathbf{V}^H\mathbf{d}$, and $\mathbf{y} = \mathbf{C}^{-1}\mathbf{x}$. To solve (4.80), we use the Lagrange multiplier technique and form the objective function h given by

$$h(\lambda,\mathbf{y}) = \|\mathbf{D}_\text{A}\mathbf{y}-\mathbf{s}\|_2^2 + \lambda\left(\|\mathbf{D}_\text{B}\mathbf{y}-\mathbf{r}\|_2^2 - \delta^2\right) \tag{4.81}$$

By setting $\partial h/\partial y_i = 0$ for $i = 1, \ldots, n$, we get the matrix equation

$$\left(\mathbf{D}_\text{A}^T\mathbf{D}_\text{A} + \lambda\mathbf{D}_\text{B}^T\mathbf{D}_\text{B}\right)\mathbf{y} = \mathbf{D}_\text{A}^T\mathbf{s} + \lambda\mathbf{D}_\text{B}^T\mathbf{r} \tag{4.82}$$

Assuming that the matrix $\left(\mathbf{D}_\text{A}^T\mathbf{D}_\text{A} + \lambda\mathbf{D}_\text{B}^T\mathbf{D}_\text{B}\right)$ is non-singular, the solution $\mathbf{y}(\lambda)$ is given by

$$y_i(\lambda) = \begin{cases} \dfrac{\alpha_i s_i + \lambda\beta_i r_i}{\alpha_i^2 + \lambda\beta_i^2}, & i = 1,\ldots,q \\ s_i/\alpha_i, & i = q+1,\ldots,n \end{cases} \tag{4.83}$$

To find the Lagrange multiplier λ, we define the function

$$\phi(\lambda) = \left\|\mathbf{D}_B\mathbf{y}(\lambda) - \mathbf{r}\right\|_2^2 = \sum_{i=1}^{q}\left(\alpha_i \frac{\beta_i s_i - \alpha_i r_i}{\alpha_i^2 + \lambda\beta_i^2}\right)^2 + \sum_{i=q+1}^{p} r_i^2 \tag{4.84}$$

and seek a solution such that $\phi(\lambda) = \delta^2$. Note that $\phi(\lambda)$ is maximum for $\lambda = 0$, and it is monotone decreasing for $\lambda > 0$. We assume that $\phi(0) > \delta^2$ for a solution to be feasible in this case. Therefore, there is a unique positive λ^* for which $\phi(\lambda^*) = \delta^2$, and the desired solution is given by

$$\mathbf{x}_{\mathrm{LSQI}} = \mathbf{C}\mathbf{y}\left(\lambda^*\right) \tag{4.85}$$

4.7.2 Equality Constraint

The LS problem with linear equality constraint arises by setting $\delta = 0$ in (4.78),

$$\min_{\mathbf{Bx}=\mathbf{d}} \left\|\mathbf{Ax} - \mathbf{b}\right\|_2 \tag{4.86}$$

We refer to (4.86) as the LSE problem which is a special case of the LSQI problem. Assume that the matrices $\mathbf{A} \in \mathbb{C}^{m\times n}$ and $\mathbf{B} \in \mathbb{C}^{p\times n}$ are full rank, and $m \ge n \ge p$ for clarity. Let the GSVD of the matrix pair $(\mathbf{A}, \mathbf{B})$ be given by (4.79) with $\mathbf{D}_B = \mathrm{diag}\left(\beta_1, \ldots, \beta_p\right)$. Utilizing the GSVD, the LSE problem is rewritten as

$$\min_{\mathbf{VD}_B\mathbf{C}^{-1}\mathbf{x}=\mathbf{d}} \left\|\mathbf{UD}_A\mathbf{C}^{-1}\mathbf{x} - \mathbf{b}\right\|_2 \tag{4.87}$$

It is interesting to note that we can find the solution of the LSE problem (4.87) directly without using the Lagrange multiplier technique. Write $\mathbf{C}^{-1}\mathbf{x} = \begin{bmatrix}\mathbf{y}\\ \mathbf{z}\end{bmatrix}$ where $\mathbf{y} \in \mathbb{C}^p$ and $\mathbf{z} \in \mathbb{C}^{n-p}$. Then, the equality constraint $\mathbf{VD}_B\mathbf{C}^{-1}\mathbf{x} = \mathbf{d}$ becomes

$$\left[\beta_1\mathbf{v}_1, \ldots, \beta_p\mathbf{v}_p\right]\mathbf{y} = \mathbf{d} \tag{4.88}$$

which can be solved for $\mathbf{y}$. Next, we expand the matrix product

$$\mathbf{UD}_A\begin{bmatrix}\mathbf{y}\\ \mathbf{z}\end{bmatrix} = \left[\alpha_1\mathbf{u}_1, \ldots, \alpha_n\mathbf{u}_n\right]\begin{bmatrix}\mathbf{y}\\ \mathbf{z}\end{bmatrix} = \mathbf{W}_1\mathbf{y} + \mathbf{W}_2\mathbf{z} \tag{4.89}$$

where $\mathbf{W}_1 = \left[\alpha_1\mathbf{u}_1, \ldots, \alpha_p\mathbf{u}_p\right] \in \mathbb{C}^{m\times p}$ and $\mathbf{W}_2 = \left[\alpha_{p+1}\mathbf{u}_{p+1}, \ldots, \alpha_n\mathbf{u}_n\right] \in \mathbb{C}^{m\times(n-p)}$. Thus, the CLS problem reduces to the unconstrained LS problem as follows

$$\min_{\mathbf{z}} \left\|\mathbf{W}_1\mathbf{y} + \mathbf{W}_2\mathbf{z} - \mathbf{b}\right\|_2 \tag{4.90}$$

which is solved for $\mathbf{z}$. Finally, the desired solution $\mathbf{x} = \mathbf{C}\begin{bmatrix}\mathbf{y}\\ \mathbf{z}\end{bmatrix}$ is expressed as

$$\mathbf{x}_{\mathrm{LSE}} = \sum_{i=1}^{p} \frac{\mathbf{v}_i^H\mathbf{d}}{\beta_i}\mathbf{c}_i + \sum_{i=p+1}^{n} \frac{\mathbf{u}_i^H\mathbf{b}}{\alpha_i}\mathbf{c}_i \tag{4.91}$$

where $\mathbf{U} = \left[\mathbf{u}_1, \ldots, \mathbf{u}_m\right]$, $\mathbf{V} = \left[\mathbf{v}_1, \ldots, \mathbf{v}_p\right]$, and $\mathbf{C} = \left[\mathbf{c}_1, \ldots, \mathbf{c}_n\right]$.

4.8 Total Least Squares Solutions

Consider the linear system $\mathbf{Ax} \approx \mathbf{b}$ where $\mathbf{A} \in \mathbb{C}^{m \times n}$ with $m > n$, $\mathbf{b} \in \mathbb{C}^m$, and $\mathbf{x} \in \mathbb{C}^n$. The LS problem to find $\mathbf{x}$ is formulated with the assumption that $\mathbf{b}$ is in error. Accordingly, we set $\mathbf{Ax} = \mathbf{b} + \mathbf{r}$, i.e., $\mathbf{b} + \mathbf{r} \in \mathscr{R}(\mathbf{A})$, the range space of $\mathbf{A}$. Then, the solution $\mathbf{x}$ is obtained by minimizing $\|\mathbf{r}\|_2$. The total least squares (TLS) problem to determine $\mathbf{x}$, on the contrary, is relevant when both $\mathbf{A}$ and $\mathbf{b}$ are in error. In this case, we set $(\mathbf{A} + \mathbf{E})\mathbf{x} = \mathbf{b} + \mathbf{r}$, i.e., $\mathbf{b} + \mathbf{r} \in \mathscr{R}(\mathbf{A} + \mathbf{E})$, where $\mathbf{E}$ and $\mathbf{r}$ are the deviations of $\mathbf{A}$ and $\mathbf{b}$, respectively. Then, the solution $\mathbf{x}$ is obtained by minimizing $\|[\mathbf{E}, \mathbf{r}]\|_F$.

The best fitting of a straight line $ax = b$ with the slope x utilizing multiple point values $\{(a_i, b_i); i = 1, \ldots, m\}$ is demonstrated by the LS and TLS techniques respectively in Figure 4.2 (a) & (b) [van Huffel and Vandewalle]. It can be shown that the root-mean-square (RMS) approximation error ε for the TLS technique is less than or equal to that for the LS technique, i.e.,

$$\varepsilon_{\text{TLS}} = \min\left(\sum_i d_i^2 \Big/ m\right)^{1/2} \le \varepsilon_{\text{LS}} = \min\left(\sum_i r_i^2 \Big/ m\right)^{1/2} \tag{4.92}$$

We now present a simple method to solve the linear system $\mathbf{Ax} = \mathbf{b}$ in the TLS sense. Let the problem be rewritten as

$$[\mathbf{A}, \mathbf{b}]\begin{bmatrix} \mathbf{x} \\ -1 \end{bmatrix} = \mathbf{0} \;\Rightarrow\; \mathbf{Gy} = \mathbf{0} \tag{4.93}$$

where $\mathbf{G} \in \mathbb{C}^{m \times (n+1)}$ with $m \ge n+1$, and $\mathbf{y} \in \mathbb{C}^{n+1}$. Denote the SVD of $\mathbf{G}$ by $\mathbf{G} = \mathbf{U}\Sigma\mathbf{V}^H$ where $\mathbf{U} = [\mathbf{u}_1, \ldots, \mathbf{u}_{n+1}]$ having $\mathbf{u}_i \in \mathbb{C}^m$, $\mathbf{V} = [\mathbf{v}_1, \ldots, \mathbf{v}_{n+1}]$ having $\mathbf{v}_i \in \mathbb{C}^{n+1}$, and $\Sigma = \text{diag}(\sigma_1, \ldots, \sigma_{n+1})$ with $\sigma_1 \ge \cdots \ge \sigma_{n+1}$. We assume that $\sigma_n > \sigma_{n+1} > 0$ for clarity and uniqueness of solution. Since $\sigma_{n+1} \ne 0$ implies $\text{rank}(\mathbf{G}) = n+1$, whereas $\text{rank}(\mathbf{A}) = n$, it is clear that (4.93) does not have an exact solution, and the trivial solution $\mathbf{y} = \mathbf{0}$ is not acceptable.

In order to obtain a solution in this case, we replace $\mathbf{G}$ by the rank-constrained matrix $\mathbf{G}_c = \mathbf{U}\Sigma_c\mathbf{V}^H$ where $\Sigma_c = \text{diag}(\sigma_1, \ldots, \sigma_n, 0)$. This amounts to adding the rank-one correction matrix $[\mathbf{E}, \mathbf{r}] = -\sigma_{n+1}\mathbf{u}_{n+1}\mathbf{v}_{n+1}^H$ to $\mathbf{G} = [\mathbf{A}, \mathbf{b}]$, and its Frobenius norm, σ_{n+1}, is the minimum TLS approximation error.

Now, the solution $\mathbf{y}$ belongs to the null space of $\mathbf{G}_c$, which is spanned by the lone right singular vector $\mathbf{v}_{n+1}$. Noting that the last component of $\mathbf{y}$ is -1, we write the TLS solution as

$$\mathbf{y}_{\text{TLS}} = \frac{-\mathbf{v}_{n+1}}{\mathbf{v}_{n+1}(n+1)} \tag{4.94}$$

where $\mathbf{v}_{n+1}(n+1)$ is the $(n+1)$th element of the vector $\mathbf{v}_{n+1}$.

Next, we consider the case when rank($\mathbf{G}$) = $l < n$. The solution $\mathbf{y}$ of (4.93) now belongs to the null space of $\mathbf{G}$, which is spanned by the right singular vectors $\{\mathbf{v}_{l+1}, \ldots, \mathbf{v}_{n+1}\}$. Once again, noting that the last component of $\mathbf{y}$ is -1, we write the TLS solution as

$$\mathbf{y}_{\text{TLS}} = \frac{-\sum_{i=l+1}^{n+1} \mathbf{v}_i^*(n+1)\mathbf{v}_i}{\sum_{i=l+1}^{n+1} |\mathbf{v}_i(n+1)|^2} \tag{4.95}$$

where $\mathbf{v}_i(n+1)$ is the $(n+1)$th element of the vector $\mathbf{v}_i$, and * denotes complex conjugation.

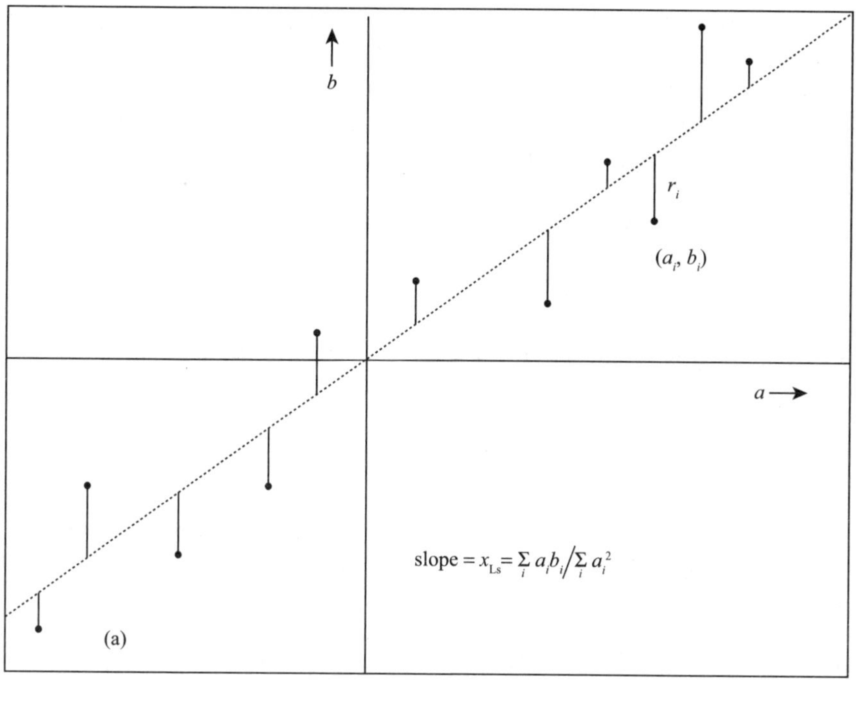

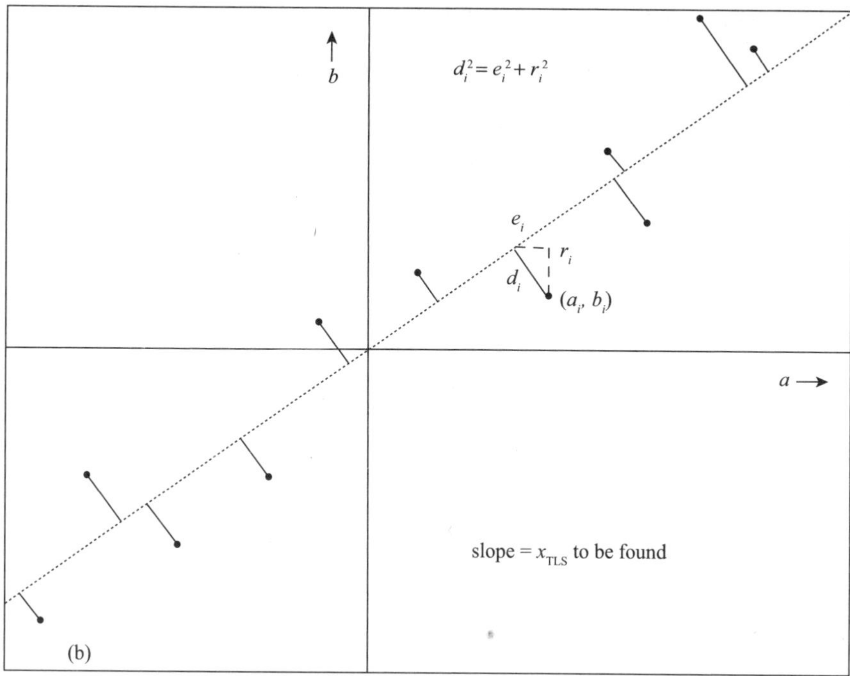

FIGURE 4.2 (a) LS technique: Minimize $\sum_i r_i^2$ (b) TLS technique: Minimize $\sum_i d_i^2$

We now present a generalization of the TLS technique to solve the multiple-input multiple-output (MIMO) linear system $\mathbf{AX} = \mathbf{B}$ where $\mathbf{A} \in \mathbb{C}^{m\times n}$ is the system matrix, $\mathbf{X} \in \mathbb{C}^{n\times p}$ is the input matrix, and $\mathbf{B} \in \mathbb{C}^{m\times p}$ is the output matrix. Assume that $\mathbf{A}$ and $\mathbf{B}$ are full rank for the following discussion. We set $(\mathbf{A}+\mathbf{E})\mathbf{X} = \mathbf{B}+\mathbf{R}$, i.e., $\mathcal{R}(\mathbf{B}+\mathbf{R}) \subseteq \mathcal{R}(\mathbf{A}+\mathbf{E})$, where $\mathbf{E}$ and $\mathbf{R}$ are the deviations of $\mathbf{A}$ and $\mathbf{B}$, respectively. The solution $\mathbf{X}$ is obtained by minimizing $\|[\mathbf{E},\mathbf{R}]\|_F$.

In order to solve the MIMO system $\mathbf{AX} = \mathbf{B}$ in the TLS sense, we rewrite the problem as

$$[\mathbf{A},\mathbf{B}]\begin{bmatrix}\mathbf{X}\\ -\mathbf{I}_p\end{bmatrix} = \mathbf{0} \;\Rightarrow\; \mathbf{GY} = \mathbf{0} \tag{4.96}$$

where $\mathbf{G} \in \mathbb{C}^{m\times(n+p)}$ with $m \geq n+p$, and $\mathbf{Y} \in \mathbb{C}^{(n+p)\times p}$. Denote the SVD of $\mathbf{G}$ by $\mathbf{G} = \mathbf{U}\Sigma\mathbf{V}^H$ where $\mathbf{U} = [\mathbf{U}_1, \mathbf{U}_2]$ having $\mathbf{U}_1 \in \mathbb{C}^{m\times n}$, $\mathbf{U}_2 \in \mathbb{C}^{m\times p}$, $\mathbf{V} = \begin{bmatrix}\mathbf{V}_{11} & \mathbf{V}_{12}\\ \mathbf{V}_{21} & \mathbf{V}_{22}\end{bmatrix}$ having $\mathbf{V}_{11} \in \mathbb{C}^{n\times n}$, $\mathbf{V}_{12} \in \mathbb{C}^{n\times p}$, $\mathbf{V}_{21} \in \mathbb{C}^{p\times n}$, $\mathbf{V}_{22} \in \mathbb{C}^{p\times p}$, and $\Sigma = \operatorname{diag}(\Sigma_1, \Sigma_2)$ with $\Sigma_1 = \operatorname{diag}(\sigma_1, \ldots, \sigma_n)$, $\Sigma_2 = \operatorname{diag}(\sigma_{n+1}, \ldots, \sigma_{n+p})$.

We assume that $\sigma_n > \sigma_{n+1} \geq \cdots \geq \sigma_{n+p} > 0$ for clarity and uniqueness of solution. Now, observe that $\operatorname{rank}(\mathbf{G}) = n+p \neq \operatorname{rank}(\mathbf{A}) = n$. Therefore, in order to obtain a solution, we need to replace $\mathbf{G}$ by the rank-constrained matrix $\mathbf{G}_c = \mathbf{U}\Sigma_c\mathbf{V}^H$ where $\Sigma_c = \operatorname{diag}(\Sigma_1, \mathbf{0})$. This can be done by adding the rank-p correction matrix $[\mathbf{E},\mathbf{R}] = -\mathbf{U}_2\Sigma_2\begin{bmatrix}\mathbf{V}_{12}\\ \mathbf{V}_{22}\end{bmatrix}^H$ to $\mathbf{G} = [\mathbf{A},\mathbf{B}]$, and its Frobenius norm, $\left(\sigma_{n+1}^2 + \cdots + \sigma_{n+p}^2\right)^{1/2}$, is the minimum TLS approximation error. Now, the column-vectors of the solution $\mathbf{Y}$ belong to the null space of $\mathbf{G}_c$, and this null space is same as the range space of the matrix $\begin{bmatrix}\mathbf{V}_{12}\\ \mathbf{V}_{22}\end{bmatrix}$. Thus, for some $p\times p$ matrix $\mathbf{S}$, the solution $\mathbf{Y}$ is given by

$$\mathbf{Y} = \begin{bmatrix}\mathbf{X}\\ -\mathbf{I}_p\end{bmatrix} = \begin{bmatrix}\mathbf{V}_{12}\\ \mathbf{V}_{22}\end{bmatrix}\mathbf{S} \tag{4.97}$$

Comparing the two sides of (4.97), we get $\mathbf{S} = -\mathbf{V}_{22}^{-1}$, and the unique solution $\mathbf{X}_{\text{TLS}}$ is given by

$$\mathbf{X}_{\text{TLS}} = -\mathbf{V}_{12}\mathbf{V}_{22}^{-1} \tag{4.98}$$

We now summarize the above results in the following theorem.

Theorem 4.29

Consider the MIMO system $\mathbf{AX} = \mathbf{B}$ given the matrices $\mathbf{A} \in \mathbb{C}^{m\times n}$ and $\mathbf{B} \in \mathbb{C}^{m\times p}$ with $m \geq n+p$. Let $\mathbf{G} = [\mathbf{A},\mathbf{B}]$ have SVD $\mathbf{U}^H\mathbf{G}\mathbf{V} = \operatorname{diag}(\sigma_1, \ldots, \sigma_{n+p}) = \Sigma$ where $\mathbf{U}$, $\mathbf{V}$, and Σ are partitioned as follows:

$$\mathbf{U} = [\underset{n}{\mathbf{U}_1} \;\; \underset{p}{\mathbf{U}_2}] \qquad \mathbf{V} = \begin{bmatrix}\mathbf{V}_{11} & \mathbf{V}_{12}\\ \mathbf{V}_{21} & \mathbf{V}_{22}\end{bmatrix}\begin{matrix}n\\ p\end{matrix} \qquad \Sigma = \begin{bmatrix}\Sigma_1 & \mathbf{0}\\ \mathbf{0} & \Sigma_2\end{bmatrix}\begin{matrix}n\\ p\end{matrix}$$

(column block sizes of $\mathbf{V}$ and Σ: n, p)

If $\sigma_n > \sigma_{n+1}$, then the matrix $[\mathbf{E},\mathbf{R}] = -\mathbf{U}_2\Sigma_2\begin{bmatrix}\mathbf{V}_{12}\\ \mathbf{V}_{22}\end{bmatrix}^H$ minimizes $\|[\mathbf{E},\mathbf{R}]\|_F$ subject to $\mathcal{R}(\mathbf{B}+\mathbf{R}) \subseteq \mathcal{R}(\mathbf{A}+\mathbf{E})$. Moreover, the matrix $\mathbf{X}_{\text{TLS}} = -\mathbf{V}_{12}\mathbf{V}_{22}^{-1}$ exists and is the unique solution to $(\mathbf{A}+\mathbf{E})\mathbf{X} = \mathbf{B}+\mathbf{R}$.

4.9 Sturm–Liouville Problem and Eigenfunction Expansion

Consider the operator equation given by

$$Ly = \lambda y \tag{4.99}$$

where L is a linear operator, $y = y(x)$ is a function defined over the interval $a \le x \le b$, and λ is a coefficient. Any non-zero function that satisfies the above equation, together with certain boundary conditions at $x = a$ and $x = b$, is called an eigenfunction of the operator L. The corresponding value of λ is called an eigenvalue.

The general Sturm–Liouville (S–L) problem is given by the second-order differential equation

$$-\frac{d}{dx}\left(p(x)\frac{dy}{dx}\right)+q(x)y=\lambda r(x)y, \quad a \le x \le b \tag{4.100}$$

together with some boundary conditions, where $p(x)$, $q(x)$ and $r(x)$ are real-valued continuous functions, $p(x)$ and $r(x)$ are positive, and λ is a parameter. When $p(x)$ is differentiable and $p(x)\,r(x)$ is twice differentiable, then the above equation with proper substitution reduces to

$$-\frac{d^2y}{dx^2}+s(x)y=\lambda y \;\Rightarrow\; Ly=\lambda y \tag{4.101}$$

where the operator L is given by $L \equiv -\dfrac{d^2}{dx^2}+s(x),\; a \le x \le b$. This is again the eigenequation (4.99).

Let λ_m and λ_n be two distinct eigenvalues of L, and ϕ_m, ϕ_n, respectively, are the corresponding eigenfunctions. Then, writing

$$-\phi_m''(x)+s(x)\phi_m(x)=\lambda_m\,\phi_m(x) \tag{4.102}$$

$$-\phi_n''(x)+s(x)\phi_n(x)=\lambda_n\,\phi_n(x) \tag{4.103}$$

next, multiplying by ϕ_n, ϕ_m, in that order, and taking difference, we get

$$\begin{aligned}(\lambda_m-\lambda_n)\phi_m(x)\phi_n(x) &= \phi_m(x)\phi_n''(x)-\phi_n(x)\phi_m''(x)\\ &= \frac{d}{dx}\left[\phi_m(x)\phi_n'(x)-\phi_n(x)\phi_m'(x)\right]\end{aligned} \tag{4.104}$$

Hence, on integration, we obtain

$$\begin{aligned}(\lambda_m-\lambda_n)\int_a^b \phi_m(x)\phi_n(x)\,dx &= \left[\phi_m(x)\phi_n'(x)-\phi_n(x)\phi_m'(x)\right]_a^b\\ &= 0\end{aligned} \tag{4.105}$$

provided ϕ_m and ϕ_n both vanish at $x = a$ and $x = b$ (or some more general condition of similar kind is satisfied). Since $\lambda_m \ne \lambda_n$, it follows that

$$\int_a^b \phi_m(x)\phi_n(x)\,dx = 0 \tag{4.106}$$

Moreover, by multiplying with a suitable constant, we can have

$$\int_a^b \phi_m^2(x)\,dx = 1 \tag{4.107}$$

Therefore, the eigenfunctions $\{\phi_m(x)\}$ constitute an orthonormal set. Suppose now that an arbitrary function $f(x)$ can be expanded in terms of the functions $\{\phi_m(x)\}$ over the interval $[a, b]$, i.e.,

$$f(x) = \sum_{m=0}^{\infty} c_m \phi_m(x) \tag{4.108}$$

then, the coefficient c_m of the above series can be computed as

$$c_m = \int_a^b f(x)\phi_m(x)dx \tag{4.109}$$

In some cases, the eigenvalues are not discrete points. Instead, they form a continuous range of values, say, for example, over $[0, \infty)$. Accordingly, the expansion of $f(x)$ takes the form

$$f(x) = \int_0^{\infty} c(\lambda)\phi_\lambda(x)d\lambda \tag{4.110}$$

which is an integral function of λ for each value of x. Our aim is to investigate in the sequel under what conditions an arbitrary function $f(x)$ can be expanded in the form of (4.108) or (4.110).

Consider the second-order differential equation involving the operator $L \equiv -\dfrac{d^2}{dx^2} + s(x)$,

$$F''(x) + [\lambda - s(x)]F(x) = f(x), \quad a \le x \le b \tag{4.111}$$

In this case, the function $F(x)$ can be expressed in terms of the solutions of (4.101). Let $\phi(x, \lambda)$ and $\psi(x, \lambda)$ be the two solutions of (4.101) such that their Wronskian $W(\phi, \psi)$ defined as $W(\phi, \psi) = \phi(x)\psi'(x) - \phi'(x)\psi(x)$ is unity. Then, $F(x, \lambda)$, a solution of (4.111), is given by

$$F(x,\lambda) = \psi(x,\lambda)\int_a^x \phi(\xi,\lambda)f(\xi)d\xi + \phi(x,\lambda)\int_x^b \psi(\xi,\lambda)f(\xi)d\xi \tag{4.112}$$

This can be easily verified by differentiating the above expression twice. Another solution of (4.111) can be written as

$$F(x) = F(x,\lambda) = \sum_{m=0}^{\infty} \frac{c_m \phi_m(x)}{\lambda - \lambda_m} \tag{4.113}$$

which can be verified by direct substitution using the expansion of $f(x)$ from (4.108). When the above solution is available, (4.113) indicates that the terms of the expansion of $f(x)$ are the residues at the poles $\lambda = \lambda_m$ of $F(x, \lambda)$.

The S–L problem given by (4.101), is called *regular* if the interval $[a, b]$ is finite and if the function $s(x)$ is summable (absolutely integrable) on the entire interval. If the interval $[a, b]$ is infinite or if $s(x)$ is not summable (or both), then the problem is called *singular*. We shall assume in all cases that $s(x)$ is a real function of x, continuous at all points interior to the interval $[a, b]$. The following theorem guarantees the existence of solution of a regular Sturm–Liouville problem.

Theorem 4.30

If $s(x)$ is a real continuous function over the interval (a, b), and $s(a)$, $s(b)$ are finite values, then the second-order differential equation

$$\frac{d^2y}{dx^2}+\left[\lambda-s(x)\right]y=0, \quad a\le x\le b$$

has a solution $\phi(x)$ such that

$$\phi(a)=\sin\alpha, \quad \phi'(a)=-\cos\alpha$$

for a given α.

Moreover, for each $x\in(a,b)$, $\phi(x)$ is an integral function of λ.

Let $\phi(x, \lambda)$ and $\psi(x, \lambda)$ be the solutions of (4.101) such that

$$\begin{aligned}\phi(a,\lambda)&=\sin\alpha, \quad \phi'(a,\lambda)=-\cos\alpha\\ \psi(b,\lambda)&=\sin\beta, \quad \psi'(b,\lambda)=-\cos\beta\end{aligned} \tag{4.114}$$

Then,

$$\begin{aligned}\frac{d}{dx}W(\phi,\psi)&=\phi(x)\psi''(x)-\psi(x)\phi''(x)\\ &=\left[s(x)-\lambda\right]\phi(x)\psi(x)-\left[s(x)-\lambda\right]\psi(x)\phi(x)=0\end{aligned} \tag{4.115}$$

Hence, $W(\phi,\psi)$ is independent of x, and it is a function of λ only. Put $W(\phi,\psi)=\omega(\lambda)$.

Let

$$F(x,\lambda)=\frac{\psi(x,\lambda)}{\omega(\lambda)}\int_a^x\phi(\xi,\lambda)f(\xi)d\xi+\frac{\phi(x,\lambda)}{\omega(\lambda)}\int_x^b\psi(\xi,\lambda)f(\xi)d\xi \tag{4.116}$$

Then, it can be verified by differentiation that $F(x, \lambda)$ satisfies (4.101) and the boundary conditions given by

$$\begin{aligned}F(a,\lambda)\cos\alpha+F'(a,\lambda)\sin\alpha&=0\\ F(b,\lambda)\cos\beta+F'(b,\lambda)\sin\beta&=0\end{aligned} \tag{4.117}$$

for all values of λ. We have assumed that $f(x)$ is continuous.

Suppose now that all zeros of $\omega(\lambda)$ are simple zeros $\{\lambda_m\}$ on the real axis. Then, the Wronskian of $\phi(x, \lambda)$ and $\psi(x, \lambda)$ is zero, which gives one solution as a constant multiple of the other, say, $\psi(x,\lambda_m)=k_m\phi(x,\lambda_m)$. Hence, $F(x, \lambda)$ has the residue

$$\lim_{\lambda\to\lambda_m}(\lambda-\lambda_m)F(x,\lambda)=\frac{k_m}{\omega'(\lambda_m)}\phi(x,\lambda_m)\int_a^b\phi(\xi,\lambda_m)f(\xi)d\xi$$

at $\lambda=\lambda_m$. Therefore, the function $f(x)$ has the expansion of the form

$$\begin{aligned}f(x)&=\sum_{m=0}^{\infty}\frac{k_m}{\omega'(\lambda_m)}\phi(x,\lambda_m)\int_a^b\phi(\xi,\lambda_m)f(\xi)d\xi\\ \Rightarrow f(x)&=\sum_{m=0}^{\infty}c_m\phi(x,\lambda_m)=\sum_{m=0}^{\infty}c_m\phi_m(x)\end{aligned} \tag{4.118}$$

This is the Sturm–Liouville (S–L) expansion of $f(x)$.

If we start with any two independent solutions $\phi_0(x, \lambda)$ and $\psi_0(x, \lambda)$ of the eigenequation (4.101), and form the functions

$$\phi(x,\lambda)=\frac{\phi_0(x)\left[\psi_0(a)\cos\alpha+\psi_0'(a)\sin\alpha\right]-\psi_0(x)\left[\phi_0(a)\cos\alpha+\phi_0'(a)\sin\alpha\right]}{\omega_0(\lambda)}$$
$$\psi(x,\lambda)=\frac{\phi_0(x)\left[\psi_0(b)\cos\beta+\psi_0'(b)\sin\beta\right]-\psi_0(x)\left[\phi_0(b)\cos\beta+\phi_0'(b)\sin\beta\right]}{\omega_0(\lambda)} \quad (4.119)$$

with $\omega_0(\lambda)=W(\phi_0,\psi_0)$, then writing

$$\frac{\phi_0(b,\lambda_m)\cos\beta-\phi_0'(b,\lambda_m)\sin\beta}{\phi_0(a,\lambda_m)\cos\alpha-\phi_0'(a,\lambda_m)\sin\alpha}=\frac{\psi_0(b,\lambda_m)\cos\beta-\psi_0'(b,\lambda_m)\sin\beta}{\psi_0(a,\lambda_m)\cos\alpha-\psi_0'(a,\lambda_m)\sin\alpha}=k_m \quad (4.120)$$

the similar analysis as above can be presented.

In the case of ordinary Fourier series, $s(x) = 0$, and the differential equation (4.101) becomes

$$\frac{d^2y}{dx^2}+\lambda y=0,\quad 0\le x\le b \quad (4.121)$$

with $a = 0$. The two solutions of (4.121) are

$$\phi_0(x,\lambda)=\cos\left(\sqrt{\lambda}x\right),\ \psi_0(x,\lambda)=\sin\left(\sqrt{\lambda}x\right),\ \text{and}\ \omega_0(\lambda)=\sqrt{\lambda} \quad (4.122)$$

Consider the case when $\alpha = 0$, $\beta = 0$. Then,

$$\phi(x,\lambda)=-\frac{\sin\left(\sqrt{\lambda}x\right)}{\sqrt{\lambda}},\ \psi(x,\lambda)=\frac{\sin\left(\sqrt{\lambda}(b-x)\right)}{\sqrt{\lambda}},\ \text{and}\ \omega(\lambda)=\frac{\sin\left(\sqrt{\lambda}b\right)}{\sqrt{\lambda}} \quad (4.123)$$

The zeros of $\omega(\lambda)$ are $\lambda_m=\left(\frac{m\pi}{b}\right)^2$, $m = 1, 2, \ldots,$

thus, we have $\omega'(\lambda_m)=\frac{b}{2\lambda_m}\cos\left(\sqrt{\lambda_m}b\right)=\frac{(-1)^m b}{2\lambda_m}$, and $k_m=\frac{\phi_0(b,\lambda_m)}{\phi_0(0,\lambda_m)}=\frac{\cos(m\pi)}{\cos(0)}=(-1)^m$.

Hence, we get the expansion of the sine series,

$$f(x)=\frac{2}{b}\sum_{m=1}^{\infty}\sin\left(\frac{m\pi x}{b}\right)\int_0^b\sin\left(\frac{m\pi\xi}{b}\right)f(\xi)d\xi \quad (4.124)$$

Similarly, considering the case when $\alpha=\frac{\pi}{2}$, $\beta=\frac{\pi}{2}$, we obtain the expansion of the cosine series,

$$f(x)=\frac{1}{b}\int_0^b f(\xi)d\xi+\frac{2}{b}\sum_{m=1}^{\infty}\cos\left(\frac{m\pi x}{b}\right)\int_0^b\cos\left(\frac{m\pi\xi}{b}\right)f(\xi)d\xi \quad (4.125)$$

For an integrable (piecewise continuous) function $f(x)$, the S–L expansion of $f(x)$ has the same convergence criterion as an ordinary Fourier series.

Theorem 4.31

Let f be a function integrable over (a, b). Then, for any $x \in (a,b)$, the S–L expansion

$$f(x) = \sum_{m=0}^{\infty} \frac{k_m}{\omega'(\lambda_m)} \phi(x, \lambda_m) \int_a^b \phi(\xi, \lambda_m) f(\xi) d\xi$$

converges to $\frac{1}{2}\left[f(x+0)+f(x-0)\right]$ when $f(x)$ is of bounded variation in the neighbourhood of x.

The general S–L equation (4.100) rewritten as

$$M\left[y(x)\right] + \lambda r(x) y(x) = 0, \quad a \le x \le b,$$

where $M\left[y(x)\right] = \left[p(x)y'(x)\right]' - q(x)y(x)$, is in self-adjoint form. The operator M and the adjoint of the operator M, denoted by $M^{\#}$, are same as shown below:

$$My = py'' + p'y' - qy,$$

$$M^{\#}y = (py)'' - (p'y)' - qy = My.$$

Some useful properties of the S–L problem emerge due to the self-adjoint nature of the operator M.

Theorem 4.32

Let λ_m and λ_n be any two distinct eigenvalues of the general S–L problem:

$$\begin{aligned} &\left[p(x)y'(x)\right]' + \left[-q(x) + \lambda r(x)\right]y(x) = 0, \quad a \le x \le b, \\ &a_1 y(a) + a_2 y'(a) = 0, \; b_1 y(b) + b_2 y'(b) = 0 \end{aligned} \tag{4.126}$$

with corresponding solutions $y_m(x)$ and $y_n(x)$. Then, $y_m(x)$ and $y_n(x)$ are orthogonal with weight function $r(x)$, i.e.,

$$\int_a^b r(x) y_m(x) y_n(x) dx = 0 \tag{4.127}$$

The orthogonality condition also holds in the following modifications of the general S–L problem:

(i) when $p(a) = 0$, and $a_1 = a_2 = 0$;
(ii) when $p(b) = 0$, and $b_1 = b_2 = 0$;
(iii) when $p(a) = p(b)$, and the boundary conditions are replaced by the conditions $y(a) = y(b), \; y'(a) = y'(b)$. (4.128)

Moreover, all eigenvalues are real in the S–L problem and its modifications (i)–(iii).

The general S–L problem with modification (i) or (ii) is singular. In the reduced S–L problem of (4.101), the corresponding condition is obtained as $s(a)$ or $s(b)$ tending to infinity, respectively. The conditions (4.128) in modification (iii) are called the periodic boundary conditions. The following theorem is on the uniqueness of eigenfunctions of the S–L problem.

Theorem 4.33

The general S–L problem (4.126), with an additional condition that either $p(a) > 0$ or $p(b) > 0$, cannot have two linearly independent eigenfunctions corresponding to the same eigenvalue.

Further, in this case, each eigenfunction can be made real-valued by multiplying it with an appropriate constant.

The above theorem is not applicable for the S–L problem with periodic boundary conditions. We have found two independent eigenfunctions for the same eigenvalue while deriving the expansions of Fourier sine or cosine series. To obtain the basic Fourier series, we consider the S–L problem on the interval $[-b, b]$,

$$y'' + \lambda y = 0,\ y(-b) = y(b),\ y'(-b) = y'(b) \tag{4.129}$$

which has the eigenvalues $\lambda_m = \left(\frac{m\pi}{b}\right)^2$, $m = 1, 2, \ldots$, and the set of eigenfunctions $\left\{1, \cos\left(\frac{m\pi x}{b}\right), \sin\left(\frac{m\pi x}{b}\right)\right\}$. In this case, we have two linearly independent eigenfunctions corresponding to the same eigenvalue λ_m, and the S–L expansion becomes

$$\begin{aligned} f(x) &= \frac{1}{2b}\int_{-b}^{b} f(\xi)\,d\xi \\ &+ \frac{1}{b}\sum_{m=1}^{\infty}\left[\cos\left(\frac{m\pi x}{b}\right)\int_{-b}^{b}\cos\left(\frac{m\pi\xi}{b}\right) f(\xi)\,d\xi + \sin\left(\frac{m\pi x}{b}\right)\int_{-b}^{b}\sin\left(\frac{m\pi\xi}{b}\right) f(\xi)\,d\xi\right]. \end{aligned} \tag{4.130}$$

We now consider two S–L problem of special interest. In each case, analysis leads to useful expansion formula. The decomposition of a function in terms of a set of orthonormal functions is called the generalized Fourier series.

4.9.1 Fourier-Bessel Series

The singular SL problem for the independent variable $x \in [0, b]$ involving the Bessel's equation

$$x^2\frac{d^2y}{dx^2} + x\frac{dy}{dx} + \left(-\upsilon^2 + \lambda x^2\right)y \quad (\upsilon = 0, 1, \ldots) \tag{4.131}$$

and the boundary condition

$$b_1 y(b) + b_2 y'(b) = 0 \tag{4.132}$$

appears in many situations. The above differential equation is written in the self-adjoint form as

$$\frac{d}{dx}\left(x\frac{dy}{dx}\right) + \left(-\frac{\upsilon^2}{x} + \lambda x\right)y = 0 \tag{4.133}$$

When the parameter λ is positive, we write $\lambda = \alpha^2 (\alpha > 0)$. The eigenfunction of the SL problem is found to be $x^{\frac{1}{2}} J_\upsilon(\alpha x)$ involving the Bessel function J_υ. No solution exists of the SL problem when λ is negative. The eigenvalue $\lambda = 0$ occurs for $\upsilon = 0$ and $b_1 = 0$, and the corresponding eigenfunction is $x^{\frac{1}{2}}$. We now state the orthogonality property of the eigenfunctions in the following theorem.

Theorem 4.34

For each value of υ = 0, 1, 2, ..., the functions $\left\{x^{\frac{1}{2}}J_{\upsilon}(\alpha_m x);\ m=1,2,\ldots\right\}$ constitute an orthogonal set on the interval [0, b] when $\alpha=\alpha_m;\ m=1,2,\ldots$ are the positive roots of the equation

(i) $J_{\upsilon}(\alpha b)=0$, or

(ii) $b_1 J_{\upsilon}(\alpha b)+b_2\alpha J_{\upsilon}'(\alpha b)=0$ with $b_1\ge 0,\ b_2>0,\ b_1$ and υ both not zero.

For $\upsilon = 0$, the functions $\left\{x^{\frac{1}{2}}J_0(\alpha_m x);\ m=1,2,\ldots\right\}$ constitute an orthogonal set on the interval [0, b] when $\alpha=\alpha_m;\ m=1,2,\ldots$ are the non-negative roots of the equation

(iii) $J_0'(\alpha b)=0$.

We generate an orthonormal set of eigenfunctions $\{\phi_m(x);\ m=1,2,\ldots\}$ on the interval [0, b] as follows

$$\phi_m(x)=\frac{x^{\frac{1}{2}}J_{\upsilon}(\alpha_m x)}{\left\|x^{\frac{1}{2}}J_{\upsilon}(\alpha_m x)\right\|}=\frac{x^{\frac{1}{2}}J_{\upsilon}(\alpha_m x)}{\rho_m} \tag{4.134}$$

and represent a function $f(x)$ by the generalized Fourier series involving those orthonormal functions as

$$f(x)=\sum_{m=1}^{\infty}\frac{J_{\upsilon}(\alpha_m x)}{\rho_m^2}\int_0^b \xi f(\xi)J_{\upsilon}(\alpha_m\xi)\,d\xi \tag{4.135}$$

Case (i)

When $\alpha=\alpha_m$; $m = 1, 2, \ldots$ are the roots of the equation $J_\upsilon(\alpha b) = 0$, we get

$$\begin{aligned}\rho_m^2&=\int_0^b x\left[J_{\upsilon}(\alpha_m x)\right]^2dx=\frac{b^2}{2}\left[J_{\upsilon}'(\alpha_m b)\right]\\&=\frac{b^2}{2}\left[J_{\upsilon+1}(\alpha_m b)\right]^2\end{aligned} \tag{4.136}$$

where the identity $xJ_{\upsilon}'(x)=\upsilon J_{\upsilon}(x)-xJ_{\upsilon+1}(x)$ is used.

Case (ii)

When $\alpha=\alpha_m$; m = 1, 2, ... are the roots of the equation $b_1J_{\upsilon}(\alpha b)+b_2\alpha J_{\upsilon}'(\alpha b)=0$ with $b_1\ge 0,\ b_2>0$, b_1 and υ both not zero, we get

$$\rho_m^2=\frac{\alpha_m^2b_2^2b^2-\upsilon^2b_2^2+b_1^2b^2}{2\alpha_m^2b_2^2}\left[J_{\upsilon}(\alpha_m b)\right]^2 \tag{4.137}$$

Case (iii)

When $\alpha=\alpha_m$; $m = 1, 2, \ldots$ are the roots of the equation $J_0'(\alpha b)=0$, we get

$$\rho_m^2=\frac{b^2}{2}\left[J_0(\alpha_m b)\right]^2 \tag{4.138}$$

Note that here $\alpha_1=0$, and since $J_0(\alpha_1 x)=1$, we have $\rho_1^2=\frac{b^2}{2}$.

4.9.2 Legendre Series

The SL problem involving the Legendre's equation

$$\left(1-x^2\right)\frac{d^2y}{dx^2}-2x\frac{dy}{dx}+\lambda y=0,\quad -1\le x\le 1 \tag{4.139}$$

is written in the self-adjoint form as

$$\frac{d}{dx}\left[\left(1-x^2\right)\frac{dy}{dx}\right]+\lambda y=0 \tag{4.140}$$

in which $p(x)=1-x^2$, $q(x)=0$, and $r(x)=1$. Since $p(x)=0$ at $x=-1$ and $x=1$, these are singular points. Accordingly, no boundary conditions are needed at the ends of the interval [–1, 1]. The singular SL problem consisting of equation (4.140) and the condition that the solution y and its derivative y' be continuous on the interval [–1, 1] is to be considered.

It turns out that an infinite set of values of the parameter λ can be found for each of which equation (4.140) has a solution. In particular, λ can have any one of the integer values

$$\lambda=m(m+1),\quad m=0,1,2,\ldots \tag{4.141}$$

and the corresponding solution is the polynomial $P_m(x)$ which the Legendre polynomial of degree m. The polynomials $\{P_m(x);\ m=0,1,2,\ldots\}$ constitute an orthogonal set on the interval [–1, 1], i.e.,

$$\int_{-1}^{1}P_m(x)P_n(x)dx=0\quad\text{for}\quad m\ne n \tag{4.142}$$

In this case, an orthonormal set of polynomials on the interval [–1, 1] is generated as $\{P_m(x)/\|P_m(x)\|;\ m=0,1,2,\ldots\}$ where $\|P_m(x)\|^2=\int_{-1}^{1}\left[P_m(x)\right]^2dx=\frac{2}{2m+1}$. Therefore, the Legendre series representation of a function $f(x)$ is given by

$$\begin{aligned} f(x)&=\sum_{m=0}^{\infty}\frac{P_m(x)}{\|P_m(x)\|^2}\int_{-1}^{1}f(\xi)P_m(\xi)d\xi \\ &=\sum_{m=0}^{\infty}\frac{2m+1}{2}P_m(x)\int_{-1}^{1}f(\xi)P_m(\xi)d\xi \end{aligned} \tag{4.143}$$

The polynomial $P_m(x)$ is an even function when m is even and an odd function when m is odd, i.e.,

$$P_m(-x)=(-1)^m P_m(x) \tag{4.144}$$

Then, it follows from the above equation that

$$\int_{0}^{1}P_{2m}(x)P_{2n}(x)dx=0\quad\text{for}\quad m\ne n \tag{4.145}$$

and

$$\int_0^1 P_{2m+1}(x)P_{2n+1}(x)dx = 0 \quad \text{for } m \neq n \tag{4.146}$$

where $m, n = 0, 1, 2, \ldots$

The SL problem consisting of the Legendre's equation on the interval [0, 1] and the boundary condition $y'(0) = 0$ has the eigenvalues $\lambda = 2m(2m+1)$; $m = 0, 1, 2, \ldots$, and the corresponding eigenfunctions are $\{P_{2m}(x);\ m = 0, 1, 2, \ldots\}$. Similarly, the SL problem consisting of the Legendre's equation on the interval [0, 1] and the condition $y(0) = 0$ has the eigenvalues $\lambda = (2m+1)(2m+2)$; $m = 0, 1, 2, \ldots$, and the corresponding eigenfunctions are $\{P_{2m+1}(x);\ m = 0, 1, 2, \ldots\}$. The representations of functions on the interval [0, 1] can be written as

$$f(x) = \sum_{m=0}^{\infty}(4m+1)P_{2m}(x)\int_0^1 f(\xi)P_{2m}(\xi)d\xi \tag{4.147}$$

when f is an even function, otherwise as

$$f(x) = \sum_{m=0}^{\infty}(4m+3)P_{2m+1}(x)\int_0^1 f(\xi)P_{2m+1}(\xi)d\xi \tag{4.148}$$

when f is an odd function. Compare the Legendre series (4.147) with the Fourier cosine series and the Legendre series (4.148) with the Fourier sine series to represent the even and odd parts, respectively, of an arbitrary function.

4.10 Signal Processing Applications IV: Parameter Estimation of Complex Exponential Signal in Noise

The problem of estimation of complex exponential signal parameters from a given set of signal values at uniform spacing when the signal samples are corrupted with noise appears in a wide variety of engineering applications.

Let the modelled signal sequence $g[n] = g(nT)$ be given by $g[n] = \sum_{i=1}^{M} b_{ci} z_i^n$ with $z_i = \exp(s_i T)$. The signal sequence satisfies the linear prediction model given by

$$g[n] = -\sum_{l=1}^{M} a_l\, g[n-l], \quad n \geq M \tag{4.149}$$

with proper initial conditions $\{g[0], g[1], \ldots, g[M-1]\}$.

The sampled data represented by $y[n] = g[n] + w[n] = \sum_{i=1}^{L} b_{ci} z_i^n$ with additive noise $w[n]$, for $n = 0, 1, \ldots, N-1$, can be fitted by the extended-order linear prediction model as follows,

$$\hat{y}[n] = -\sum_{l=1}^{L} a_l\, y[n-l], \quad n \geq L \geq M \tag{4.150}$$

with proper initial conditions $\{y[0], y[1], \ldots, y[L-1]\}$. The coefficients a_l form the characteristic polynomial of the extended-order prediction-error filter $A(z)$,

$$\begin{aligned} A(z) &= 1 + \sum_{l=1}^{L} a_l z^{-l} \\ &= \prod_{l=1}^{L} \left(1 - z_l z^{-1}\right), \quad L \geq M \end{aligned} \tag{4.151}$$

whose M signal roots are located at $z_i = \exp(s_i T)$, and $(L-M)$ additional roots are used to model the additive noise.

4.10.1 Forward–Backward Linear Prediction Method

Let the sequence $\{y[n];\ n = 0, 1, \ldots, N-1\}$ be the sum of L complex exponentials satisfying the L-order forward and backward linear prediction equations as shown below:

$$y[n] = -\sum_{l=1}^{L} a_l\, y[n-l] + e_f[n], \quad L \leq n \leq N-1 \tag{4.152}$$

and

$$y[n] = -\sum_{l=1}^{L} a_l^*\, y[n+l] + e_b[n], \quad 0 \leq n \leq N-L-1 \tag{4.153}$$

where $e_f[n]$ and $e_b[n]$ are the forward and backward prediction errors, respectively. We rewrite the backward prediction equation as

$$y^*[n] = -\sum_{l=1}^{L} a_l\, y^*[n+l] + e_b^*[n], \quad 0 \leq n \leq N-L-1 \tag{4.154}$$

Combining (4.152) and (4.154), we get the forward–backward prediction equations in matrix form as

$$\begin{bmatrix} y[L-1] & y[L-2] & \cdots & y[0] \\ y[L] & y[L-1] & \cdots & y[1] \\ \vdots & \vdots & & \vdots \\ y[N-2] & y[N-3] & \cdots & y[N-L-1] \\ y^*[1] & y^*[2] & \cdots & y^*[L] \\ y^*[2] & y^*[3] & \cdots & y^*[L+1] \\ \vdots & \vdots & & \vdots \\ y^*[N-L] & y^*[N-L+1] & \cdots & y^*[N-1] \end{bmatrix} \begin{bmatrix} a_1 \\ a_2 \\ \vdots \\ a_L \end{bmatrix} = - \begin{bmatrix} y[L] \\ y[L+1] \\ \vdots \\ y[N-1] \\ y^*[0] \\ y^*[1] \\ \vdots \\ y^*[N-L-1] \end{bmatrix} \tag{4.155}$$

or, $\quad \bar{\mathbf{Y}}\mathbf{a} = -\bar{\mathbf{y}}$

where the forward and backward prediction errors are neglected.

Solving (4.155) for the coefficient vector $\mathbf{a}$ in the least squares sense, we obtain

$$\mathbf{a} = \left(\bar{\mathbf{Y}}^H\bar{\mathbf{Y}}\right)^{-1}\bar{\mathbf{Y}}^H\bar{\mathbf{y}}$$
$$= \left(\bar{\mathbf{Y}}^H\bar{\mathbf{Y}}\right)^{-1}\mathbf{r} \tag{4.156}$$

Then, the characteristic equation $A(z) = 0$ of the prediction error filter is formed with the coefficients as

$$z^L + a_1 z^{L-1} + a_2 z^{L-2} + \cdots + a_{L-1}z + a_L = 0 \tag{4.157}$$

and the roots are computed for the z_i-values.

When the sequence $y[n]$ is not corrupted with noise, the matrix $\bar{\mathbf{Y}}^H\bar{\mathbf{Y}}$ is found to be singular. In fact, with $y[n] = g[n] = \sum_{i=1}^{M} b_{ci} z_i^n$, the rank of $\bar{\mathbf{Y}}$ is M, because each column of $\bar{\mathbf{Y}}$ can be written as a linear combination of the linearly independent vectors $\{\bar{\mathbf{z}}_i\,;\ i = 1, 2, \ldots, M\}$ given by

$$\bar{\mathbf{z}}_i = \left[1, z_i, z_i^2, \ldots, z_i^{N-L-1}, z_i^{-L}, z_i^{-(L+1)}, \ldots, z_i^{-(N-1)}\right]^T \tag{4.158}$$

In order to have linear independence of the vectors $\{\bar{\mathbf{z}}_i\}$, it is necessary that the length of the vector, $2(N-L) \geq M$. Thus, the permissible range of the extended order is given by

$$M \leq L \leq N - \frac{M}{2} \tag{4.159}$$

In the noise-free case, the rank of $\bar{\mathbf{Y}}^H\bar{\mathbf{Y}}$ will be M. We compute the eigenvalue/eigenvector decomposition of the Hermitian matrix $\bar{\mathbf{Y}}^H\bar{\mathbf{Y}}$ as

$$\bar{\mathbf{Y}}^H\bar{\mathbf{Y}} = \mathbf{U}\Lambda\mathbf{U}^H \tag{4.160}$$

where $\mathbf{U}$ is a unitary matrix, $\Lambda = \text{diag}(\lambda_1, \lambda_2, \ldots, \lambda_L)$ with $\lambda_1 \geq \lambda_2 \geq \cdots \geq \lambda_M > 0$ and $\lambda_{M+1} = \lambda_{M+2} = \cdots = \lambda_L = 0$. Thus, the solution for the coefficient vector $\mathbf{a}$ is obtained as

$$\mathbf{a} = \sum_{i=1}^{M} \frac{\left(\mathbf{u}_i^H\mathbf{r}\right)}{\lambda_i}\mathbf{u}_i \tag{4.161}$$

where $\mathbf{u}_i$ are the columns of $\mathbf{U}$, and $\mathbf{r} = \bar{\mathbf{Y}}^H\bar{\mathbf{y}}$.

4.10.2 Null-Space Solution Method

Let the sequence $\{y[n];\ n = 0, 1, \ldots, N-1\}$ be the sum of L complex exponentials satisfying the L-order linear prediction equations for $n \geq L$. Then, the prediction coefficients are given by the matrix equation as shown below

$$\begin{bmatrix} y[L] & y[L-1] & \cdots & y[0] \\ y[L+1] & y[L] & \cdots & y[1] \\ \vdots & \vdots & & \vdots \\ y[N-1] & y[N-2] & \cdots & y[N-L-1] \end{bmatrix} \begin{bmatrix} 1 \\ a_1 \\ \vdots \\ a_L \end{bmatrix} = \mathbf{0}$$

$$\text{or,} \qquad \mathbf{Y}\bar{\mathbf{a}} = \mathbf{0} \tag{4.162}$$

where the prediction error filter is given by $z^L + a_1 z^{L-1} + a_2 z^{L-2} + \cdots + a_{L-1}z + a_L = 0$, whose roots are the z_i-values.

Note that in the noise-free case, when $y[n]=g[n]=\sum_{i=1}^{M} b_{ci} z_i^n$, the rank of the matrix $\mathbf{Y}$ is M, because each column of $\mathbf{Y}$ can be written as a linear combination of the linearly independent vectors $\{\mathbf{z}_i\,;\ i=1,2,\ldots,M\}$ given by

$$\mathbf{z}_i=\left[1,\, z_i,\, z_i^2,\ldots,\, z_i^{N-L-1}\right]^T \tag{4.163}$$

The permissible range of the extended order is now given by

$$M \le L \le N-M \tag{4.164}$$

Premultiplying both sides of (4.162) by $\mathbf{Y}^H$, we get

$$\mathbf{Y}^H \mathbf{Y} \bar{\mathbf{a}} = \mathbf{0} \tag{4.165}$$

where the Hermitian matrix $\mathbf{Y}^H\mathbf{Y}$ is of rank M, and the coefficient vector $\mathbf{a}$ belongs to the null-space of the matrix $\mathbf{Y}^H\mathbf{Y}$ of dimension $(L+1)^2$.

When $L=M$, the eigenvalue/eigenvector decomposition of $\mathbf{Y}^H\mathbf{Y}$ is given by

$$\mathbf{Y}^H \mathbf{Y} = \mathbf{V}\bar{\Lambda}\mathbf{V}^H \tag{4.166}$$

where $\mathbf{V}$ is a unitary matrix, $\bar{\Lambda}=\operatorname{diag}(\lambda_1,\lambda_2,\ldots,\lambda_{M+1})$ with $\lambda_1 \ge \lambda_2 \ge \cdots \ge \lambda_M > 0$ and $\lambda_{M+1}=0$. Thus, the solution for the coefficient vector $\bar{\mathbf{a}}$ is obtained as

$$\bar{\mathbf{a}} = \frac{\mathbf{v}_{M+1}}{\mathbf{v}_{M+1}(1)} \tag{4.167}$$

where $\mathbf{v}_{M+1}$ is the eigenvector corresponding to the eigenvalue λ_{M+1}, and $\mathbf{v}_{M+1}(1)$ is the first element of the eigenvector. This normalization is required because the first element of the vector $\bar{\mathbf{a}}$ is unity.

For other values of $L > M$, the null-space solution for the vector $\bar{\mathbf{a}}$ is given by

$$\bar{\mathbf{a}} = \frac{\sum_{i=M+1}^{L+1} \mathbf{v}_i^*(1)\,\mathbf{v}_i}{\sum_{j=M+1}^{L+1} \left|\mathbf{v}_j(1)\right|^2} \tag{4.168}$$

where the eigenvectors $\{\mathbf{v}_i\,;\ i=M+1,\ldots,L+1\}$ corresponding to the eigenvalues $\lambda_{M+1}=\cdots=\lambda_{L+1}=0$ span the null-space of $\mathbf{Y}^H\mathbf{Y}$, and $\mathbf{v}_i(1)$ is the first element of the eigenvector $\mathbf{v}_i$.

The derivation of (4.168) is left as an exercise to the reader. Note that the vector $\bar{\mathbf{a}}$ can be written as $\bar{\mathbf{a}}=\sum_{i=M+1}^{L+1}\gamma_i \mathbf{v}_i$, where the coefficients γ_i are to be determined by minimizing $\|\bar{\mathbf{a}}\|^2$ subject to the constraint $\bar{\mathbf{a}}(1)=1$. This is the minimum norm solution for the desired vector $\bar{\mathbf{a}}$.

4.10.3 Matrix Pencil Method

Let the sequence $\{y[n];\ n=0,1,\ldots,N-1\}$ comprising of M complex exponentials in additive noise be available for processing. We define two matrices $\mathbf{Y}_0$ and $\mathbf{Y}_1$, both of dimension $(N-L)\times L$, as shown below

$$\mathbf{Y}_0 = \begin{bmatrix} y[L-1] & y[L-2] & \dots & y[0] \\ y[L] & y[L-1] & \dots & y[1] \\ \vdots & \vdots & & \vdots \\ y[N-2] & y[N-3] & \dots & y[N-L-1] \end{bmatrix} \tag{4.169}$$

$$\mathbf{Y}_1 = \begin{bmatrix} y[L] & y[L-1] & \dots & y[1] \\ y[L+1] & y[L] & \dots & y[2] \\ \vdots & \vdots & & \vdots \\ y[N-1] & y[N-2] & \dots & y[N-L] \end{bmatrix} \tag{4.170}$$

and form the matrix pencil $(\mathbf{Y}_1 - z\mathbf{Y}_0)$ with a complex parameter z.

In the noise-free case, when $y[n] = g[n] = \sum_{i=1}^{M} b_{ci} z_i^n$, the matrices $\mathbf{Y}_0$ and $\mathbf{Y}_1$ can be decomposed as follows

$$\mathbf{Y}_0 = \bar{\mathbf{Z}}_V \mathbf{B}_c \bar{\mathbf{Z}}_R \tag{4.171}$$

$$\mathbf{Y}_1 = \bar{\mathbf{Z}}_V \mathbf{B}_c \mathbf{Z} \bar{\mathbf{Z}}_R \tag{4.172}$$

where

$$\bar{\mathbf{Z}}_V = \begin{bmatrix} 1 & \dots & 1 \\ z_1 & \dots & z_M \\ \vdots & & \vdots \\ z_1^{N-L-1} & \dots & z_M^{N-L-1} \end{bmatrix}, \ \mathbf{B}_c = \operatorname{diag}\left(b_{c1}, b_{c2}, \dots, b_{cM}\right),$$

$$\bar{\mathbf{Z}}_R = \begin{bmatrix} z_1^{L-1} & z_1^{L-2} & \dots & 1 \\ \vdots & \vdots & & \vdots \\ z_M^{L-1} & z_M^{L-2} & \dots & 1 \end{bmatrix}, \mathbf{Z} = \operatorname{diag}(z_1, z_2, \dots, z_M).$$

It is easy to verify now that the rank of every constituent matrix in the above decompositions is M, and as a result, the rank of $\mathbf{Y}_0$ or $\mathbf{Y}_1$ is M, provided the extended order L is chosen such that $M \le L \le N - M$. Moreover, the rank of the matrix pencil $(\mathbf{Y}_1 - z\mathbf{Y}_0)$ is M for any arbitrary value of z. However, it is clear from the matrix decompositions that the rank of $(\mathbf{Y}_1 - z\mathbf{Y}_0)$ is $(M–1)$ when $z = z_i$, z_i being one of the distinct modes of the underlying signal. Therefore, we should determine the rank-reducing numbers $z = z_i$ of the matrix pencil.

Premultiplying the matrix pencil by $\mathbf{Y}_0^H$, we get the modified matrix pencil $\left(\mathbf{Y}_0^H \mathbf{Y}_1 - z\mathbf{Y}_0^H \mathbf{Y}_0\right)$. Now the rank-reducing numbers of the modified matrix pencil can be determined by computing the generalized eigenvalues of the matrix pair $\left(\mathbf{Y}_0^H \mathbf{Y}_1, \mathbf{Y}_0^H \mathbf{Y}_0\right)$,

$$z = \lambda\left(\mathbf{Y}_0^H \mathbf{Y}_1, \mathbf{Y}_0^H \mathbf{Y}_0\right) = z_i, \quad i = 1, 2, \dots, M \tag{4.173}$$

Alternatively, we may first calculate the Moore–Penrose inverse $\left(\mathbf{Y}_0^H \mathbf{Y}_0\right)^+$, and then compute the eigenvalues of the matrix $\left(\mathbf{Y}_0^H \mathbf{Y}_0\right)^+ \mathbf{Y}_0^H \mathbf{Y}_1$,

$$z = \lambda\left(\left(\mathbf{Y}_0^H \mathbf{Y}_0\right)^+ \mathbf{Y}_0^H \mathbf{Y}_1\right) = z_i, \quad i = 1, 2, \dots, M \tag{4.174}$$

We obtain the eigenvalue/eigenvector decomposition of the Hermitian matrix $\mathbf{Y}_0^H\mathbf{Y}_0$ as

$$\mathbf{Y}_0^H\mathbf{Y}_0 = \mathbf{W}\Lambda\mathbf{W}^H \tag{4.175}$$

where $\mathbf{W}$ is a unitary matrix, $\Lambda = \text{diag}(\lambda_1, \lambda_2, \ldots, \lambda_L)$ with $\lambda_1 \geq \lambda_2 \geq \cdots \geq \lambda_M > 0$ and $\lambda_{M+1} = \lambda_{M+2} = \cdots = \lambda_L = 0$. The Moore–Penrose inverse $(\mathbf{Y}_0^H\mathbf{Y}_0)$ is computed as

$$(\mathbf{Y}_0^H\mathbf{Y}_0)^+ = \mathbf{W}\Lambda^+\mathbf{W}^H \tag{4.176}$$

where $\Lambda^+ = \text{diag}\left(1/\lambda_1, \ldots, 1/\lambda_M, 0, \ldots, 0\right)$.

4.10.4 Transient Signal Analysis III

The multi-component complex exponential signal model is often used for representation of transient signals. When the transient signal happens to be the impulse response of a linear system, the estimation of parameters of the complex signal model is equivalent to the extraction of the poles and residues at the poles of the system. In this way, the problem of system identification is related to the parameter estimation of the complex exponential signals.

The techniques developed for parameter estimation of the complex exponential signals can also be applied for parameter estimation of the complex sinusoidal signals. For instance, the methods developed here are equally applicable to the problem of direction-of-arrival (DOA) estimation of radio waves. In the DOA estimation, the signal measured at the sensor array is represented by the weighted sum of complex sinusoidal signals whose spatial frequencies contain the DOA information. In this case, the independent variable is the spatial distance, instead of time which is the independent variable in the system identification problem.

We discuss some methods for parameter estimation of the complex exponential signals here. It is to be pointed out that all these methods are developed based on the geometrical properties of the complex exponential signals. When the signal is corrupted with noise, the techniques do not provide satisfactory results because the consideration of noise is not built in the developed theory. We use an extended order modelling is such cases to model simultaneously the signal and the noise, and a subspace decomposition technique is employed to separate the signal subspace from the noise subspace. It turns out that with a proper choice of extended model order and a suitable subspace decomposition technique, the methods developed based on the geometrical properties perform comparably with the optimal methods developed based on the statistical properties of the noisy signal.

4.10.5 Simulation Study: Example 1 (FBLP Method)

The test signal to be considered for simulation study is the damped sinusoidal signal $g(t) = \exp(-0.5t)\sin 2t + \exp(-0.8t)\sin 5t$, corresponding to the complex conjugate pole-pairs $s_{1,2} = -0.5 \pm j2.0$ and $s_{3,4} = -0.8 \pm j5.0$ with number of components $M = 4$, sampled at $N = 40$ uniformly spaced points $\{nT; n = 0, 1, \ldots, N-1\}$ with the sampling interval $T = 0.08$. The zero-mean white complex Gaussian noise is added to the signal samples setting the signal-to-noise ratio (SNR) at 10 dB. To estimate the complex modes $z_i = \exp(s_i T)$ of the signal, the forward–backward linear prediction (FBLP) method with the extended model order $L = 24$ is used.

There are two possible procedures that can be adopted here. In the first procedure, we compute the pseudoinverse of the system matrix by the least squares (LS) technique, and the vector of prediction coefficients is calculated by using the pseudoinverse. In the second procedure, we compute the rank-constrained generalized inverse of the system matrix by the SVD technique, and the coefficient vector is calculated accordingly. Note that the presence of additive noise in data makes the extended-order system matrix full-rank, however, the effective rank of the matrix still remains same as the number of signal components $M = 4$. The effective rank is indicated in the number of significant singular values of the matrix, which are found to be some order higher in magnitude than the other singular values that are related to noise.

We estimate the z_i-values for 40 realizations using independent noise sequences employing, in turn, the full-rank pseudoinverse and the rank-constrained generalized inverse computed by the SVD technique. The estimated z_i-values are plotted in Figures 4.3 and 4.4, where the signal modes are shown by circles. The advantage of applying the SVD technique is apparent.

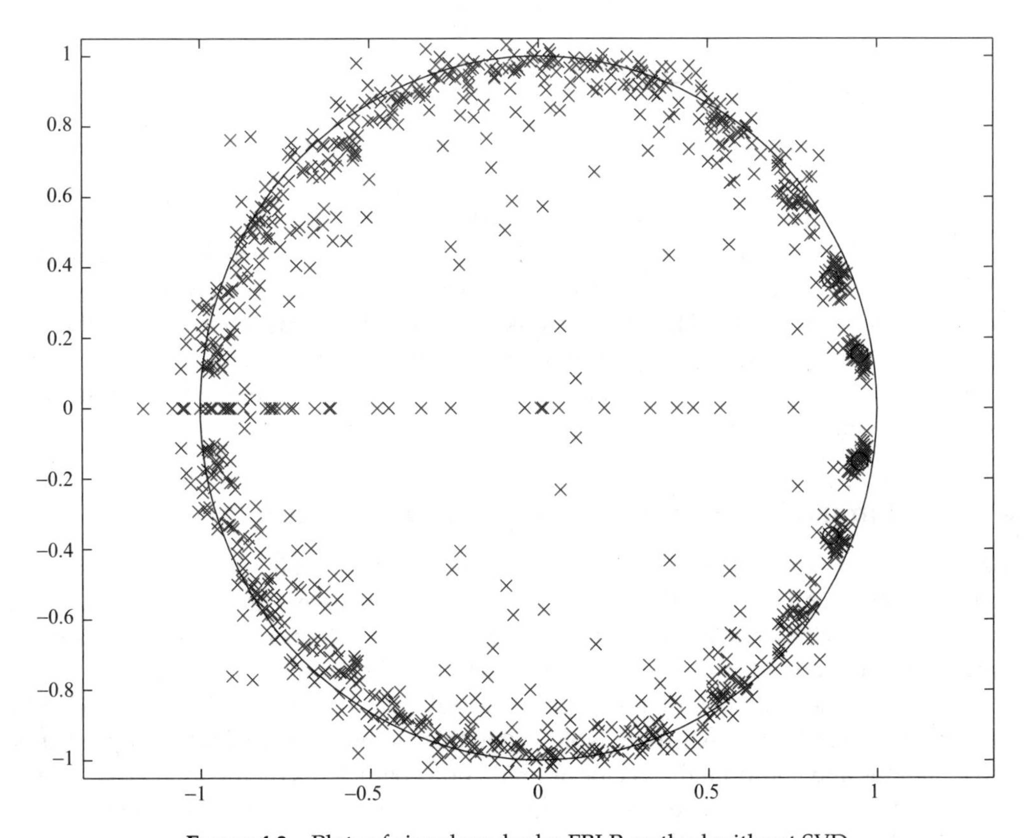

FIGURE 4.3 Plots of signal modes by FBLP method without SVD

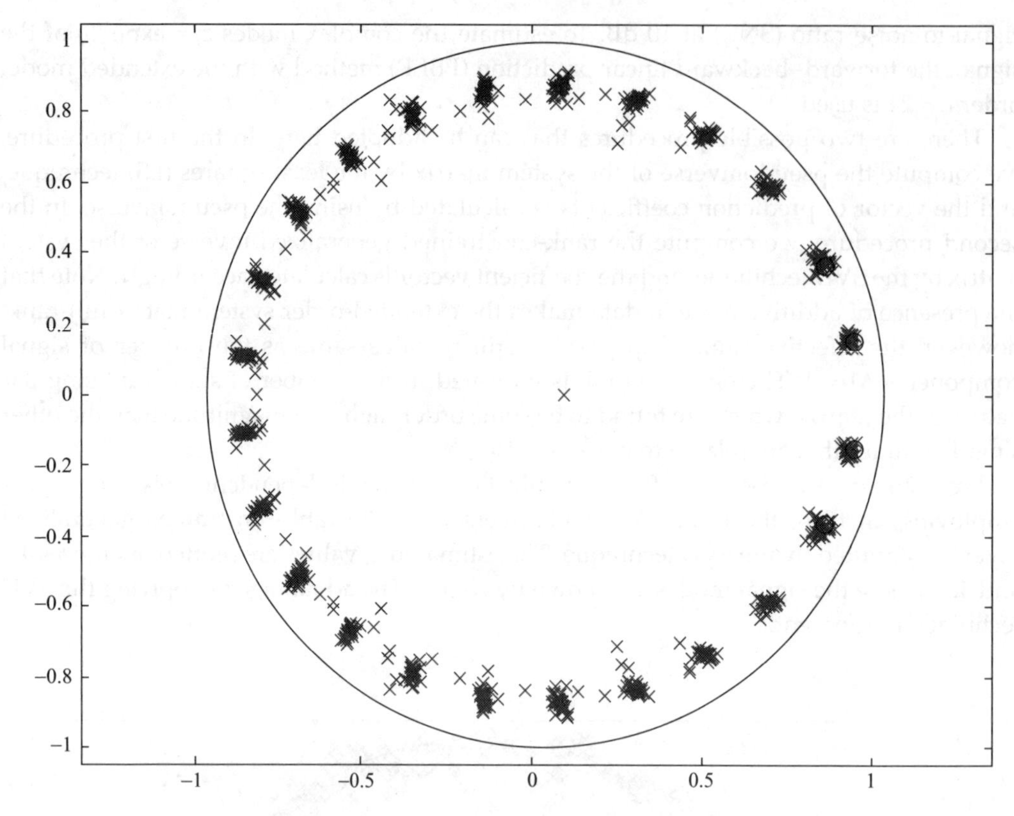

FIGURE 4.4 Plots of signal modes by FBLP method with SVD

It is to be mentioned that the extended order $L = 24$ of the FBLP method has been set by trial for best performance. Note that in this case, system matrix has the dimension $2(N–L) \times L = 32 \times 24$, which provides most accurate results when we compute the pseudoinverse (pinv) by MATLAB or the rank-constrained generalized inverse by the SVD technique. Usually, the accuracy of parameter estimation improves with increase of model order. However, the optimal range of model order (L) relative to the number of points (N) in the data-set may depend on a specific problem.

4.10.6 Simulation Study: Example 2 (NSS Method)

The signal $g(t) = \exp(-0.5t)\sin 2t + \exp(-0.8t)\sin 5t$ is sampled at $N = 40$ with the sampling interval $T = 0.08$ unit of time. The complex frequency parameters are $s_{1,2} = -0.5 \pm j2.0$ and $s_{3,4} = -0.8 \pm j5.0$ with number of signal components $M = 4$. The zero-mean white complex Gaussian noise is added to the signal samples setting SNR=20 dB.

To estimate the z_i-modes of the signal, the null-space solution (NSS) method with extended model order $L = 16$ is used. Note that in this method, we use the linear prediction (LP) equations and compute the vector of prediction coefficients by the total least squares (TLS) technique. It is to be pointed out that the formulation of the NSS method looks similar to the Pisarenko's Method as discussed in Chapter 1, particularly the procedure of eigenvalue decomposition (EVD) of the normal matrices in both methods. However, it should be noted that the Pisarenko's method uses the LP equations of autocorrelation functions of pure sinusoids. In the case of damped sinusoids, the autocorrelation functions (ACFs) are time-dependent, and we are getting the sum of ACFs over a moving window in the normal matrix here. In fact in the present example, one should use the SVD technique on the data matrix, instead of the EVD technique on the normal matrix for better accuracy of estimation.

We estimate the z_i-values for 40 realizations using independent noise sequences employing, in turn, the EVD technique and the SVD technique as explained above. The estimated z_i-values are plotted in Figures 4.5 and 4.6, where the signal modes are shown by circles. There seems to be some advantage of applying the SVD technique for parameter estimation. However, it should be noted that this advantage is coming solely from the better conditioning of the data matrix over the normal matrix. In each method, the rank of the system matrix is fixed at $M = 4$ in computation of the null-space of the matrix. Therefore, the computations are equally affected by noise in both the methods.

It is to be mentioned that the extended order $L = 16$ of the NSS method has been set by trial for best performance. Note that in this case, the dimensions of the system matrices are $L \times L = 16 \times 16$ and $(N - L) \times L = 24 \times 16$, which provide most accurate eigenvectors and

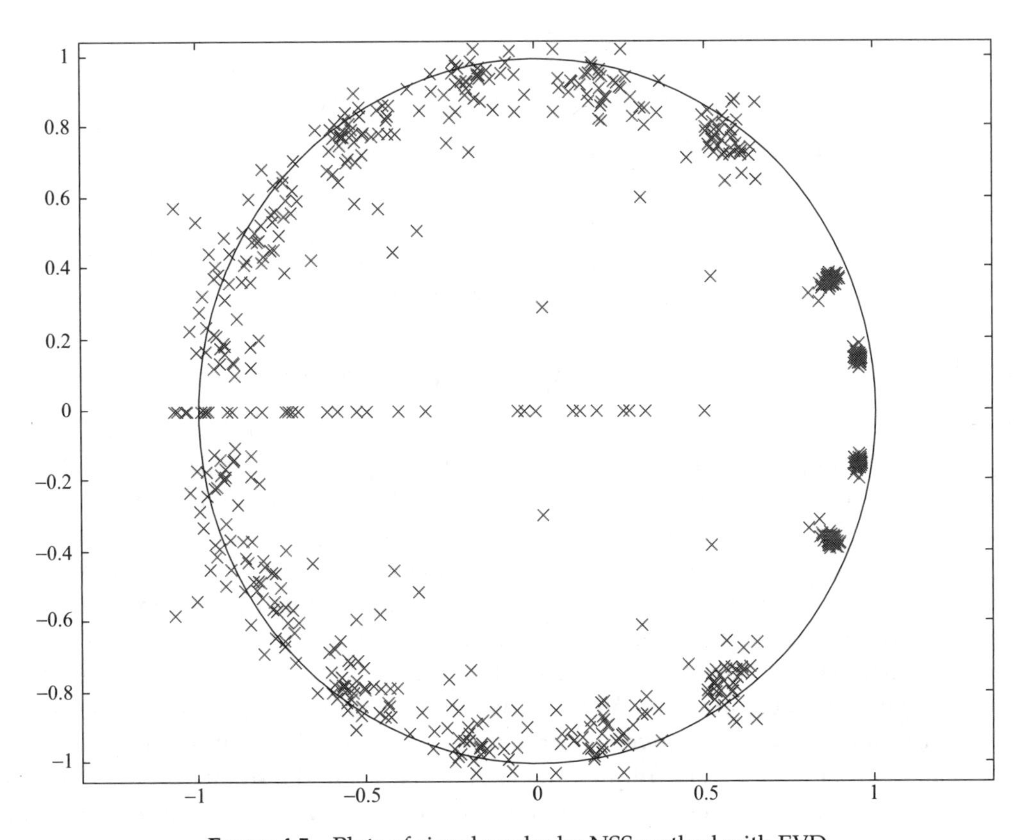

FIGURE 4.5 Plots of signal modes by NSS method with EVD

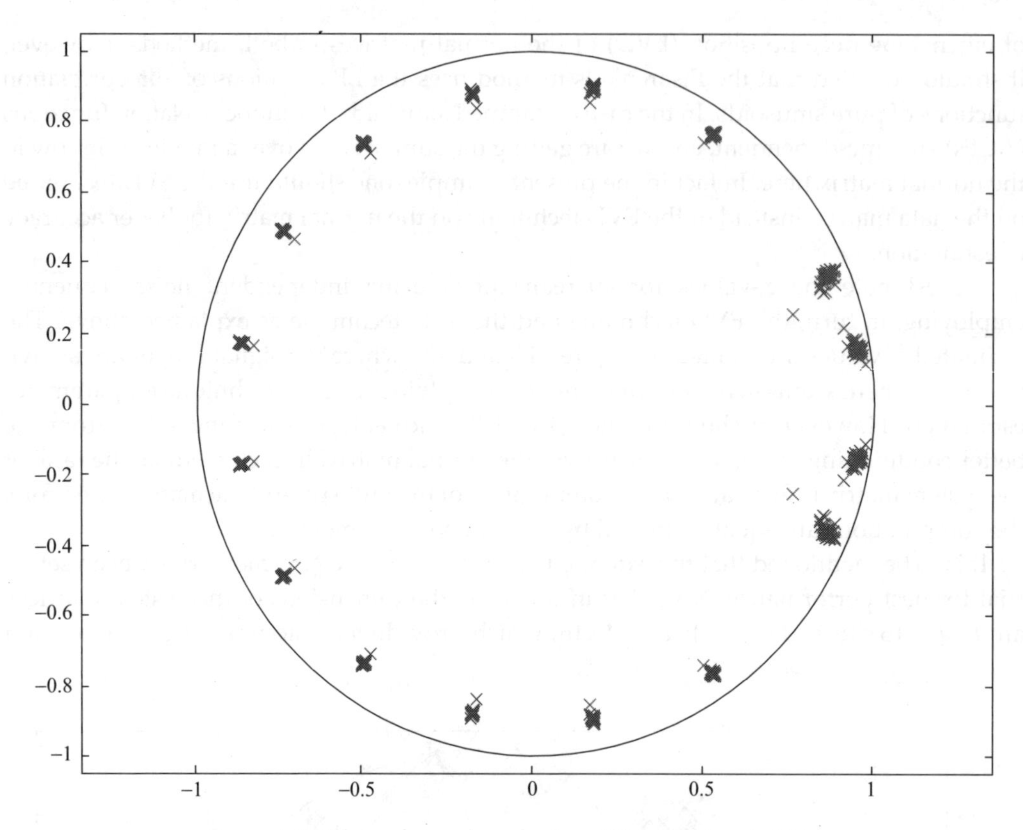

FIGURE 4.6 Plots of signal modes by NSS method with SVD

singular vectors of the corresponding matrices, respectively. Once again, the choice of optimal L for a given N may depend on a specific problem.

4.10.7 Simulation Study: Example 3 (MP Method)

We consider the transient signal $g(t) = \exp(-0.5t)\sin 2t + \exp(-0.8t)\sin 5t$ corresponding to the complex conjugate pole-pairs $s_{1,2} = -0.5 \pm j2.0$ and $s_{3,4} = -0.8 \pm j5.0$, $M = 4$, sampled at $N = 40$ equispaced points with sampling interval $T = 0.08$ units of time. The zero-mean white complex Gaussian noise is mixed with the signal setting SNR=20 dB.

In the matrix pencil (MP) method, we form a matrix pencil with two data matrices of similar dimensions whose corresponding elements are related by one time delay. Let these matrices be referred to as the time-lag matrix and the time-lead matrix. Here, we are interested to find the rank-reducing number of the matrix pencil. To convert this problem into an eigenvalue decomposition problem, we compute either the pseudoinverse or the rank-constrained generalized inverse of the time-lag matrix. Then, premultiplying the time-lead matrix by the inverse of the time-lag matrix, we find a square product matrix whose non-zero eigenvalues are the modes of the underlying complex exponential signal.

The extended-order data matrices are full-rank when noise is superimposed on the signal sequence. The pseudoinverse of the time-lag matrix will be full-rank in this case, which in turn, will ensure that the square product matrix also be full-rank. Thus, the eigenvalues of the matrix will provide the signal and noise modes. However, when we compute the rank-constrained generalized inverse of the time-lag matrix by the SVD technique, the rank of the product matrix will be same as the number of signal components $M = 4$, and the non-zero eigenvalues of the matrix will now give only the signal modes. Thus, the SVD technique reduces computation of eigenvalue decomposition, and at the same time, removes ambiguity between the signal modes and the noise modes.

We estimate the z_i-values for 40 realizations using independent noise sequences, calculating in turn, the full-rank pseudoinverse and the rank-constrained generalized inverse computed by the SVD technique of the time-lag data matrix. The estimated z_i-values are plotted in Figures 4.7 and 4.8, where the signal modes are shown by circles. The advantage of applying the SVD technique is to reduce the dimensionality of the problem. However, the accuracy of estimation of signal modes depends primarily on the extended model order, which allows not only modelling the signal component but also the noise component in the data.

It is to be mentioned that the extended order $L = 16$ of the MP method has been set by trial for best performance. Note that in this case, the data matrices have the dimension

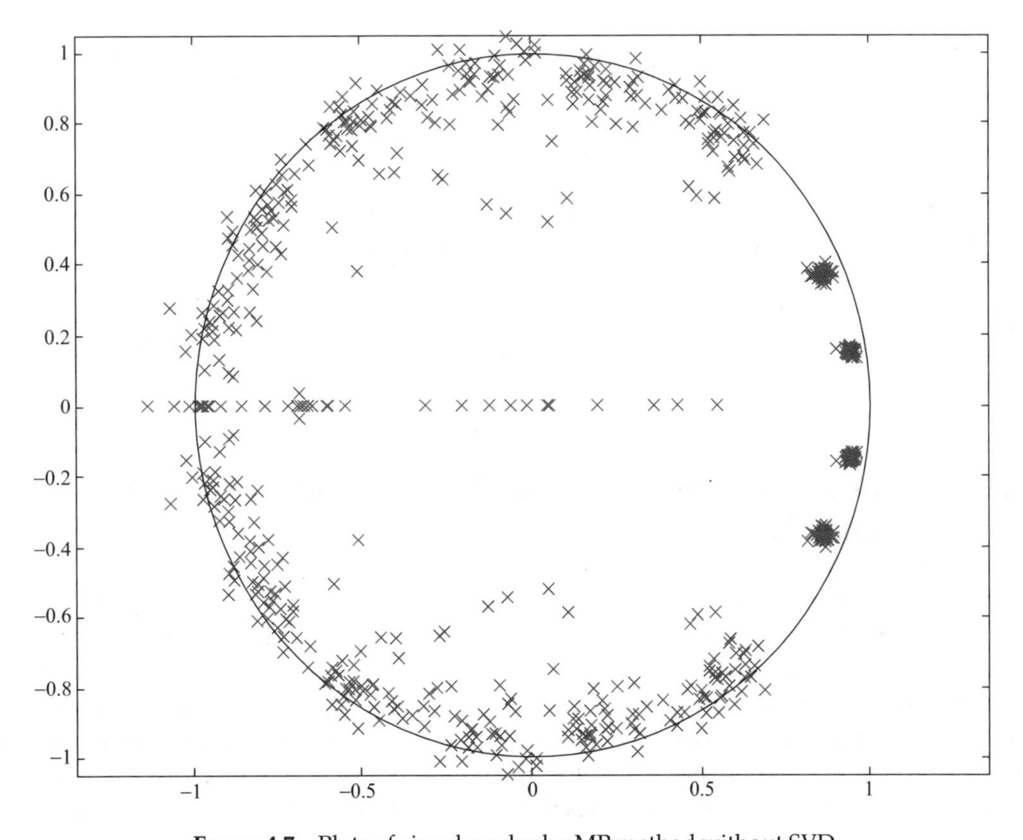

FIGURE 4.7 Plots of signal modes by MP method without SVD

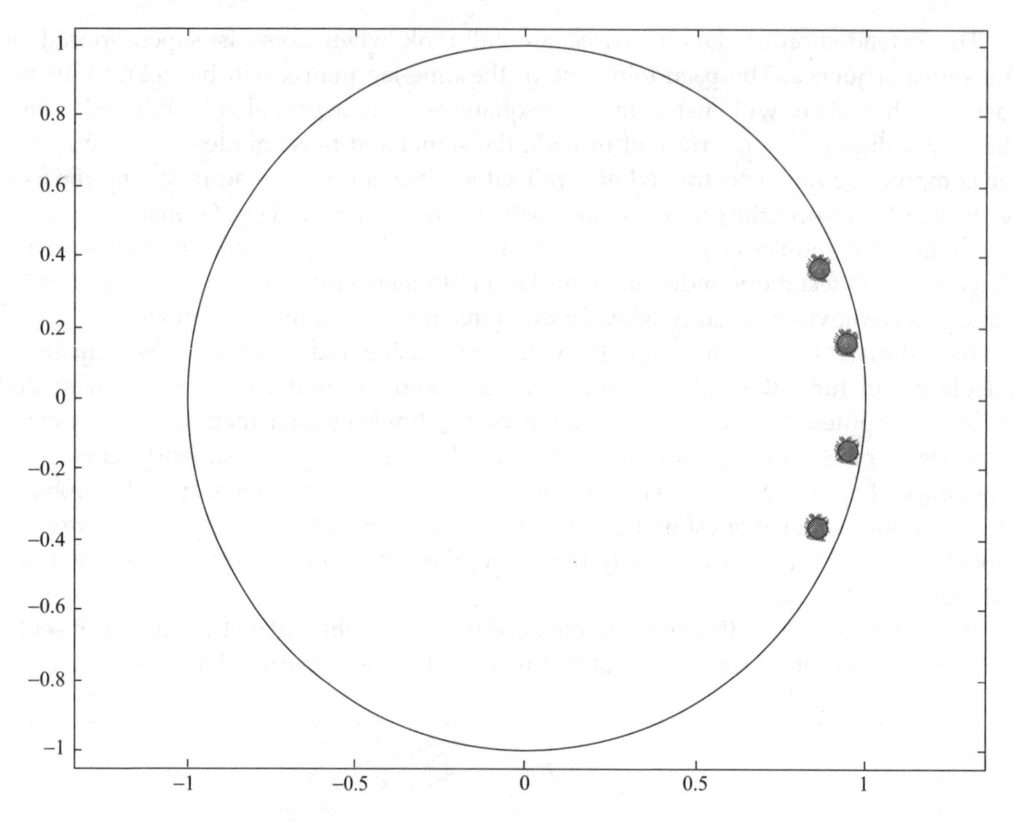

FIGURE 4.8 Plots of signal modes by MP method with SVD

$(N-L)\times L = 24\times 16$, which provides most accurate results when we compute the pseudoinverse (pinv) by MATLAB or the rank-constrained generalized inverse by the SVD technique of the time-lag matrix. Some empirical results are available in the literature relating the accuracy of parameter estimation and the extended model order. However, the optimal range of model order (L) relative to the number of points (N) in the data-set may depend on a specific problem.

4.11 Signal Processing Applications V: Multicomponent Signal Analysis

A multicomponent signal is characterized by the activities on the time-frequency plane at two or more places. In other words, there are two or more frequency-activities at a particular time with the separation of the mid-frequencies set by more than the bandwidths of such activities or components. Similarly, there may be two or more time-events at a specific frequency with the difference of the mid-time instants set by more than the time-spreads of such events or components. Thus, the separation between two or more components throughout the time-frequency plane can be considered as the characteristic property of a multicomponent signal.

A monocomponent signal looks like a single mountain ridge on the time-frequency plane, and this ridge defines the instantaneous frequency and bandwidth of the signal at a given time. A monocomponent signal is usually written in the form $g(t) = A(t)\exp\left[j\phi(t)\right]$ where its trajectory is given by the instantaneous circular frequency $\omega(t) = \phi'(t)$ and the amplitude envelope $A(t)$.

4.11.1 Discrete Energy Separation Algorithm

Let the discrete-time real signal $g[n]$ be given by $g[n] = A[n]\cos(\phi[n])$. The Teager energy of the signal $g[n]$ is calculated by the operator Ψ expressed as

$$\Psi(g[n]) = g^2[n] - g[n-1]g[n+1] \tag{4.177}$$

From the energy operators of the signal $g[n]$ and the difference signals $r[n] = g[n] - g[n-1]$ and $r[n+1] = g[n+1] - g[n]$, the instantaneous circular frequency $\omega[n]$ and the amplitude envelope $|A[n]|$ are calculated using the discrete energy separation algorithm (DESA) as follows,

$$\omega[n] \simeq \cos^{-1}\left(1 - \frac{\Psi(r[n]) + \Psi(r[n+1])}{4\Psi(g[n])}\right) \tag{4.178}$$

$$|A[n]| \simeq \sqrt{\frac{\Psi(g[n])}{1 - \left(1 - \frac{\Psi(r[n]) + \Psi(r[n+1])}{4\Psi(g[n])}\right)^2}} \tag{4.179}$$

It is assumed that $0 \le \omega[n] \le \pi$, i.e., the above technique can estimate frequencies up to the half of sampling frequency.

Suppose now that a multicomponent signal $g[n] = \sum_{i=1}^{m} A_i[n]\cos(\phi_i[n])$ is given for analysis. In order to extract the amplitude and frequency modulation laws of each component by employing the DESA technique, it is required that the individual components of the signal be separated. We shall demonstrate that a specific eigenfunction decomposition of the signal can be used to separate the signal components.

4.11.2 Signal Decomposition via Fourier-Bessel Series Expansion

Consider the zero-order Fourier-Bessel (FB) series expansion of a signal $g(t)$ over some finite interval $(0, b)$ expressed as

$$g(t) = \sum_{k=1}^{q} c_k J_0\left(\frac{\lambda_k}{b}t\right) \tag{4.180}$$

where $\lambda = \lambda_k;\ k = 1,2,\ldots$ are the ascending order positive roots of $J_0(\lambda) = 0$, and $J_0(\cdot)$ are the zero-order Bessel functions. The roots of the Bessel function $J_0(\lambda)$ can be computed using the Newton–Raphson method.

The sequence of the Bessel function $\left\{J_0\left(\frac{\lambda_k}{b}t\right)\right\}$ forms an orthogonal set on the interval $0 \le t \le b$ with respect to the weight t, i.e.,

$$\int_0^b tJ_0\left(\frac{\lambda_k}{b}t\right)J_0\left(\frac{\lambda_l}{b}t\right)dt = 0 \quad \text{for } k \ne l \tag{4.181}$$

Using the orthogonality of the set $\left\{J_0\left(\frac{\lambda_k}{b}t\right)\right\}$, the FB coefficients c_k are computed by using the following relation

$$c_k = \frac{2\int_0^b tg(t)J_0\left(\frac{\lambda_k}{b}t\right)dt}{b^2\left[J_1(\lambda_k)\right]^2}, \quad 1 \le k \dot{\le} q \tag{4.182}$$

where $J_1(\lambda_k)$ are the first-order Bessel functions. The integral in the above equation is known as the finite Hankel transform (FHT). Many fast numerical computation methods are available to calculate the FHT and the corresponding FB coefficients.

The spectral property of the FB coefficients reveals that the FB series expansion decomposes a time-signal along the frequency axis. Let the signal be the sinusoidal signal $g(t) = A\cos(\omega t)$. The FB coefficients c_k for the signal is given by

$$c_k = \frac{2A\lambda_k}{J_1(\lambda_k)}\left(\frac{\cos(\omega b - \theta)}{\sqrt{\left(\lambda_k^2 - \omega^2 b^2\right)^2 + \omega^2 b^2}}\right) \tag{4.183}$$

where $\theta = \sin^{-1}\left(\frac{\omega b}{\sqrt{\left(\lambda_k^2 - \omega^2 b^2\right)^2 + \omega^2 b^2}}\right)$.

The peak magnitude of c_k is

$$c_{k,\text{peak}} = \frac{2A\sin(\omega b)}{J_1(\omega b)},$$

which occurs at the order k where the root $\lambda_k \approx \omega b$. The magnitude of c_k decays rapidly away from the order where the peak appears, and the value becomes insignificant at far-away orders.

Now, consider the amplitude and frequency modulated (AFM) sinusoidal signal $g(t)$ given by

$$\begin{aligned} g(t) &= A\left[1+\mu\cos(\upsilon_a t)\right]\cos\left[\omega t + \beta\sin\left(\upsilon_f t\right)\right] \\ &= A\left[1+\mu\cos(\upsilon_a t)\right]\sum_{p=-\infty}^{\infty} J_p(\beta)\cos\left[\left(\omega + p\upsilon_f\right)t\right] \end{aligned} \tag{4.184}$$

The largest $|p|$ for appreciable value of $J_p(\beta)$ is given by $P = \mathrm{Int}\left[1+\beta+\ln(1+2\beta)\right]$ where $\mathrm{Int}[\cdot]$ denotes nearest integer. Note that $J_{-p}(\beta) = (-1)^p J_p(\beta)$. Thus, the signal can be expressed after simplification as

$$\begin{aligned} g(t) &= A \sum_{p=-P}^{P} J_p(\beta)\cos\left[\left(\omega + p\upsilon_f\right)t\right] \\ &+ \frac{\mu A}{2} \sum_{p=-P}^{P} J_p(\beta)\cos\left[\left(\omega + \upsilon_a + p\upsilon_f\right)t\right] + \frac{\mu A}{2} \sum_{p=-P}^{P} J_p(\beta)\cos\left[\left(\omega - \upsilon_a + p\upsilon_f\right)t\right] \end{aligned} \tag{4.185}$$

The above expression shows that the AFM sinusoidal signal can be expanded as a group of sinusoidal signals with circular frequencies clustered around the mid-frequency ω. Therefore, the FB coefficients computed for the AFM signal will show a cluster of orders k with high magnitudes, and the values become insignificant at far-away orders.

Consider now a multicomponent AFM sinusoidal signal given by

$$g(t) = \sum_{i=1}^{M} A_i\left[1 + \mu_i \cos(\upsilon_{ai} t)\right]\cos\left[\omega_i t + \beta_i \sin\left(\upsilon_{fi} t\right)\right] \tag{4.186}$$

We assume that the circular frequencies ω_i are well separated, and the difference between successive ω_i's is more than the bandwidths of the corresponding AFM components. The FB coefficients computed for the multicomponent AFM signal will then show M clusters of orders with high magnitudes and insignificant values in between two successive clusters. The reconstruction of the individual components of the composite signal can be achieved by summing the FB series for the separated clusters of the FB coefficients. Thus, the decomposition of a multicomponent signal into its individual components can be accomplished by the FB series expansion.

4.11.3 Simulation Study: Example 4 (Multicomponent AFM Signal)

We consider the discrete-time multicomponent AFM signal $g[n]$ sampled over N points,

$$g[n] = \sum_{i=1}^{M} A_i\left[1 + \mu_i \cos(\upsilon_{ai} n)\right]\cos\left[\omega_i n + \beta_i \sin\left(\upsilon_{fi} n\right)\right], \quad n = 0, \ldots, N-1 \tag{4.187}$$

with the set of parameters as given below:
$M = 3,\ A_1 = 1,\ \omega_1 = 2\pi(0.05),\ \mu_1 = 0.5,\ \upsilon_{a1} = 2\pi(0.005),\ \beta_1 = 0.4,\ \upsilon_{f1} = 2\pi(0.001),$
$A_2 = 0.8,\ \omega_2 = 2\pi(0.15),\ \mu_2 = 0.6,\ \upsilon_{a2} = 2\pi(0.0025),\ \beta_2 = 0.7,\ \upsilon_{f2} = 2\pi(0.0005),$
$A_3 = 0.6,\ \omega_3 = 2\pi(0.25),\ \mu_3 = 0.4,\ \upsilon_{a3} = 2\pi(0.005),\ \beta_3 = 0.8,\ \upsilon_{f3} = 2\pi(0.001),$ and $N = 500$.

The signal sequence $g[n]$ is plotted in Figure 4.9. We compute the FB coefficients c_k as

$$c_k = \frac{\sum_{n=0}^{N-1} (n/N)\, g[n]\, J_0(\lambda_k n/N)}{(N/2)\left[J_1(\lambda_k)\right]^2}, \quad 1 \le k \le N \tag{4.188}$$

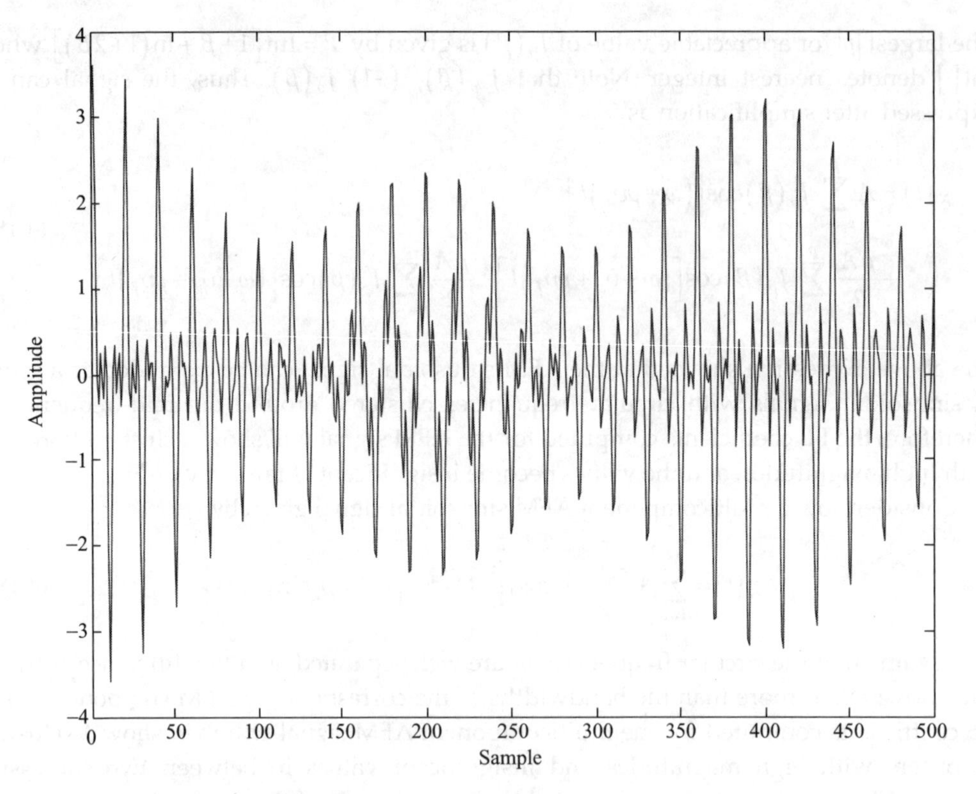

FIGURE 4.9 Multicomponent AFM signal

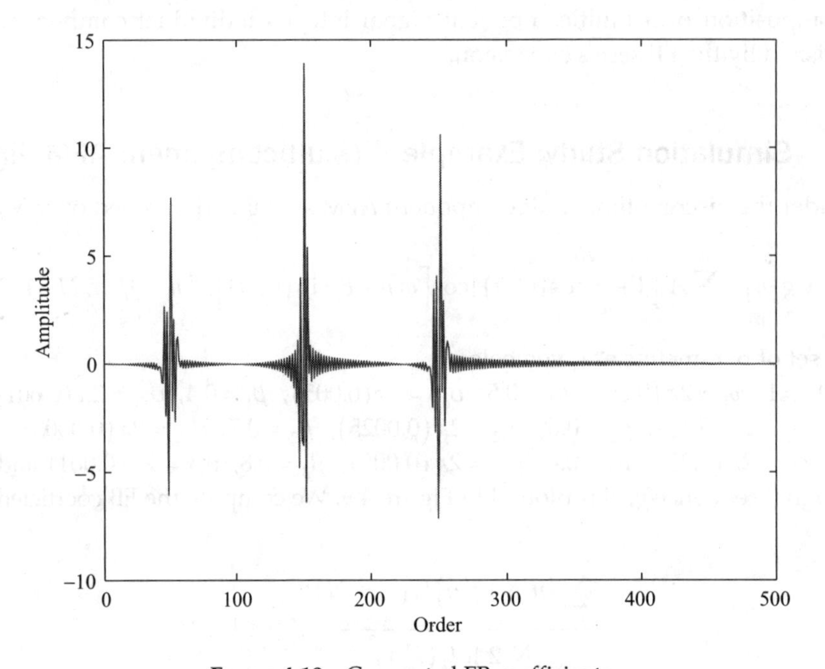

FIGURE 4.10 Computed FB coefficients

and the coefficient versus order plot is shown in Figure 4.10. The first root $\lambda_1 = 2.4048$ of $J_0(\lambda_k)$ is fitted and the other roots are approximated by $\lambda_{k+1} = \lambda_k + \pi$. The Bessel functions (besselj) and the summation (sum) are computed by MATLAB.

The plot of the FB coefficients clearly shows three clusters corresponding to three AFM components of the signal. Now, by using the computed FB coefficients over one cluster, one signal component can be reconstructed at a time. We reconstruct the first component by summing the FB series over the range of order 21–95, the second component over 111–220 and the third component over 225–365. The plots of the reconstructed component signals are shown in Figures 4.11–4.13, where the original component signals are also shown for comparison.

Thus, the separation of components can be achieved by the FB series expansion of the composite signal. Once we separate the individual components, the DESA technique can be used to extract the amplitude and frequency modulation laws of each component. We leave this part of the simulation study as an exercise to the reader.

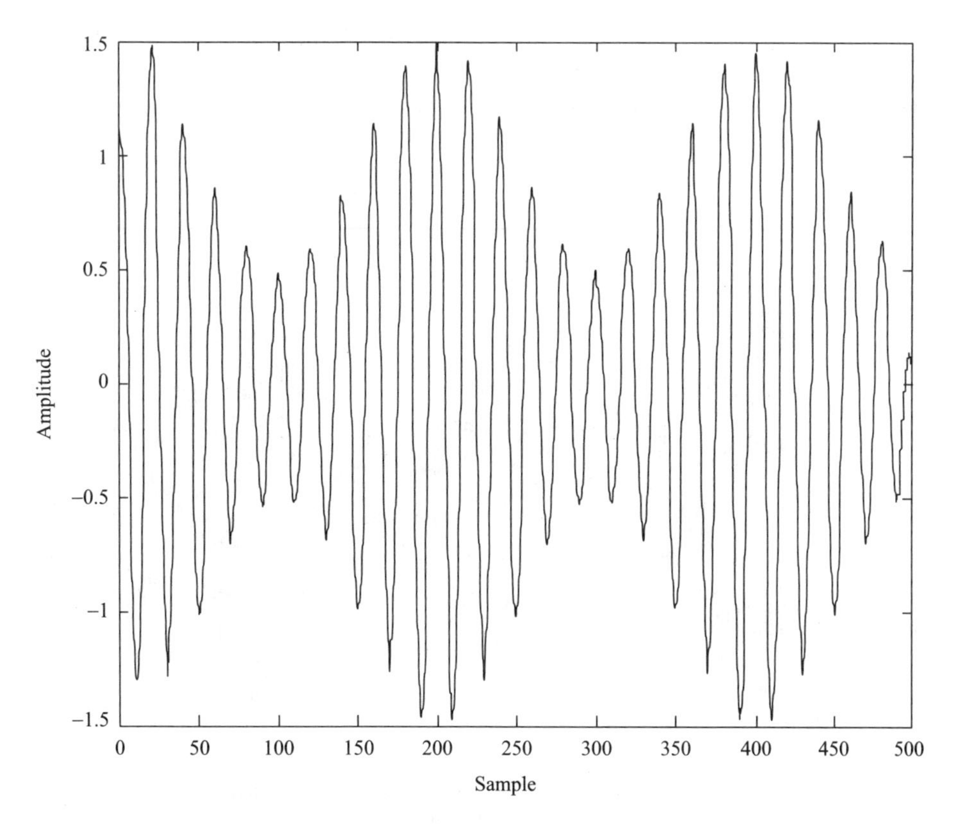

FIGURE 4.11 First AFM component; reconstructed (solid line), original (dotted line)

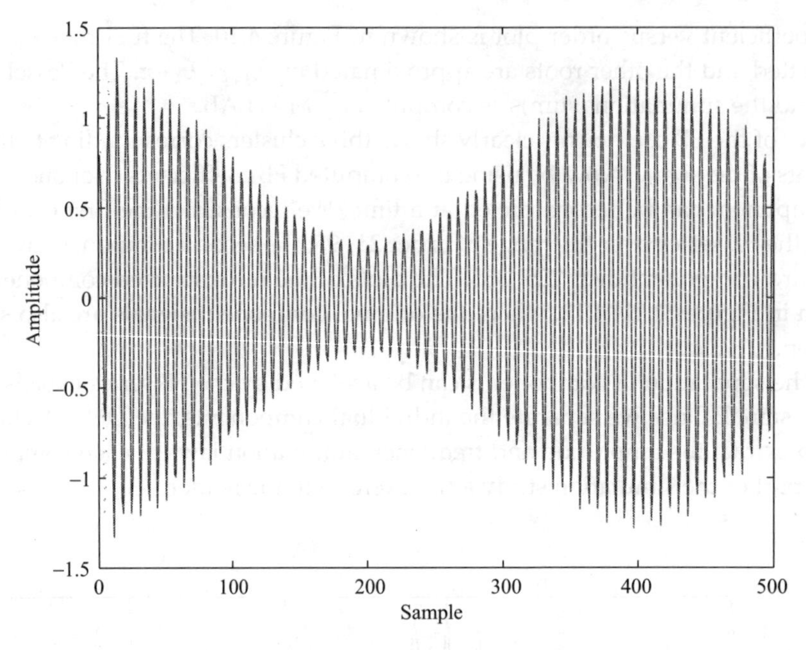

FIGURE 4.12 Second AFM component; reconstructed (solid line), original (dotted line)

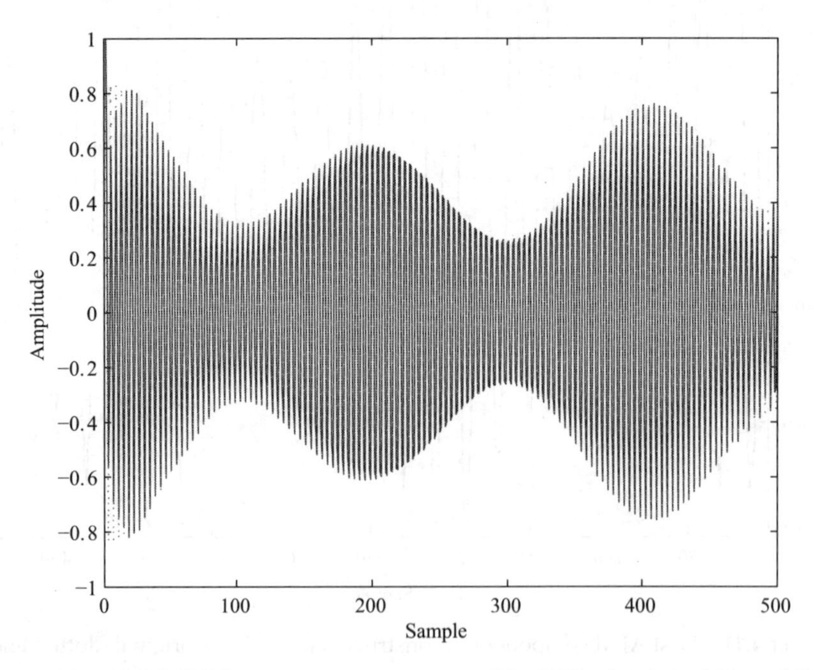

FIGURE 4.13 Third AFM component; reconstructed (solid line), original (dotted line)

Problems

Some sample problems are given below. Additional problems can be obtained from standard texts on matrix analysis, SVD and eigenfunction expansion.

P4.1 Compute the eigenvalues of

$$g(\mathbf{A}) = \left[\mathbf{A}^3 + \mathbf{A}^2 - \mathbf{A} + \mathbf{I}\right]\exp(\mathbf{A})$$

when $\mathbf{A} = \begin{bmatrix} 1 & 0 & -2 \\ 0 & 1 & 0 \\ 1 & 0 & -2 \end{bmatrix}$.

P4.2 Starting with the SVD of a general matrix **A**, find the matrix decomposition when
(i) **A** is normal ($A^H A = AA^H$), and
(ii) **A** is Hermitian ($A = A^H$).
Then, show that the Hermitian matrices **A** and **B** are simultaneously diagonalizable by the same unitary matrix **U** if they commute, i.e., **AB** = **BA**.

P4.3 Show that the solution of the homogenous equation **Ax** = **0**, where $\mathbf{A} \in \mathbb{C}^{m \times n}$, $m \geq n$ and $\text{rank}(\mathbf{A}) = l < n$ is given by

$$\mathbf{x} = \sum_{i=1}^{n-l} c_i \mathbf{v}_{l+i} \text{ with } c_i = \frac{\mathbf{v}_{l+i}^*(1)}{\sum_{k=1}^{n-l} \left|\mathbf{v}_{l+k}(1)\right|^2}$$

where $\mathbf{v}_i$ are the right singular vectors of the matrix **A**. For computations of the coefficients c_i, minimize $\|\mathbf{x}\|^2$ subject to the constraint $\mathbf{x}^*(1) = 1$.

P4.4 Consider the matrix equation **Ax** = **b**. Let the SVD of **A** be $\mathbf{U}\Sigma\mathbf{V}^H$. Show that the minimum square error-norm while computing the least squares solution $\mathbf{x}_{\text{LS}}$ is given by $\|\mathbf{A}\mathbf{x}_{\text{LS}} - \mathbf{b}\|^2 = \|\mathbf{U}_2^H \mathbf{b}\|^2$, where $\mathbf{U}_2$ comprises of the left singular vectors of **A**, which span the null-space of $\mathbf{A}^H$.

P4.5 Consider the perturbed matrix equation $(\mathbf{A} + \mathbf{E})\mathbf{x} = \mathbf{0}$, and show that any simultaneous SVDs of the form $\mathbf{A} = \mathbf{U}\Sigma\mathbf{V}^H$ and $\mathbf{E} = \mathbf{U}\Lambda\mathbf{V}^H$ are possible if and only if the matrices $\mathbf{A}\mathbf{E}^H$ and $\mathbf{E}^H\mathbf{A}$ are both normal. Then, state the conditions when the solutions of the original matrix equation **Ax** = **0** and the perturbed matrix equation will be the same.

P4.6 The signal consisting of m complex sinusoids of the form $x[n] = \sum_{i=1}^{m} \alpha_i \exp\left[j(2\pi f_i n + \phi_i)\right]$, where the phases ϕ_i are statistically independent random variables uniformly distributed on $[0, 2\pi)$, can be shown to be wide-sense stationary with autocorrelation function

$r_{xx}[k] = \sum_{i=1}^{m} p_i \exp(j2\pi f_i k)$, where $p_i = \alpha_i^2$ is the power of the ith sinusoid.

The signal autocorrelation matrix $\mathbf{R}_{xx}$ of dimension $(m+1)\times(m+1)$ may be represented as

$$\mathbf{R}_{xx} = \sum_{i=1}^{m} p_i \mathbf{s}_i \mathbf{s}_i^H \text{ where } \mathbf{s}_i = \left[1, e^{j2\pi f_i}, e^{j4\pi f_i}, \ldots, e^{j2\pi m f_i}\right]^T.$$

Moreover, it can be shown that the polynomial $A(z)=1+\sum_{l=1}^{m} a_l z^{-l}$ has m roots on the unit circle, which correspond to the frequencies of the sinusoids, and the coefficient vector $\mathbf{a}=[1, a_1, \ldots, a_m]$ is given by $\mathbf{R}_{xx}\mathbf{a}=\mathbf{0}$.

Using the eigenvalue decomposition of the computed autocorrelation matrix

$$\mathbf{R}_{yy}=\mathbf{R}_{xx}+\sigma_w^2\mathbf{I}$$

for the observed sequence $y[n]=x[n]+w[n]$, where $w[n]$ is white noise of variance σ_w^2, find an algorithm to estimate the frequencies of the signal. Show all the derivations.

Bibliography

The list is indicative, not exhaustive. Similar references can be searched in the library.

Berezanskii, J. M. 1968. *Expansions in Eigenfunctions of Self-adjoint Operators*. Providence, RI: American Mathematical Society.

Cadzow, J. A., and M.-M. Wu. 1987. 'Analysis of Transient Data in Noise.' *IEE Proceedings F, Commun., Radar & Signal Process.*, 134: 69–78.

Churchill, R. V., and J. W. Brown. 1978. *Fourier Series and Boundary Value Problems*. 3rd ed., New York: McGraw-Hill.

Cohen, L. 1992. 'What is a Multicomponent Signal?' *Proc. IEEE Int. Conf.* ASSP, ICASSP-92, 5: 113–16.

Cullum, J., and R.A. Willoughby, eds. 1986. *Large Scale Eigenvalue Problems*. Amsterdam: North-Holland.

De Moor, B., J. Staar, and J. Vandewalle. 1988. 'Oriented Energy and Oriented Signal-to-Signal Ratio Concepts in the Analysis of Vector Sequences and Time Series.' In *SVD and Signal Processing: Algorithms, Applications and Architectures*, edited by E. F. Deprettere, 209–232. Amsterdam: Elsevier Science.

Dewilde, P., and E. F. Deprettere. 1988. 'Singular Value Decomposition: An Introduction.' In *SVD and Signal Processing: Algorithms, Applications and Architectures*, edited by E. F. Deprettere, 3–41. Amsterdam: Elsevier Science.

Golub, G. H., and C. F. Van Loan. 1996. *Matrix Computations*. 3rd ed., Baltimore: The Johns Hopkins University Press.

Henderson, T. L., 1981. 'Geometric Methods for Determining System Poles from Transient Response.' *IEEE Trans. Acoust. Speech Signal Process.* ASSP-29: 982–88.

Horn, R. A., and C. R. Johnson. 2013. *Matrix Analysis*. 2nd ed., Cambridge: Cambridge University Press.

Hua, Y., and T. K. Sarkar. 1990. 'Matrix Pencil Method for Estimating Parameters of Exponentially Damped/Undamped Sinusoids in Noise.' *IEEE Trans. Acoust. Speech Signal Process.* ASSP-38: 814–24.

Jain, V. K., T. K. Sarkar, and D. D. Weiner. 1983. 'Rational Modeling by the Pencil-of-Function Method.' *IEEE Trans. Acoust. Speech Signal Process.* ASSP-31: 564–73.

Kay, S. M. 1988. *Modern Spectral Estimation: Theory and Application*. Englewood Cliffs, NJ: Prentice Hall.

Klema, V. C., and A. J. Laub. 1980. 'The Singular Value Decomposition: Its Computation and Some Applications.' *IEEE Trans. Automatic Control*. AC-25: 164–76.

Kumaresan, R., and D. W. Tufts. 1982. 'Estimating the parameters of exponentially damped sinusoids and pole-zero modeling in noise.' *IEEE Trans. Acoust. Speech Signal Process*. ASSP-30: 833–40.

Kung, S.-Y. 1978. 'A New Identification and Model Reduction Algorithm via Singular Value Decomposition.' Proc. 12th Asilomar Conf. Circuits, Syst. Comput., 705–714, Pacific Grove, CA.

Ouibrahim, H., D. D. Weiner, and T. K. Sarkar. 1986. 'A Generalized Approach to Direction Finding.' Proc. IEEE Military Comm. Conf. - MILCOM 1986, 3: 41.4.1–41.4.5.

Pachori, R. B. 2007. 'Methods Based on Fourier-Bessel Representation for Analysis of Nonstationary Signals.' PhD diss., Dept. of Elec. Eng., I.I.T. Kanpur.

Pachori, R. B., and P. Sircar. 2007. 'A New Technique to Reduce Cross Terms in the Wigner Distribution.' *Digital Signal Process* 17 (2): 466–74.

Ralston, A., and P. Rabinowitz. 1978. *A First Course in Numerical Analysis*. 2nd ed., Singapore: McGraw-Hill.

Roy, R., and T. Kailath. 1988. 'ESPIRIT – Estimation of Signal Parameters via Rotational Invariance Techniques.' In *SVD and Signal Processing: Algorithms, Applications and Architectures*, edited by E. F. Deprettere, 233–265. Amsterdam: Elsevier Science.

Schott, J. R. 2005. *Matrix Analysis for Statistics*. 2nd ed., Hoboken, NJ: John Wiley.

Schroeder, J. 1993. 'Signal Processing via Fourier-Bessel Series Expansion.' *Digital Signal Process.*, 3 (2): 112–24.

Sircar, P., and S. Mukhopadhyay. 1995. 'Accumulated Moment Method for Estimating Parameters of the Complex Exponential Signal Models in Noise.' *Signal Processing* 45: 231–43.

Sircar, P., and A.C. Ranade. 1992. 'Nonuniform Sampling and Study of Transient Systems Response.' *IEE Proceedings*, part F, 139: 49–55.

Stewart, G.W. 1973. *Introduction to Matrix Computation*. New York: Academic Press.

Titchmarsh, E.C. 1962. *Eigenfunction Expansions: Associated with Second-order Differential Equations*. 2nd ed., London: Oxford University Press.

Tufts, D.W., and R. Kumaresan. 1982. 'Estimation of Frequencies of Multiple Sinusoids: Making Linear Prediction Perform like Maximum Likelihood.' *Proceedings IEEE*, 70 (9): 975–89.

van der Veen, A. -J., E. F. Deprettere, and A. L. Swindlehurst. 1993. 'Subspace-Based Signal Analysis Using Singular Value Decomposition.' *Proceedings IEEE*, 81: 1277–1308.

van Huffel, S., and J. Vandewalle. 1988. 'The Total Least Squares Technique: Computation, Properties and Applications.' In *SVD and Signal Processing: Algorithms, Applications and Architectures*, edited by E. F. Deprettere, 189–207. Amsterdam: Elsevier Science.

Watkins, D. S. 2002. *Fundamentals of Matrix Computations*. 2nd ed., New York: John Wiley.

Zhang, F. 2011. *Matrix Theory: Basic Results and Techniques*. 2nd ed., New York: Springer.

5 Optimization

5.1 Unconstrained Extrema

Let $f:\mathbb{R}^n \to \mathbb{R}$ be a function of n real variables $\{x_1,\ldots,x_n\}$. We denote the multivariable real-valued scalar function as $f(\mathbf{x})$ where $\mathbf{x}=[x_1,\ldots,x_n]^T \in \mathbb{R}^n$.

The function $f(\mathbf{x})$, $\mathbf{x}\in\mathbb{R}^n$, has a relative (local) maximum at $\mathbf{x}^* \in \mathbb{R}^n$ if and only if there is a positive real number δ such that for any $\mathbf{x}$ satisfying $\|\mathbf{x}-\mathbf{x}^*\|<\delta$, the condition $f(\mathbf{x})\le f(\mathbf{x}^*)$ is satisfied. Similarly, for a relative (local) minimum, the condition $f(\mathbf{x}^*)\le f(\mathbf{x})$ is to be satisfied. The subspace $\mathscr{D} \equiv \{\mathbf{x}:\|\mathbf{x}-\mathbf{x}^*\|<\delta\}$ is called the spherical δ-neighbourhood of $\mathbf{x}^*$. If the inequalities hold for all $\mathbf{x}\in\mathbb{R}^n$, then $f(\mathbf{x})$ has an absolute (global) maximum or minimum (extremum) at $\mathbf{x}^* \in \mathbb{R}^n$. An absolute or global extremum may be chosen out of possibly many relative or local extrema. Thus, a global extremum is also a local extremum, and the condition for a point to be a local extremum naturally applies for it to be a global extremum. In addition, there may be some condition for the point to qualify as a global extremum.

Assume that $f(x)$, $x\in\mathbb{R}$, is continuous and has continuous first and second derivatives $\frac{df}{dx}$ and $\frac{d^2 f}{dx^2}$ for all $x\in\mathbb{R}$. Then, a necessary condition for a point $x=x^*$ to be a local extremum is that

$$\left.\frac{df(x)}{dx}\right|_{x=x^*}=0, \tag{5.1}$$

and if in addition, it is given that

$$\left.\frac{d^2 f(x)}{dx^2}\right|_{x=x^*}\neq 0 \tag{5.2}$$

then, it is a sufficient condition for $x=x^*$ to be an extremum. Moreover, for x^* to be a local maximum, $\left.\frac{d^2 f(x)}{dx^2}\right|_{x=x^*}$ is negative, and for x^* to be a minimum, $\left.\frac{d^2 f(x)}{dx^2}\right|_{x=x^*}$ is positive.

Consider the function $f_1(x)=2x^2+4x+3$ for $x\in[-\infty,\infty]$. This function has an absolute or global minimum at $x=-1$. Next, consider the function $f_2(x)=x^2(3-x)$ for

$x \in [-\infty, \infty]$. This function has a relative or local minimum at $x = 0$ and a relative or local maximum at $x = 2$.

We now state the necessary and sufficient conditions for an extremum of a function f of n variables in the following two theorems.

Theorem 5.1

Let $f : \mathbb{R}^n \to \mathbb{R}$ be a function of n variables $\mathbf{x} \in \mathbb{R}^n$. A necessary condition for a point $\mathbf{x}^* \in \mathbb{R}^n$ to be a maximum or minimum is that the gradient of f evaluated at $\mathbf{x}^*$

$$\nabla f(\mathbf{x}^*) = \left[\left. \frac{\partial f(\mathbf{x})}{\partial x_i} \right|_{\mathbf{x}=\mathbf{x}^*} ; \ i = 1, \ldots, n \right] = \mathbf{0} \tag{5.3}$$

where all partial derivatives exist and are continuous at $\mathbf{x} = \mathbf{x}^*$.

A point $\mathbf{x}^*$ which satisfies the condition (5.3) is called a stationary point. However, a stationary point may not be an extremum. Therefore, (5.3) alone is not a sufficient condition.

Theorem 5.2

Let $f : \mathbb{R}^n \to \mathbb{R}$ be a function of n variables $\mathbf{x} \in \mathbb{R}^n$. A sufficient condition for a point $\mathbf{x}^* \in \mathbb{R}^n$ to be an extremum is that

(a) the gradient vector (stationarity)

$$\nabla f(\mathbf{x}^*) = \left[\left. \frac{\partial f(\mathbf{x})}{\partial x_i} \right|_{\mathbf{x}=\mathbf{x}^*} ; \ i = 1, \ldots, n \right] = \mathbf{0}$$

(b) the Hessian matrix

$$\nabla^2 f(\mathbf{x}^*) = \left[\left. \frac{\partial^2 f}{\partial x_i \partial x_j} \right|_{\mathbf{x}=\mathbf{x}^*} ; \ i = 1, \ldots, n; \ j = 1, \ldots, n \right]$$

is negative definite for a maximum and positive definite for a minimum, where all partial derivatives exist and are continuous at $\mathbf{x} = \mathbf{x}^*$.

Note that if the Hessian $\nabla^2 f(\mathbf{x}^*)$ becomes a null matrix together with stationarity for $\mathbf{x} = \mathbf{x}^*$, then f has a point of inflection at $\mathbf{x}^*$.

We now define a convex or concave function whose extremum is to be found.

A subspace $\mathcal{S} \subseteq \mathbb{R}^n$ is said to be a convex subspace if and only if for any $\mathbf{x}, \mathbf{y} \in \mathcal{S}$ and $\lambda \in [0, 1]$, the point

$$\lambda \mathbf{x} + (1 - \lambda)\mathbf{y} \in \mathcal{S} \tag{5.4}$$

A function $f : \mathbb{R}^n \to \mathbb{R}$ defined on a convex subspace $\mathcal{S} \subseteq \mathbb{R}^n$ is convex if for any $\mathbf{x}, \mathbf{y} \in \mathcal{S}$ and $\lambda \in [0, 1]$, the function satisfies

$$f(\lambda \mathbf{x} + (1 - \lambda)\mathbf{y}) \le \lambda f(\mathbf{x}) + (1 - \lambda) f(\mathbf{y}) \tag{5.5}$$

A function f is concave if for the same conditions given above, the function satisfies

$$f\left(\lambda\mathbf{x}+(1-\lambda)\mathbf{y}\right)\geq\lambda f\left(\mathbf{x}\right)+(1-\lambda)f\left(\mathbf{y}\right) \tag{5.6}$$

Theorem 5.3

Let $f:\mathbb{R}^n\to\mathbb{R}$ be a continuously twice differentiable convex (or concave) function of $\mathbf{x}\in\mathbb{R}^n$. Then, a necessary and sufficient condition for $\mathbf{x}=\mathbf{x}^*$ to be a global minimum (maximum) of f is that $\nabla f\left(\mathbf{x}^*\right)=\mathbf{0}$.

Note that the condition on the Hessian as required in Theorem 5.2 is inherently satisfied when f is a convex (or concave) function. Conversely, the condition of positive definiteness (negative definiteness) of the Hessian matrix amounts to assuming that f is convex (or concave) in the neighbourhood of $\mathbf{x}^*$.

We now define a unimodal function whose extremum is to be found.

A one-variable function $f(x)$ is unimodal on the real interval $[a, b]$ if f has a minimum at $x^*\in[a,b]$, and for any $x_1,x_2\in[a,b]$ with $x_1<x_2$, we have

$$\begin{aligned} x_2\leq x^* \quad &\text{implies} \quad f(x_1)>f(x_2)\\ x_1\geq x^* \quad &\text{implies} \quad f(x_1)<f(x_2) \end{aligned} \tag{5.7}$$

A unimodal function on $[a, b]$ has a unique local minimum. An equivalent definition of a unimodal function $f(x)$ which has a unique local maximum at $x^*\in[a,b]$ can be stated in a similar way as above.

A multi-variable function $f(\mathbf{x})$ is unimodal on the subspace $\mathscr{S}\in\mathbb{R}^n$ if f has a minimum at $\mathbf{x}^*\in\mathscr{S}$, and f is monotonically decreasing along any direction contained in $\mathscr{S}$ towards $\mathbf{x}^*$ and monotonically increasing along any direction away from $\mathbf{x}^*$. An equivalent definition of a unimodal function $f(\mathbf{x})$ which has a local maximum can be stated in a similar way.

Theorem 5.4

Let $f:\mathbb{R}^n\to\mathbb{R}$ be a unimodal function on $\mathscr{S}\in\mathbb{R}^n$. Then, f has a unique local extremum.

A quadratic polynomial is an example of a unimodal function.

By the principle of duality, a point $\mathbf{x}^*$ which minimizes the function $f(\mathbf{x})$ also maximizes the function $-f(\mathbf{x})$. Moreover, if $f(\mathbf{x})$ is convex, then $-f(\mathbf{x})$ is concave and vice versa. Thus, every minimization problem can be formulated as an equivalent maximization problem, and conversely. Therefore in the sequel, we shall consider a maximization problem or a minimization problem but not both, because the equivalent dual problem can be formulated in a straightforward way.

5.2 Extrema of Functions with Equality Constraints

We want to find the extremum of a function $f:\mathbb{R}^n\to\mathbb{R}$ subject to the equality constraint $\mathbf{g}(\mathbf{x})=\mathbf{0}$ where $\mathbf{g}:\mathbb{R}^n\to\mathbb{R}^m$, $m<n$.

Consider the case with $n = 2$ and $m = 1$. Assuming all partial derivatives exist, we obtain by using the chain rule

$$\frac{df(\mathbf{x})}{dx_1} = \frac{\partial f(\mathbf{x})}{\partial x_1} + \frac{\partial f(\mathbf{x})}{\partial x_2}\frac{dx_2}{dx_1} \tag{5.8}$$

and

$$\frac{dg(\mathbf{x})}{dx_1} = \frac{\partial g(\mathbf{x})}{\partial x_1} + \frac{\partial g(\mathbf{x})}{\partial x_2}\frac{dx_2}{dx_1} \tag{5.9}$$

Since $\mathbf{x}$ is to be chosen to satisfy $g(\mathbf{x}) = 0$, a constant, we have

$$\frac{\partial g(\mathbf{x})}{\partial x_1} + \frac{\partial g(\mathbf{x})}{\partial x_2}\frac{dx_2}{dx_1} = 0 \tag{5.10}$$

Moreover, if $\mathbf{x}$ is a candidate for an extremum, then $\mathbf{x}$ must also satisfy

$$\frac{\partial f(\mathbf{x})}{\partial x_1} + \frac{\partial f(\mathbf{x})}{\partial x_2}\frac{dx_2}{dx_1} = 0 \tag{5.11}$$

It follows from (5.10) and (5.11) that

$$\frac{dx_2}{dx_1} = -\frac{\dfrac{\partial f(\mathbf{x})}{\partial x_1}}{\dfrac{\partial f(\mathbf{x})}{\partial x_2}} = -\frac{\dfrac{\partial g(\mathbf{x})}{\partial x_1}}{\dfrac{\partial g(\mathbf{x})}{\partial x_2}} = \lambda, \text{ say} \tag{5.12}$$

where λ is the proportionality constant to be evaluated for the problem.

It turns out that if we define the composite function (Lagrangian) as

$$F(\mathbf{x},\lambda) \equiv f(\mathbf{x}) + \lambda g(\mathbf{x}) \tag{5.13}$$

then the constrained optimization problem becomes an unconstrained optimization problem, because (5.12) is equivalent to $\frac{\partial F}{\partial \mathbf{x}} = \mathbf{0}$ and $g(\mathbf{x}) = 0$ is equivalent to $\frac{\partial F}{\partial \lambda} = 0$.

5.2.1 Lagrange Multiplier Approach

Let $f(\mathbf{x})$, $\mathbf{x}^T = [x_1, \ldots, x_n]$, be the real-valued function to be maximized subject to the equality constraints $g_i(\mathbf{x}) = 0$, $i = 1, \ldots, m < n$. We form the Lagrange function

$$F(\mathbf{x},\boldsymbol{\lambda}) \equiv f(\mathbf{x}) + \sum_i \lambda_i g_i(\mathbf{x}) \tag{5.14}$$

It can be shown that the constrained optimization problem is transformed into an equivalent problem of unconstrained optimization through the formation of the Lagrange function as above.

Assuming all partial derivatives exist and are continuous at the candidate optimum point $\mathbf{x} = \mathbf{x}^*$, we must have

$$df \equiv \sum_j \frac{\partial f}{\partial x_j} dx_j = 0 \tag{5.15}$$

and

$$dg_i \equiv \sum_j \frac{\partial g_i}{\partial x_j} dx_j = 0 \quad \text{for } i = 1, \ldots, m \tag{5.16}$$

at $\mathbf{x} = \mathbf{x}^*$. Combining (5.15) and (5.16) with properly selected multipliers λ_i, we obtain

$$\begin{aligned} dF &\equiv df + \sum_i \lambda_i dg_i \\ &\equiv \sum_j \frac{\partial F}{\partial x_j} dx_j = 0 \end{aligned} \tag{5.17}$$

Now, since there are m number of independent equality constraints, there must be $(n-m)$ number of independent variables x_j. Suppose that $\{x_{m+1}, \ldots, x_n\}$ are the independent variables. Let the multipliers λ_i be chosen such that at $\mathbf{x} = \mathbf{x}^*$,

$$\frac{\partial F}{\partial x_i} = 0 \quad \text{for } i = 1, \ldots, m \tag{5.18}$$

and the optimum Lagrange multiplier vector $\boldsymbol{\lambda} = \boldsymbol{\lambda}^*$ is obtained.

Then, from (5.17) and (5.18), we get

$$\frac{\partial F}{\partial x_{m+1}} dx_{m+1} + \cdots + \frac{\partial F}{\partial x_n} dx_n = 0 \tag{5.19}$$

Since $\{x_{m+1}, \ldots, x_n\}$ are the independent variables, (5.19) yields to

$$\frac{\partial F}{\partial x_{m+1}} = \cdots = \frac{\partial F}{\partial x_n} = 0 \tag{5.20}$$

Thus, combining (5.18) and (5.20), we obtain

$$\frac{\partial F}{\partial x_j} \equiv \frac{\partial f}{\partial x_j} + \sum_i \lambda_i \frac{\partial g_i}{\partial x_j} = 0 \quad \text{for } j = 1, \ldots, n \tag{5.21}$$

where all partial derivatives are evaluated at $(\mathbf{x}, \boldsymbol{\lambda}) = (\mathbf{x}^*, \boldsymbol{\lambda}^*)$. Note that (5.21) provides n necessary conditions for the existence of a constrained extremum at $\mathbf{x}^*$.

Further, note that the equality constraints can be expressed as

$$\frac{\partial F}{\partial \lambda_i} \equiv g_i = 0 \quad \text{for } i = 1, \ldots, m \tag{5.22}$$

evaluated at $\mathbf{x}^*$.

Thus, the necessary condition and constraint condition are given in the symmetrical form as

$\frac{\partial F}{\partial x_j} = 0;\ j = 1, \ldots, n$ and $\frac{\partial F}{\partial \lambda_i} = 0;\ i = 1, \ldots, m$, respectively, at the optimum value $(\mathbf{x}^*, \boldsymbol{\lambda}^*)$.

We shall now derive the sufficient condition for a point $\mathbf{x}=\mathbf{x}^*$ to be a local maximum of the function $f(\mathbf{x})$ subject to the equality constraints $g_i(\mathbf{x})=0$ for $i=1,\ldots,m$.

Assuming all partial derivatives exist and are continuous at the candidate local maximum point $\mathbf{x}=\mathbf{x}^*$, we must have the following sufficient conditions satisfied

$$df \equiv \left[\nabla f\left(\mathbf{x}^*\right)\right]^T \mathbf{dx}=0 \tag{5.23}$$

and

$$d^2 f \equiv \mathbf{dx}^T\left[\nabla^2 f\left(\mathbf{x}^*\right)\right]\mathbf{dx}+\left[\nabla f\left(\mathbf{x}^*\right)\right]^T \mathbf{d}^2\mathbf{x}<0 \tag{5.24}$$

where $\mathbf{dx}\equiv\left[dx_1,\ldots,dx_n\right]^T$ and $d^2 f=d(df)$.

Moreover, the vectors $\mathbf{dx}$ and $\mathbf{d}^2\mathbf{x}$ must satisfy

$$dg_i \equiv \left[\nabla g_i\left(\mathbf{x}^*\right)\right]^T \mathbf{dx}=0 \tag{5.25}$$

$$d^2 g_i \equiv \mathbf{dx}^T\left[\nabla^2 g_i\left(\mathbf{x}^*\right)\right]\mathbf{dx}+\left[\nabla g_i\left(\mathbf{x}^*\right)\right]^T \mathbf{d}^2\mathbf{x}=0 \tag{5.26}$$

at $\mathbf{x}=\mathbf{x}^*$, where the gradient vector and the Hessian matrix are defined as usual.

We now derive an alternative sufficient condition involving the Lagrange function $F(\mathbf{x},\boldsymbol{\lambda})$. Let $(\mathbf{x},\boldsymbol{\lambda})=\left(\mathbf{x}^*,\boldsymbol{\lambda}^*\right)$ be the optimum value for a constrained local maximum of $f(\mathbf{x})$. Consider a second-order Taylor expansion of $F\left(\mathbf{x},\boldsymbol{\lambda}^*\right)$ in the δ-neighbourhood of $\mathbf{x}=\mathbf{x}^*$ as given by

$$F\left(\mathbf{x},\boldsymbol{\lambda}^*\right)=F\left(\mathbf{x}^*,\boldsymbol{\lambda}^*\right)+\mathbf{d}^T\left[\nabla_x F\left(\mathbf{x}^*,\boldsymbol{\lambda}^*\right)\right]+\frac{1}{2}\mathbf{d}^T\left[\nabla_x^2 F\left(\xi_0,\boldsymbol{\lambda}^*\right)\right]\mathbf{d} \tag{5.27}$$

where $\mathbf{d}=\mathbf{x}-\mathbf{x}^*$, $\|\mathbf{d}\|<\delta$ for positive real δ, and $\xi_0=\mathbf{x}^*+\theta_0\,\mathbf{d}$ $\left(0\le\theta_0\le 1\right)$.

Assume that the point $\mathbf{x}=\mathbf{x}^*+\mathbf{d}$ satisfies the constraint equations $\mathbf{g}_i(\mathbf{x})=0$; $i=1,\ldots,m$. Then, $F\left(\mathbf{x},\boldsymbol{\lambda}^*\right)=f(\mathbf{x})$, and from (5.27), we get

$$\begin{aligned} f\left(\mathbf{x}^*+\mathbf{d}\right)&=f\left(\mathbf{x}^*\right)+\mathbf{d}^T\left[\nabla f\left(\mathbf{x}^*\right)+\sum_i \lambda_i^*\nabla g_i\left(\mathbf{x}^*\right)\right]+\frac{1}{2}\mathbf{d}^T\left[\nabla_x^2 F\left(\xi_0,\boldsymbol{\lambda}^*\right)\right]\mathbf{d} \\ &=f\left(\mathbf{x}^*\right)+\frac{1}{2}\mathbf{d}^T\left[\nabla_x^2 F\left(\xi_0,\boldsymbol{\lambda}^*\right)\right]\mathbf{d} \end{aligned} \tag{5.28}$$

where we have used (5.21). Now, from a first-order Taylor expansion of $g_i(\mathbf{x})$, we get

$$g_i\left(\mathbf{x}^*+\mathbf{d}\right)=g_i\left(\mathbf{x}^*\right)+\mathbf{d}^T\nabla g_i\left(\xi_i\right) \tag{5.29}$$

where $\xi_i=\mathbf{x}^*+\theta_i\,\mathbf{d}$ $\left(0\le\theta_i\le 1\right)$.

Since $g_i\left(\mathbf{x}^*+\mathbf{d}\right)=g_i\left(\mathbf{x}^*\right)=0$ for $i=1,\ldots,m$, (5.29) yields to

$$\mathbf{d}^T\nabla g_i\left(\xi_i\right)=0 \tag{5.30}$$

Thus, from (5.28) and (5.30), we find a sufficient condition for the existence of a constrained local maximum of $f(\mathbf{x})$ at $\mathbf{x}=\mathbf{x}^*$ as follows

$$\mathbf{d}^T\left[\nabla_x^2 F\left(\xi_0, \lambda^*\right)\right]\mathbf{d}<0 \tag{5.31}$$

for all $\mathbf{d}$ satisfying $\mathbf{d}^T\nabla g_i(\xi_i)=0$ with $\|\mathbf{d}\|<\delta$, δ being a small positive real number.

Since the elements of $\nabla_x^2 F(\mathbf{x},\lambda)$ and $\nabla g_i(\mathbf{x})$ are continuous functions of $\mathbf{x}$, we state the required sufficient condition as

$$\mathbf{d}^T\left[\nabla_x^2 F\left(\mathbf{x}^*, \lambda^*\right)\right]\mathbf{d}<0 \tag{5.32}$$

for all $\mathbf{d}$ satisfying $\mathbf{d}^T\nabla g_i(\xi_i)=0;\ i=1,\ldots,m$. Note that the condition (5.32) is to be satisfied in addition to the condition $\nabla_x F\left(\mathbf{x}^*, \lambda^*\right)=0$ as given by (5.21).

We now state the necessary and sufficient conditions for a constrained local extremum of a function f of n variables in the following two theorems.

Theorem 5.5

Let $f:\mathbb{R}^n\to\mathbb{R}$ be the function to be maximized subject to the equality constraint $\mathbf{g}(\mathbf{x})=\mathbf{0}$ where $\mathbf{g}:\mathbb{R}^n\to\mathbb{R}^m$, $m<n$. Let the Lagrange function be defined as

$$F(\mathbf{x},\lambda)=f(\mathbf{x})+\lambda^T\mathbf{g}(\mathbf{x}) \tag{5.33}$$

where $\lambda\in\mathbb{R}^m$. Then, a necessary condition for a constrained local or global maximum (or minimum) at a point $\mathbf{x}^*\in\mathbb{R}^n$ is that the gradient vector of F satisfies

$$\nabla_x F\left(\mathbf{x}^*, \lambda^*\right)=\frac{\partial f}{\partial \mathbf{x}}+\frac{\partial \mathbf{g}^T}{\partial \mathbf{x}}\lambda\bigg|_{(\mathbf{x},\lambda)=(\mathbf{x}^*,\lambda^*)}=\mathbf{0} \tag{5.34}$$

where λ^* is the optimum value of λ.

Further, the constraint condition $\mathbf{g}(\mathbf{x})=\mathbf{0}$ is given by

$$\nabla_\lambda F\left(\mathbf{x}^*\right)=\mathbf{g}\big|_{\mathbf{x}=\mathbf{x}^*}=\mathbf{0} \tag{5.35}$$

Theorem 5.6

Let $f:\mathbb{R}^n\to\mathbb{R}$ be the function to be maximized subject to the equality constraint $\mathbf{g}(\mathbf{x})=\mathbf{0}$ where $\mathbf{g}:\mathbb{R}^n\to\mathbb{R}^m$, $m<n$. Let the Lagrange function be defined as $F(\mathbf{x},\lambda)=f(\mathbf{x})+\lambda^T\mathbf{g}(\mathbf{x})$ where $\lambda\in\mathbb{R}^m$. Then, a sufficient condition for a constrained local maximum at a point $\mathbf{x}^*\in\mathbb{R}^n$ is that

(a) the gradient vectors satisfy (stationarity)

$\nabla_x F\left(\mathbf{x}^*, \lambda^*\right)=\mathbf{0}$ and $\nabla_\lambda F\left(\mathbf{x}^*\right)=\mathbf{0}$

where λ^* is the optimum value of λ;

(b) the quadratic term $\mathbf{d}^T\left[\nabla_x^2 F\left(\mathbf{x}^*, \lambda^*\right)\right]\mathbf{d}$ is negative (positive for a minimizing problem) for all $\mathbf{d}$ satisfying $\mathbf{d}^T\nabla g_i\left(\mathbf{x}^*\right)=0$, where

$$\nabla_x^2 F\left(\mathbf{x}^*,\lambda^*\right)=\frac{\partial}{\partial \mathbf{x}}\left(\frac{\partial f}{\partial \mathbf{x}}\right)^T+\frac{\partial}{\partial \mathbf{x}}\left(\frac{\partial \mathbf{g}^T}{\partial \mathbf{x}}\lambda\right)^T\Bigg|_{(\mathbf{x},\lambda)=\left(\mathbf{x}^*,\lambda^*\right)} \tag{5.36}$$

The sufficient condition (b) of the above theorem which states that the Hessian matrix is negative definite along the surface **g(x) = 0**, cannot be used immediately. However, it can be shown that there is an equivalent condition which is in a form that can be tested. The following theorem is stated with this alternative sufficient condition.

Theorem 5.7

Let the constrained optimization problem be given as:

Maximize $f(\mathbf{x})$, $\mathbf{x}\in\mathbb{R}^n$ subject to $g_i(\mathbf{x})=0$ for $i=1,\ldots,m<n$.

Let the Lagrange function be defined as $F(\mathbf{x},\lambda)=f(\mathbf{x})+\sum_i \lambda_i g_i(\mathbf{x})$.

Then, a sufficient condition for a local maximum at $\mathbf{x}=\mathbf{x}^*$ is that

(a) $\nabla_x F\left(\mathbf{x}^*, \lambda^*\right)=\mathbf{0}$ and $\nabla_\lambda F\left(\mathbf{x}^*\right)=\mathbf{0}$

where λ^* is the optimum value of λ;

(b) every root of the polynomial equation

$$P(\mu)=\begin{vmatrix}\mathbf{H}-\mu\mathbf{I} & \mathbf{G}\\ \mathbf{G}^T & \mathbf{0}\end{vmatrix}=0 \tag{5.37}$$

is negative (positive for a minimizing problem), where $\mathbf{H}=\nabla_x^2 F\left(\mathbf{x}^*, \lambda^*\right)$ and the ith column of $\mathbf{G}$ is $\nabla g_i\left(\mathbf{x}^*\right)$.

Note that since $\mathbf{H}$ is a matrix of quadratic form, it can be assumed to be symmetric. Then, it can be shown that every root of $P(\mu)$ is real.

We now state the following theorem on the existence and uniqueness of the Lagrange multiplier.

Theorem 5.8

Let the constrained optimization problem be given as:

Maximize f(**x**) subject to **g(x) = 0**.

Let the Lagrange function be defined as $F(\mathbf{x},\lambda)=f(\mathbf{x})+\lambda^T\mathbf{g}(\mathbf{x})$.

Then, a necessary and sufficient condition for the existence of the optimum value λ^* in the stationarity condition $\nabla_x F(\mathbf{x}^*, \lambda^*) = \mathbf{0}$ is that the $n \times m$ matrix $\mathbf{G} = \frac{\partial \mathbf{g}^T(\mathbf{x}^*)}{\partial \mathbf{x}}$ and the $n \times (m+1)$ matrix $\left[\mathbf{G} \;\; \nabla f(\mathbf{x}^*)\right]$ have the same rank at the maximum point $\mathbf{x}^*$.

Further, the optimum value λ^* is unique if and only if the same rank of the two matrices is m.

Let us now find an interpretation of the Lagrange multiplier.

Suppose that we find the extremum point $\mathbf{x} = \mathbf{x}^*$ of the objective function $f(\mathbf{x})$ subject to the constraint equation $g(\mathbf{x}) = b$, where b is a positive real number. We want to investigate how the optimum point $\mathbf{x}^*$ changes when the value of b varies by a small amount.

We define the Lagrange function with dependence on b as

$$F(\mathbf{x}, \lambda; b) = f(\mathbf{x}) + \lambda\left[b - g(\mathbf{x})\right] \tag{5.38}$$

and write the following set of necessary equations to be satisfied at $\mathbf{x}^*$ and λ^*,

$$\begin{aligned} &\frac{\partial F}{\partial x_j} = \frac{\partial f}{\partial x_j} - \lambda \frac{\partial g}{\partial x_j} = 0 \\ &\Rightarrow \frac{\partial g}{\partial x_j} = \frac{1}{\lambda}\frac{\partial f}{\partial x_j} \quad \text{for } j = 1, \ldots, n \end{aligned} \tag{5.39}$$

and

$$\frac{\partial F}{\partial \lambda} = b - \mathrm{g} = 0 \tag{5.40}$$

Now from the above equations, we get

$$db = dg = \sum_j \frac{\partial g}{\partial x_j} dx_j = \frac{1}{\lambda}\sum_j \frac{\partial f}{\partial x_j} dx_j = \frac{df}{\lambda} \tag{5.41}$$

which provides the required expression for λ as

$$\lambda = \frac{df}{db} \tag{5.42}$$

Thus, the Lagrange multiplier λ provides a measure of the rate of change of the extremum value of an objective function f with respect to the constraint-setting parameter b. We shall use this interpretation of the Lagrange multiplier when we consider optimization of a function subject to inequality constraint in the next section.

5.3 Extrema of Functions with Inequality Constraints

We want to find the extremum of the function $f: \mathbb{R}^n \rightarrow \mathbb{R}$ subject to the inequality constraint $\mathbf{g}(\mathbf{x}) \leq \mathbf{0}$ where $\mathbf{g}: \mathbb{R}^n \rightarrow \mathbb{R}^m$, $m < n$.

Consider the case with $n = 2$ and $m = 1$.

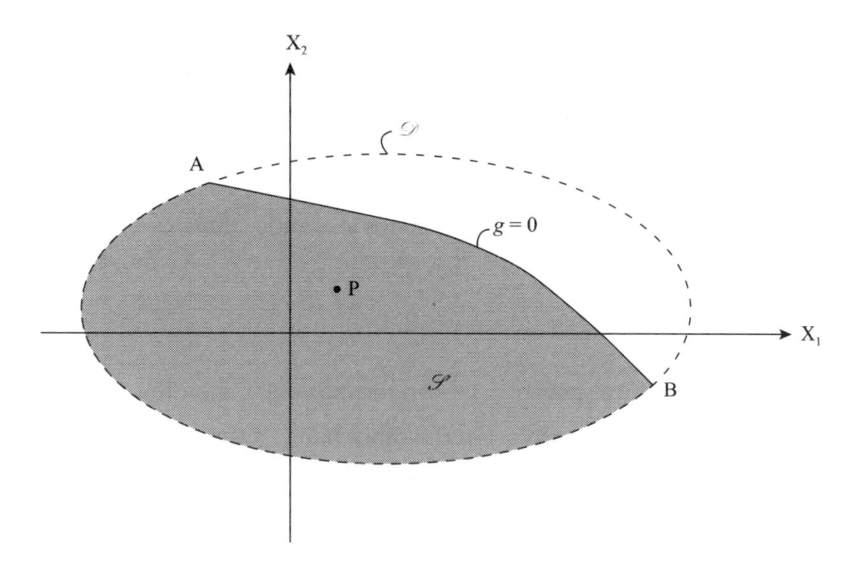

FIGURE 5.1 Feasible region $\mathscr{S}$ with inequality constraint $g(\mathbf{x}) \le 0$

Assume that both f and $g = 0$ are defined over an open region $\mathscr{D} \subseteq \mathbb{R}^2$ as shown in Figure 5.1. For the optimization problem with equality constraint $g = 0$, the admissible points which are candidates for a constrained extremum lie on the curve AB. However, for the optimization problem with inequality constraint $g \le 0$, we have a region of feasible solutions. This feasible region $\mathscr{S}$ is shown by the shaded portion of the above figure, which includes all points on the curve AB and below it [Panik].

Let us now find the necessary conditions for the point $\mathbf{x} = \mathbf{x}^*$ to be a solution of the optimization problem: Maximize $f(\mathbf{x})$ subject to $g(\mathbf{x}) \le 0$.

Introducing a real-valued slack variable s, we rewrite the constraint equation as $g(\mathbf{x}) + s^2 = 0$. The augmented Lagrange function is defined as

$$F(\mathbf{x}, \lambda, s) = f(\mathbf{x}) + \lambda\left[g(\mathbf{x}) + s^2\right] \tag{5.43}$$

Then, the necessary conditions satisfied at $\mathbf{x} = \mathbf{x}^*$ are

$$\frac{\partial F}{\partial \mathbf{x}} = \frac{\partial f}{\partial \mathbf{x}} + \lambda \frac{\partial g}{\partial \mathbf{x}} = 0 \tag{5.44}$$

$$\frac{\partial F}{\partial \lambda} = g + s^2 = 0 \quad \Rightarrow g \le 0 \tag{5.45}$$

$$\frac{\partial F}{\partial s} = 2\lambda s = 0 \quad \Rightarrow \lambda g = 0 \tag{5.46}$$

where the last equation is obtained by multiplying both sides with $(-s/2)$ and substituting $g = -s^2$.

There is one necessary condition related to the sign of the Lagrange multiplier that can be found in this case. Note that the inequality constraint $g \leq 0$ can be written as $g = -b$ where the constraint-setting parameter b is a positive real number which is varied to scan the feasible region $\mathscr{S}$. From (5.42), we find that λ is now given by $\lambda = -\frac{df}{db}$, and as b increases and the region of permissible solutions expands, the maximum value of the objective function cannot decrease. Thus, the optimum value $\lambda = \lambda^*$ must satisfy

$$\lambda \leq 0 \tag{5.47}$$

In this regard, if the extremum point $\mathbf{x}^*$ lies on the curve AB, then the constraint $g = 0$ is active and the optimum λ is negative. On the other hand, if the extremum point $\mathbf{x}^*$ is an interior point of the feasible region, such as P, then the constraint $g = 0$ is inactive and the required λ is zero.

A point $\mathbf{x}^* \in \mathscr{S}$ is an interior point of $\mathscr{S}$ if and only if there is a positive real number ε such that if $\mathbf{x}$ satisfies $\left\|\mathbf{x} - \mathbf{x}^*\right\| < \varepsilon$, then $\mathbf{x} \in \mathscr{S}$.

A subspace $\mathscr{D} \subseteq \mathbb{R}^n$ is termed open if it contains only interior points.

5.3.1 Kuhn–Tucker Necessary Conditions for Inequality Constraints

We now state a necessary condition for a candidate point $\mathbf{x} = \mathbf{x}^*$ to be a local extremum of an objective function $f(\mathbf{x})$, $\mathbf{x} \in \mathbb{R}^n$, subject to a prescribed inequality constraint. There are three distinct types of inequality constraints that may arise:

(I) $\mathbf{g}(\mathbf{x}) \geq \mathbf{0}$, $\mathbf{g} : \mathbb{R}^n \rightarrow \mathbb{R}^m$ and $\mathbf{x}$ unrestricted

(II) $\mathbf{g}(\mathbf{x}) = \mathbf{0}$, $\mathbf{g} : \mathbb{R}^n \rightarrow \mathbb{R}^m$ and $\mathbf{x} \geq \mathbf{0}$

(III) $\mathbf{g}(\mathbf{x}) \geq \mathbf{0}$, $\mathbf{g} : \mathbb{R}^n \rightarrow \mathbb{R}^m$ and $\mathbf{x} \geq \mathbf{0}$

Note that an inequality constraint $\mathbf{g}(\mathbf{x}) \leq \mathbf{0}$ can be formulated as $-\mathbf{g}(\mathbf{x}) \geq \mathbf{0}$, and thus, the given constraint belongs to the same type as mentioned in (I) or (III).

Theorem 5.9

Let the real-valued function $f(\mathbf{x})$, $\mathbf{x} \in \mathbb{R}^n$, and the constraint equation $\mathbf{g}(\mathbf{x}) \geq \mathbf{0}$, $\mathbf{g} : \mathbb{R}^n \rightarrow \mathbb{R}^m$, be defined over an open subspace $\mathscr{D} \subseteq \mathbb{R}^n$.

Let the Lagrange function be defined as $F(\mathbf{x}, \boldsymbol{\lambda}) = f(\mathbf{x}) + \boldsymbol{\lambda}^T \mathbf{g}(\mathbf{x})$.

Then, a necessary condition for a constrained local or global maximum of f subject to $\mathbf{g} \geq \mathbf{0}$ at a point $\mathbf{x}^* \in \mathscr{D}$ is given by

(a) $\nabla_x F(\mathbf{x}^*, \boldsymbol{\lambda}^*) = 0$

(b) $\mathbf{g}^T(\mathbf{x}^*)\boldsymbol{\lambda}^* = 0$

(c) $\nabla_\lambda F(\mathbf{x}^*) = \mathbf{g}(\mathbf{x}^*) \geq \mathbf{0}$

(d) $\boldsymbol{\lambda}^* \geq \mathbf{0}$

where $\boldsymbol{\lambda}^*$ is the optimum value of $\boldsymbol{\lambda}$.

Theorem 5.10

Let the real-valued function $f(\mathbf{x}),\ \mathbf{x} \in \mathbb{R}^n$, and the constraint equation $\mathbf{g}(\mathbf{x}) = \mathbf{0},\ \mathbf{g}: \mathbb{R}^n \to \mathbb{R}^m$, be defined over an open subspace $\mathscr{D} \subseteq \mathbb{R}^n$.

Let the Lagrange function be defined as $F(\mathbf{x},\boldsymbol{\lambda}) = f(\mathbf{x}) + \boldsymbol{\lambda}^T \mathbf{g}(\mathbf{x})$.

Then, a necessary condition for a constrained local or global maximum of *f* subject to $\mathbf{g} = \mathbf{0}$ and $\mathbf{x} \geq \mathbf{0}$ at a point $\mathbf{x}^* \in \mathscr{D}$ is given by

(a) $\nabla_x F(\mathbf{x}^*, \boldsymbol{\lambda}^*) \leq \mathbf{0}$

(b) $\left[\nabla_x F(\mathbf{x}^*, \boldsymbol{\lambda}^*)\right]^T \mathbf{x}^* = 0$

(c) $\nabla_\lambda F(\mathbf{x}^*) = \mathbf{g}(\mathbf{x}^*) = \mathbf{0}$

(d) $\mathbf{x}^* \geq \mathbf{0}$

where $\boldsymbol{\lambda}^*$ is the optimum value of $\boldsymbol{\lambda}$.

Theorem 5.11

Let the real-valued function $f(\mathbf{x}),\ \mathbf{x} \in \mathbb{R}^n$, and the constraint equation $\mathbf{g}(\mathbf{x}) \geq \mathbf{0},\ \mathbf{g}: \mathbb{R}^n \to \mathbb{R}^m$, be defined over an open subspace $\mathscr{D} \subseteq \mathbb{R}^n$.

Let the Lagrange function be defined as $F(\mathbf{x},\boldsymbol{\lambda}) = f(\mathbf{x}) + \boldsymbol{\lambda}^T \mathbf{g}(\mathbf{x})$.

Then, a necessary condition for a constrained local or global maximum of *f* subject to $\mathbf{g} \geq \mathbf{0}$ and $\mathbf{x} \geq \mathbf{0}$ at a point $\mathbf{x}^* \in \mathscr{D}$ is given by

(a) $\nabla_x F(\mathbf{x}^*, \boldsymbol{\lambda}^*) \leq \mathbf{0}$

(b) $\left[\nabla_x F(\mathbf{x}^*, \boldsymbol{\lambda}^*)\right]^T \mathbf{x}^* = \mathbf{0}$

(c) $\mathbf{g}^T(\mathbf{x}^*)\boldsymbol{\lambda}^* = 0$

(d) $\nabla_\lambda F(\mathbf{x}^*) = \mathbf{g}(\mathbf{x}^*) \geq \mathbf{0}$

(e) $\mathbf{x}^* \geq \mathbf{0}$

(f) $\boldsymbol{\lambda}^* \geq \mathbf{0}$

where $\boldsymbol{\lambda}^*$ is the optimum value of $\boldsymbol{\lambda}$.

5.3.2 Example 1 (Kuhn–Tucker)

We need to maximize $f(\mathbf{x}) = x_1$ subject to $g(\mathbf{x}) \equiv (1-x_1)^3 - x_2 \geq 0$ and $x_1, x_2 \geq 0$.

The feasible region of solution, $\mathscr{S}$, is shaded in Figure 5.2. The optimum point is $x_1^* = 1,\ x_2^* = 0$ giving $f(\mathbf{x}^*) = 1$.

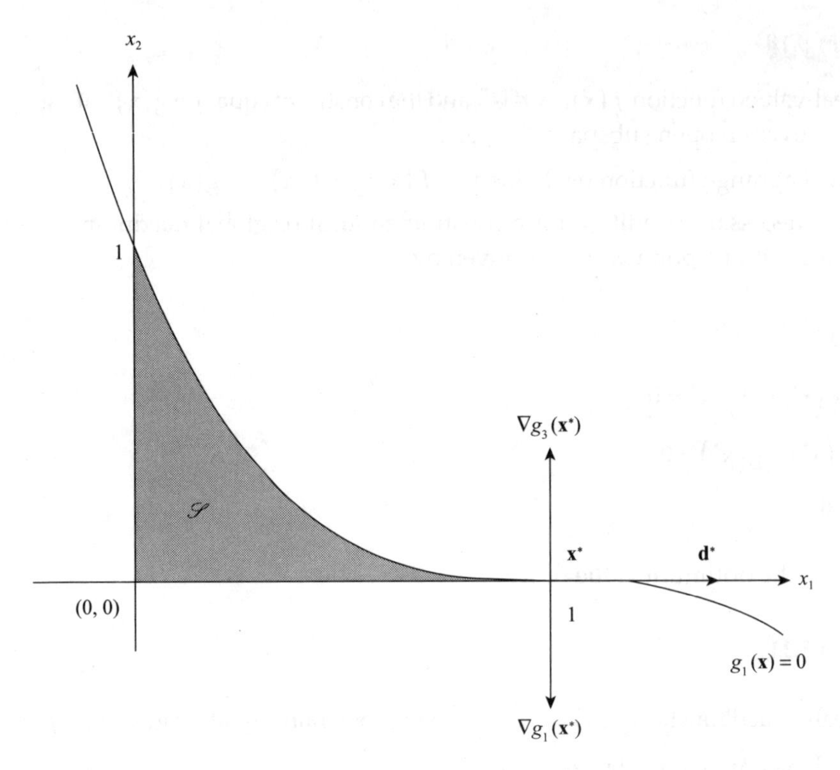

FIGURE 5.2 Direction $\mathbf{d}^*$ satisfies constraint requirements, yet is inadmissible

The Lagrange function is defined as

$$F(\mathbf{x}, \lambda) \equiv x_1 + \lambda\left[(1-x_1)^3 - x_2\right],$$

and the necessary conditions for a maximum at $\mathbf{x}^* = [1 \quad 0]^T$, according to Theorem 5.11, are given by

$$\frac{\partial F}{\partial x_1}(\mathbf{x}^*, \lambda^*) \equiv 1 - 3\lambda^*(1-x_1^*)^2 = 1, \text{ should be } \leq 0$$

$$\frac{\partial F}{\partial x_2}(\mathbf{x}^*, \lambda^*) \equiv -\lambda^* \leq 0 \Rightarrow \lambda^* \geq 0$$

$$\left[\nabla_x F(\mathbf{x}^*, \lambda^*)\right]^T \mathbf{x}^* \equiv \begin{bmatrix} 1 & -\lambda^* \end{bmatrix} \begin{bmatrix} x_1^* \\ x_2^* \end{bmatrix} = 1, \text{ should be } = 0$$

$$g(\mathbf{x}^*)\lambda^* \equiv \left[\left(1-x_1^*\right)^3 - x_2^*\right]\lambda^* = 0$$

Obviously, two of the Kuhn–Tucker necessary conditions are violated at the optimum point.

Note that there are three inequality constraints in this problem:

$$g_1(\mathbf{x}) \equiv (1-x_1)^3 - x_2 \geq 0,\ g_2(\mathbf{x}) \equiv x_1 \geq 0, \text{ and } g_3(\mathbf{x}) \equiv x_2 \geq 0.$$

The active constraints at the optimum point $\mathbf{x}^*$ constitute the active set $I_a = \{1, 3\}$. It turns out that the active constraints do not satisfy the constraint qualifications at the optimum point. This regularity requirement on the active constraint set is explained below.

At an optimal point $\mathbf{x}^*$, any admissible direction $\mathbf{d}$ must satisfy

1. $\left[\nabla g_i(\mathbf{x}^*)\right]^T \mathbf{d} \geq 0,\ i \in I_a$, and
2. $\left[\nabla f(\mathbf{x}^*)\right]^T \mathbf{d} \leq 0$, simultaneously.

In this case for $\mathbf{x}^* = [1\quad 0]^T$, we require that

$$\left[\nabla g_1(\mathbf{x}^*)\right]^T \mathbf{d} = [0\quad -1]\begin{bmatrix} d_1 \\ d_2 \end{bmatrix} = -d_2 \geq 0$$

$$\left[\nabla g_3(\mathbf{x}^*)\right]^T \mathbf{d} = [0\quad 1]\begin{bmatrix} d_1 \\ d_2 \end{bmatrix} = d_2 \geq 0$$

for all admissible $\mathbf{d}$.

If we let $\mathbf{d}^* = [1\quad 0]^T$, then $\mathbf{d}^*$ satisfies the above equations. However, $\mathbf{d}^*$ is not an admissible direction because the vector does not enter the feasible region $\mathscr{S}$, as shown in the above figure.

5.3.3 Constraint Qualification

Let $\mathbf{x}^* \in \mathscr{S} = \left\{\mathbf{x} \middle| \mathbf{g}(\mathbf{x}) \geq \mathbf{0}, \mathbf{x} \in \mathbb{R}^n\right\}$ with $\mathbf{g}(\mathbf{x})$ partially differentiable having continuous derivatives at $\mathbf{x}^*$. Then, the constraint qualification holds locally at $\mathbf{x}^*$ if for each $\mathbf{d} \neq \mathbf{0}$ satisfying $\left[\nabla g_i(\mathbf{x}^*)\right]^T \mathbf{d} \geq 0$ for all active constraints $i \in I_a$, there is a continuous differentiable arc $\mathbf{x} = \mathbf{a}(t),\ \mathbf{a}(0) = \mathbf{x}^*$, such that for each small positive t, $\mathbf{g}(\mathbf{x}) = \mathbf{g}(\mathbf{a}(t)) \geq \mathbf{0}$ and $\mathbf{d} = \mathbf{a}'(0)$.

The constraint qualification is sufficient to guarantee the following regularity condition

$$\left\{\mathbf{d} \middle| \left[\nabla f(\mathbf{x}^*)\right]^T \mathbf{d} > 0,\ \left[\nabla g_i(\mathbf{x}^*)\right]^T \mathbf{d} \geq 0,\ i \in I_a \right\} = \phi \tag{5.48}$$

that is, no such $\mathbf{d}$ can be found to satisfy both the conditions simultaneously. The regularity condition is necessary and sufficient to ensure the existence of an optimal $\boldsymbol{\lambda}^* \geq \mathbf{0}$.

Theorem 5.12

Let the real-valued functions $f(\mathbf{x})$ and $\mathbf{g}(\mathbf{x}) \geq \mathbf{0}$, $\mathbf{x} \in \mathbb{R}^n$ be partially differentiable having continuous derivatives at $\mathbf{x}^*$. Let $f(\mathbf{x})$ subject to $\mathbf{g}(\mathbf{x}) \geq \mathbf{0}$ have a local maximum at $\mathbf{x}^*$. Then, there exists an optimal $\boldsymbol{\lambda}^* \geq \mathbf{0}$ such that

$$\nabla f(\mathbf{x}^*) = \sum_i \lambda_i^* \left(-\nabla g_i(\mathbf{x}^*)\right), \quad i \in I_a \tag{5.49}$$

if and only if the regularity condition

$$\left\{ \mathbf{d} \middle| \left[\nabla f(\mathbf{x}^*)\right]^T \mathbf{d} > 0, \left[\nabla g_i(\mathbf{x}^*)\right]^T \mathbf{d} \geq 0, \ i \in I_a \right\} = \phi$$

is satisfied, where I_a is the active set of constraints at the optimum point.

The Fiacco–McCormick regularity requirement states that for $\mathbf{x}^* \in \mathscr{S} = \{\mathbf{x} | \mathbf{g}(\mathbf{x}) \geq \mathbf{0}, \mathbf{x} \in \mathbb{R}^n\}$ with $\mathbf{g}(\mathbf{x})$ differential throughout some neighbourhood of $\mathbf{x}^*$, if the set of vectors $\{\nabla g_i(\mathbf{x}^*),\ i \in I_a\}$ evaluated at $\mathbf{x}^*$ is linearly independent, then the Kuhn–Tucker constraint qualification is satisfied at $\mathbf{x}^*$. Refer to Theorem 5.8 here for comparison.

5.3.4 Example 2 (Regularity Condition)

Consider now the modified problem of minimizing $f(\mathbf{x}) = x_2$ subject to $g(\mathbf{x}) \equiv (1 - x_1)^3 - x_2 \geq 0$ and $x_1, x_2 \geq 0$. The optimal solution is $x_1^* \in [0, 1]$, $x_2^* = 0$. We formulate the dual problem of maximizing $-f(\mathbf{x}) = -x_2$, and the Lagrange function is written as

$$F(\mathbf{x}, \lambda) \equiv -x_2 + \lambda\left[(1 - x_1)^3 - x_2\right].$$

the necessary conditions according to Theorem 5.11 and evaluated at $\mathbf{x}^* = [1 \quad 0]^T$ are given by

$$\frac{\partial F}{\partial x_1}(\mathbf{x}^*, \lambda^*) \equiv -3\lambda^*(1 - x_1^*)^2 = 0 \leq 0$$

$$\frac{\partial F}{\partial x_2}(\mathbf{x}^*, \lambda^*) \equiv -1 - \lambda^* \leq 0 \Rightarrow \lambda^* \geq -1, \text{ satisfies } \lambda^* \geq 0$$

$$\left[\nabla_x F(\mathbf{x}^*, \lambda^*)\right]^T \mathbf{x}^* \equiv \begin{bmatrix} 0 & -1 - \lambda^* \end{bmatrix} \begin{bmatrix} x_1^* \\ x_2^* \end{bmatrix} = 0$$

$$g(\mathbf{x}^*)\lambda^* \equiv \left[(1 - x_1^*)^3 - x_2^*\right]\lambda^* = 0$$

Thus, all the necessary conditions are satisfied at the optimal point $\mathbf{x}^* = [1 \quad 0]^T$, $\lambda^* > 0$. Similarly, the necessary conditions are also satisfied at all other points where $\lambda^* = 0$.

Note that the constraint qualifications are still not satisfied in the modified problem. However, the regularity condition is satisfied now, whereas in the previous problem, the regularity condition is violated for the objective function $f(\mathbf{x}) = x_1$.

5.3.5 Kuhn–Tucker Necessary Conditions for Mixed Constraints

We now state two theorems describing the necessary conditions for a candidate point $\mathbf{x} = \mathbf{x}^*$ to be a local extremum of an objective function $f(\mathbf{x})$, $\mathbf{x} \in \mathbb{R}^n$, subject to a combination of prescribed equality and inequality constraints.

Theorem 5.13

Let the real-valued function $f(\mathbf{x})$, $\mathbf{x} \in \mathbb{R}^n$, and the constraint equations $\mathbf{g}(\mathbf{x}) \geq \mathbf{0}$, $\mathbf{g}: \mathbb{R}^n \to \mathbb{R}^m$ and $\mathbf{h}(\mathbf{x}) = \mathbf{0}$, $\mathbf{h}: \mathbb{R}^n \to \mathbb{R}^p$ be defined over an open subspace $\mathscr{D} \subseteq \mathbb{R}^n$.

Let the Lagrange function $F(\mathbf{x}, \boldsymbol{\lambda}, \boldsymbol{\mu})$ be defined as

$$F(\mathbf{x}, \boldsymbol{\lambda}, \boldsymbol{\mu}) = f(\mathbf{x}) + \boldsymbol{\lambda}^T \mathbf{g}(\mathbf{x}) + \boldsymbol{\mu}^T \mathbf{h}(\mathbf{x}) \tag{5.50}$$

Then, a necessary condition for a constrained local or global maximum of *f* subject to $\mathbf{g} \geq \mathbf{0}$ and $\mathbf{h} = \mathbf{0}$ at a point $\mathbf{x}^* \in \mathscr{D}$ is given by

(a) $\nabla_x F(\mathbf{x}^*, \boldsymbol{\lambda}^*, \boldsymbol{\mu}^*) = \mathbf{0}$

(b) $\mathbf{g}^T(\mathbf{x}^*)\boldsymbol{\lambda}^* = 0$

(c) $\nabla_\lambda F(\mathbf{x}^*) = \mathbf{g}(\mathbf{x}^*) \geq \mathbf{0}$

(d) $\nabla_\mu F(\mathbf{x}^*) = \mathbf{h}(\mathbf{x}^*) = \mathbf{0}$

(e) $\boldsymbol{\lambda}^* \geq \mathbf{0}$

where $\boldsymbol{\lambda}^*$ and $\boldsymbol{\mu}^*$ are the optimum values of $\boldsymbol{\lambda}$ and $\boldsymbol{\mu}$ respectively, and $\boldsymbol{\mu}^*$ is unrestricted in sign.

Theorem 5.14

Let the real-valued function $f(\mathbf{x})$, $\mathbf{x} \in \mathbb{R}^n$, and the constraint equations $\mathbf{g}(\mathbf{x}) \geq \mathbf{0}$, $\mathbf{g}: \mathbb{R}^n \to \mathbb{R}^m$, and $\mathbf{h}(\mathbf{x}) = \mathbf{0}$, $\mathbf{h}: \mathbb{R}^n \to \mathbb{R}^p$ be defined over an open subspace $\mathscr{D} \subseteq \mathbb{R}^n$.

Let the Lagrange function be defined as $F(\mathbf{x}, \boldsymbol{\lambda}, \boldsymbol{\mu}) = f(\mathbf{x}) + \boldsymbol{\lambda}^T \mathbf{g}(\mathbf{x}) + \boldsymbol{\mu}^T \mathbf{h}(\mathbf{x})$.

Then, a necessary condition for a constrained local or global maximum of *f* subject to $\mathbf{g} \geq \mathbf{0}$, $\mathbf{h} = \mathbf{0}$ and $\mathbf{x} \geq \mathbf{0}$ at a point $\mathbf{x}^* \in \mathscr{D}$ is given by

(a) $\nabla_x F(\mathbf{x}^*, \boldsymbol{\lambda}^*, \boldsymbol{\mu}^*) \leq \mathbf{0}$

(b) $\left[\nabla_x F(\mathbf{x}^*, \boldsymbol{\lambda}^*, \boldsymbol{\mu}^*)\right]^T \mathbf{x}^* = 0$

(c) $\mathbf{g}^T(\mathbf{x}^*)\boldsymbol{\lambda}^* = 0$

(d) $\nabla_\lambda F(\mathbf{x}^*) = \mathbf{g}(\mathbf{x}^*) \geq \mathbf{0}$

(e) $\nabla_\mu F(\mathbf{x}^*) = \mathbf{h}(\mathbf{x}^*) = \mathbf{0}$

(f) $\mathbf{x}^* \geq \mathbf{0}$

(g) $\boldsymbol{\lambda}^* \geq \mathbf{0}$

where $\boldsymbol{\lambda}^*$ and $\boldsymbol{\mu}^*$ are the optimum values of $\boldsymbol{\lambda}$ and $\boldsymbol{\mu}$ respectively, and $\boldsymbol{\mu}^*$ is unrestricted in sign.

The regularity condition of the maximizing problem of $f(\mathbf{x})$ with a combination of equality and inequality constraints becomes

$$\left\{\mathbf{d} \middle| \left[\nabla f(\mathbf{x}^*)\right]^T \mathbf{d} > 0,\ \left[\nabla g_i(\mathbf{x}^*)\right]^T \mathbf{d} \geq 0,\ i \in I_a,\ \left[\nabla h_k(\mathbf{x}^*)\right]^T \mathbf{d} = 0,\ k = 1, \ldots, p\right\} = \phi \quad (5.51)$$

where I_a is the active set. The regularity condition is necessary and sufficient to ensure the existence of some optimal $\boldsymbol{\lambda}^*$ and $\boldsymbol{\mu}^*$. Note that the inequality constraint $\mathbf{x}^* \geq \mathbf{0}$ can be included in the constraint $\mathbf{g}(\mathbf{x}^*) \geq \mathbf{0}$.

5.4 Lagrangian Saddle Point and Kuhn–Tucker Theory

The Lagrange function is defined as $F(\mathbf{x}, \boldsymbol{\lambda}) \equiv f(\mathbf{x}) + \boldsymbol{\lambda}^T \mathbf{g}(\mathbf{x}),\ \mathbf{x} \in \mathbb{R}^n,\ \boldsymbol{\lambda} \in \mathbb{R}^m$, corresponding to the constrained maximization problem

$$\max_{\mathbf{x}} \left\{ f(\mathbf{x}) \mid g_i(\mathbf{x}) \geq 0 \text{ for } i = 1, \ldots, m \right\} \quad (5.52)$$

We shall now study the saddle point property of the Lagrange function. A saddle point of $F(x, y)$ at $(x, y) = (0, 0)$ is shown in Figure 5.3.

A point $(\mathbf{x}^*, \boldsymbol{\lambda}^*)$ is termed a local saddle point of $F(\mathbf{x}, \boldsymbol{\lambda})$ if the following relations are satisfied,

$$F(\mathbf{x}, \boldsymbol{\lambda}^*) \leq F(\mathbf{x}^*, \boldsymbol{\lambda}^*) \leq F(\mathbf{x}^*, \boldsymbol{\lambda}) \quad (5.53)$$

for all $(\mathbf{x}, \boldsymbol{\lambda}) \in \mathscr{D} \subseteq \mathbb{R}^{n+m}$, $\mathscr{D}$ being an open neighbourhood of $(\mathbf{x}^*, \boldsymbol{\lambda}^*)$, and $\mathbf{x}$ unrestricted, $\boldsymbol{\lambda} \geq \mathbf{0}$.

A point $(\mathbf{x}^*, \boldsymbol{\lambda}^*)$ is termed a global saddle point of $F(\mathbf{x}, \boldsymbol{\lambda})$ if the relations (5.53) are satisfied for all $(\mathbf{x}, \boldsymbol{\lambda}) \in \mathbb{R}^{n+m}$, $\mathbf{x}$ unrestricted, $\boldsymbol{\lambda} \geq \mathbf{0}$.

The above definitions imply that $F(\mathbf{x}, \boldsymbol{\lambda})$ simultaneously attains a maximum with respect to $\mathbf{x}$, and a minimum with respect to $\boldsymbol{\lambda}$. Thus,

$$\begin{aligned} F(\mathbf{x}^*, \boldsymbol{\lambda}^*) &= \max_{\mathbf{x}} \left\{ \min_{\boldsymbol{\lambda} \geq \mathbf{0}} F(\mathbf{x}, \boldsymbol{\lambda}) \right\} \\ &= \min_{\boldsymbol{\lambda} \geq \mathbf{0}} \left\{ \max_{\mathbf{x}} F(\mathbf{x}, \boldsymbol{\lambda}) \right\} \end{aligned} \quad (5.54)$$

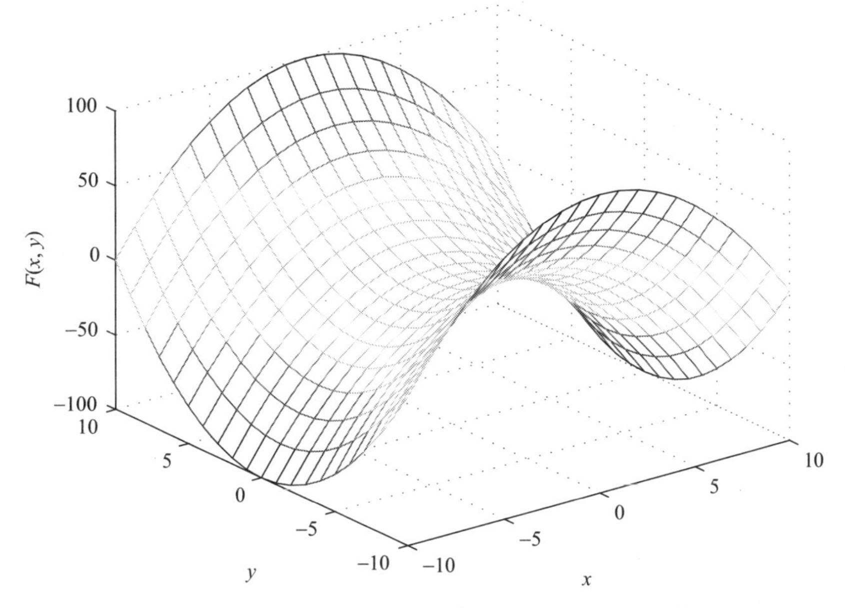

FIGURE 5.3 A saddle point of $F(x, y)$ at $(x, y) = (0, 0)$

The saddle point problem refers to finding a point $(\mathbf{x}^*, \boldsymbol{\lambda}^*)$ such that the above properties are satisfied for the Lagrange function $F(\mathbf{x}, \boldsymbol{\lambda})$. Note that for an equality constraint $\mathbf{g}(\mathbf{x}) = \mathbf{0}$, we get a degenerate form of saddle point at $(\mathbf{x}^*, \boldsymbol{\lambda}^*)$,

$$F(\mathbf{x}, \boldsymbol{\lambda}^*) \le F(\mathbf{x}^*, \boldsymbol{\lambda}^*) = F(\mathbf{x}^*, \boldsymbol{\lambda}) \tag{5.55}$$

We now state the properties of the Lagrangian saddle point in the following theorems.

Theorem 5.15

Let $\nabla_x F(\mathbf{x}, \boldsymbol{\lambda})$ and $\nabla_\lambda F(\mathbf{x}, \boldsymbol{\lambda})$ be defined for all $(\mathbf{x}, \boldsymbol{\lambda})$ in an open neighbourhood of $(\mathbf{x}^*, \boldsymbol{\lambda}^*)$, where $\mathbf{x} \ge \mathbf{0}$, $\boldsymbol{\lambda} \ge \mathbf{0}$. Then, a necessary condition for $F(\mathbf{x}, \boldsymbol{\lambda})$ to have a local saddle point at $(\mathbf{x}^*, \boldsymbol{\lambda}^*)$ is

(a) $\nabla_x F(\mathbf{x}^*, \boldsymbol{\lambda}^*) \le \mathbf{0}$, $\left[\nabla_x F(\mathbf{x}^*, \boldsymbol{\lambda}^*)\right]^T \mathbf{x}^* = 0$

(b) $\nabla_\lambda F(\mathbf{x}^*, \boldsymbol{\lambda}^*) \ge \mathbf{0}$, $\left[\nabla_\lambda F(\mathbf{x}^*, \boldsymbol{\lambda}^*)\right]^T \boldsymbol{\lambda}^* = 0$

Theorem 5.16

Let $\nabla_x F(\mathbf{x}, \boldsymbol{\lambda})$ and $\nabla_\lambda F(\mathbf{x}, \boldsymbol{\lambda})$ be defined for all $(\mathbf{x}, \boldsymbol{\lambda})$ in an open neighbourhood of $(\mathbf{x}^*, \boldsymbol{\lambda}^*)$, where $\mathbf{x} \ge \mathbf{0}$, $\boldsymbol{\lambda} \ge \mathbf{0}$. Then, a necessary and sufficient condition for $F(\mathbf{x}, \boldsymbol{\lambda})$ to have a local saddle point at $(\mathbf{x}^*, \boldsymbol{\lambda}^*)$ is

(a) $\nabla_x F(\mathbf{x}^*, \boldsymbol{\lambda}^*) \le \mathbf{0}, \ [\nabla_x F(\mathbf{x}^*, \boldsymbol{\lambda}^*)]^T \mathbf{x}^* = 0$

(b) $\nabla_\lambda F(\mathbf{x}^*, \boldsymbol{\lambda}^*) \ge \mathbf{0}, \ [\nabla_\lambda F(\mathbf{x}^*, \boldsymbol{\lambda}^*)]^T \boldsymbol{\lambda}^* = 0$

and

(c) $F(\mathbf{x}, \boldsymbol{\lambda}^*) \le F(\mathbf{x}^*, \boldsymbol{\lambda}^*) + [\nabla_x F(\mathbf{x}^*, \boldsymbol{\lambda}^*)]^T (\mathbf{x} - \mathbf{x}^*)$

(d) $F(\mathbf{x}^*, \boldsymbol{\lambda}) \ge F(\mathbf{x}^*, \boldsymbol{\lambda}^*) + [\nabla_\lambda F(\mathbf{x}^*, \boldsymbol{\lambda}^*)]^T (\boldsymbol{\lambda} - \boldsymbol{\lambda}^*)$

If the Lagrangian $F(\mathbf{x}, \boldsymbol{\lambda})$ is concave in $\mathbf{x}$ with $\boldsymbol{\lambda} = \boldsymbol{\lambda}^*$, and convex in $\boldsymbol{\lambda}$ with $\mathbf{x} = \mathbf{x}^*$, then the conditions (c) and (d) of Theorem 5.16 are always satisfied about $(\mathbf{x}^*, \boldsymbol{\lambda}^*)$. Accordingly, the property of a saddle point can be restated as given below.

Theorem 5.17

Let $F(\mathbf{x}, \boldsymbol{\lambda}^*)$ be concave in $\mathbf{x}$, and $F(\mathbf{x}^*, \boldsymbol{\lambda})$ convex in $\boldsymbol{\lambda}$, with $\nabla_x F(\mathbf{x}, \boldsymbol{\lambda})$ and $\nabla_\lambda F(\mathbf{x}, \boldsymbol{\lambda})$ defined for all $(\mathbf{x}, \boldsymbol{\lambda})$ in an open neighbourhood of $(\mathbf{x}^*, \boldsymbol{\lambda}^*)$, where $\mathbf{x} \ge \mathbf{0}, \ \boldsymbol{\lambda} \ge \mathbf{0}$. Then, $(\mathbf{x}^*, \boldsymbol{\lambda}^*)$ is a saddle point of $F(\mathbf{x}, \boldsymbol{\lambda})$ if and only if

(a) $\nabla_x F(\mathbf{x}^*, \boldsymbol{\lambda}^*) \le \mathbf{0}, \ [\nabla_x F(\mathbf{x}^*, \boldsymbol{\lambda}^*)]^T \mathbf{x}^* = 0$

(b) $\nabla_\lambda F(\mathbf{x}^*, \boldsymbol{\lambda}^*) \ge \mathbf{0}, \ [\nabla_\lambda F(\mathbf{x}^*, \boldsymbol{\lambda}^*)]^T \boldsymbol{\lambda}^* = 0$

We now relate the local saddle point problem with the local maximum problem in the following theorems.

Theorem 5.18

Let $F(\mathbf{x}, \boldsymbol{\lambda})$ be defined for all $(\mathbf{x}, \boldsymbol{\lambda})$ in an open neighbourhood of $(\mathbf{x}^*, \boldsymbol{\lambda}^*)$, where $\mathbf{x}$ is unrestricted and $\boldsymbol{\lambda} \ge \mathbf{0}$. If $(\mathbf{x}^*, \boldsymbol{\lambda}^*)$ is a local saddle point of $F(\mathbf{x}, \boldsymbol{\lambda})$, then $\mathbf{x}^*$ solves the local maximum problem (sufficient condition).

Theorem 5.19

Let $F(\mathbf{x}, \boldsymbol{\lambda})$ be defined for all $(\mathbf{x}, \boldsymbol{\lambda})$ in an open neighbourhood of $(\mathbf{x}^*, \boldsymbol{\lambda}^*)$, where $\mathbf{x}$ is unrestricted and $\boldsymbol{\lambda} \ge \mathbf{0}$. Then, a necessary and sufficient condition for $F(\mathbf{x}, \boldsymbol{\lambda})$ to have a local saddle point at $(\mathbf{x}^*, \boldsymbol{\lambda}^*)$ is that

(a) $F(\mathbf{x}, \boldsymbol{\lambda}^*)$ attains a local maximum at $\mathbf{x}^*$

(b) $\mathbf{g}(\mathbf{x}^*) \ge \mathbf{0}$

(c) $\mathbf{g}^T(\mathbf{x}^*)\boldsymbol{\lambda}^* = 0$

When the Lagrangian is concave and a constraint qualification is imposed on the constraint set, it is possible to relate the global maximum problem at $\mathbf{x}^*$ and the global saddle point problem at $(\mathbf{x}^*, \boldsymbol{\lambda}^*)$ for some $\boldsymbol{\lambda}^*$. This is stated in the following theorem.

Theorem 5.20 (Panik)

Let the real-valued scalar function $f(\mathbf{x})$ and the vector function $\mathbf{g}(\mathbf{x}) \geq \mathbf{0},\ \mathbf{x} \in \mathbb{R}^n$ be concave and differentiable over the convex region $\mathcal{S} = \{\mathbf{x} \mid \mathbf{g}(\mathbf{x}) \geq \mathbf{0}, \mathbf{x} \in \mathbb{R}^n\}$ with the property that for some $\mathbf{x} \in \mathcal{S}$, $\mathbf{g}(\mathbf{x}) > \mathbf{0}$ (Slater's constraint qualification). Then, $\mathbf{x}^*$ solves the global maximum problem if and only if there exists a $\boldsymbol{\lambda}^* \geq \mathbf{0}$ such that $(\mathbf{x}^*, \boldsymbol{\lambda}^*)$ solves the saddle point problem with the Kuhn–Tucker conditions

$$\nabla f(\mathbf{x}^*) + \nabla \mathbf{g}(\mathbf{x}^*)\boldsymbol{\lambda}^* = \mathbf{0},\ \mathbf{g}(\mathbf{x}^*) \geq \mathbf{0},\ \mathbf{g}^T(\mathbf{x}^*)\boldsymbol{\lambda}^* = 0.$$

The Slater's constraint qualification is applicable when $\mathbf{g}(\mathbf{x})$ is concave over a convex feasible region $\mathcal{S} = \{\mathbf{x} \mid \mathbf{g}(\mathbf{x}) \geq \mathbf{0}, \mathbf{x} \in \mathbb{R}^n\}$. The constraint equation $\mathbf{g}(\mathbf{x}) \geq \mathbf{0}$ satisfies the constraint qualification if there exists an $\mathbf{x} \in \mathcal{S}$ such that $\mathbf{g}(\mathbf{x}) > \mathbf{0}$, that is, $\mathcal{S}$ has a non-empty interior. The Slater's constraint qualification is easily verifiable.

We now state the equivalence theorem which ensures that the solution of a constrained maximum problem is conveniently expressed as the solution of a saddle point problem of the associated Lagrangian.

Theorem 5.21 (Kuhn–Tucker theorem)

Let the real-valued scalar function $f(\mathbf{x})$ and the vector function $\mathbf{g}(\mathbf{x}) \geq \mathbf{0},\ \mathbf{x} \in \mathbb{R}^n$ be concave and differentiable over the convex region $\mathcal{S} = \{\mathbf{x} \mid \mathbf{g}(\mathbf{x}) \geq \mathbf{0}, \mathbf{x} \in \mathbb{R}^n\}$ with the property that for some $\mathbf{x} \in \mathcal{S}$, $\mathbf{g}(\mathbf{x}) > \mathbf{0}$. Then, $\mathbf{x}^*$ solves the global maximum problem if and only if there exists a $\boldsymbol{\lambda}^* \geq \mathbf{0}$ such that $(\mathbf{x}^*, \boldsymbol{\lambda}^*)$ solves the global saddle-point problem (necessary and sufficient condition).

According to the above theorem, the saddle point problem and constrained maximum problem are formally equivalent. Therefore in order to solve the constrained maximum problem, a numerical technique can be developed which, under certain conditions, ensures convergence to a saddle point of $F(\mathbf{x}, \boldsymbol{\lambda})$. Thus, the constrained maximum of $f(\mathbf{x})$ can be found.

5.5 One–Dimensional Optimization

In order to find the extremum of a function f of n variables $(x_1, \ldots, x_n)$, we shall use some iterative method which requires the solution of an optimization problem in one dimension at each step of iteration. This one-dimensional optimization problem is given as

$$\min_{\alpha \geq 0} \{s(\alpha) = f(\mathbf{x}_0 + \alpha\,\mathbf{d})\} \tag{5.56}$$

where $\mathbf{x}_0$ is the initial point and $\mathbf{d}$ is an appropriate direction of displacement. In general, the direction $\mathbf{d}$ is a direction of descent, that is

$$\left[\nabla f(\mathbf{x}_0)\right]^T \mathbf{d} = \left.\frac{ds}{d\alpha}\right|_{\alpha=0} < 0 \tag{5.57}$$

5.5.1 The Newton–Raphson Method

Assuming that the function $s(\alpha)$ is twice continuously differentiable, we search for a minimum of $s(\alpha)$ by an iterative approach to reach a stationary point α^* where the condition

$$\left.\frac{ds}{d\alpha}\right|_{\alpha=\alpha^*} = s'(\alpha^*) = 0 \tag{5.58}$$

is satisfied.

Let α_k be the point reached at kth iteration. Then, at $(k+1)$th iteration, the function $s'(\alpha)$ is approximated by its tangent $z(\alpha)$ at α_k,

$$z(\alpha) = s'(\alpha_k) + (\alpha - \alpha_k) s''(\alpha_k) \tag{5.59}$$

Setting the above equation equal to zero, we get $\alpha = \alpha_{k+1}$ as the point to start the next iteration,

$$\alpha_{k+1} = \alpha_k - \frac{s'(\alpha_k)}{s''(\alpha_k)} \tag{5.60}$$

This iterative approach is known as the Newton–Raphson (NR) method, and the method is illustrated in Figure 5.4.

If $s(\alpha)$ is a quadratic function of the form

$$s(\alpha) = a\alpha^2 + b\alpha + c, \quad a > 0 \tag{5.61}$$

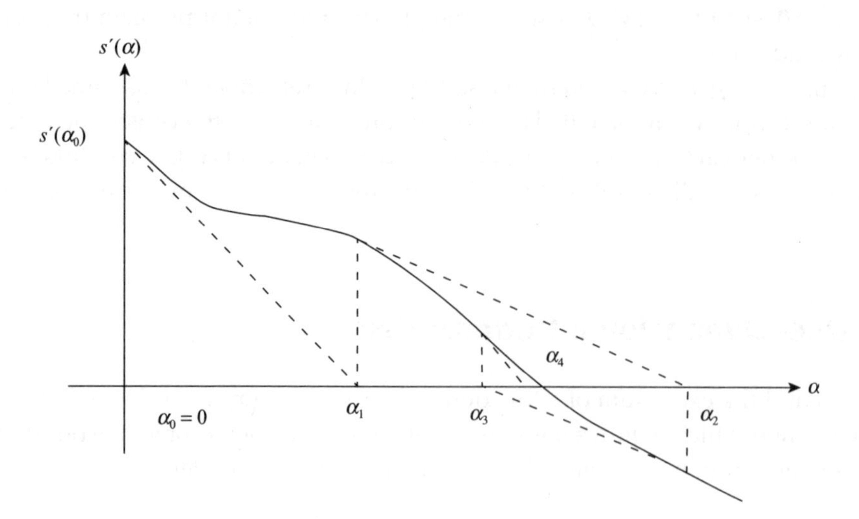

FIGURE 5.4 The Newton–Raphson method

then, for any starting point α_0, we have

$$s'(\alpha_0) = 2a\alpha_0 + b,\ s''(\alpha_0) = 2a$$

Thus,

$$\alpha_1 = \alpha_0 - \frac{2a\alpha_0 + b}{2a} = -\frac{b}{2a} \tag{5.62}$$

which is the optimum point α^* reached in a single step.

Any function $s(\alpha)$ which is twice continuously differentiable approximates a quadratic function near the optimum point. Thus, the NR method has the property of finite convergence for a sufficiently regular function.

There are two drawbacks of the NR method, namely,

1. The first and second order derivatives need to be computed at every point of iteration, which makes the computational cost of the method quite high.
2. The global convergence of the method is not guaranteed for any starting point. In fact, it is a better strategy to get a point close to the optimum by another procedure, like direct search, and then, apply the NR method to take advantage of its quadratic convergence.

5.5.2 The Secant Method

One major disadvantage of the NR method is the necessity of computing the first and second derivatives at every point.

Approximating the second derivative $s''(\alpha)$ at α_k by

$$\frac{s'(\alpha_k) - s'(\alpha_{k-1})}{\alpha_k - \alpha_{k-1}} \tag{5.63}$$

we rewrite the above equation as

$$\alpha_{k+1} = \alpha_k - s'(\alpha_k)\frac{\alpha_k - \alpha_{k-1}}{s'(\alpha_k) - s'(\alpha_{k-1})} \tag{5.64}$$

This iterative approach is known as the Secant method. Note that in search of the root of the equation $s'(\alpha) = 0$, this method approximates the function $s'(\alpha)$ at α_k by the straight line joining the points $(\alpha_{k-1}, s'(\alpha_{k-1}))$ and $(\alpha_k, s'(\alpha_k))$. The Secant method is illustrated in Figure 5.5.

The global convergence of the Secant method is not guaranteed. The starting points α_0 and α_1 must be chosen sufficiently close to the optimum point.

5.5.3 Quadratic Interpolation Method

This method does not require the computation of first or second derivatives of the function s.

Let three distinct points of α be chosen as $\alpha_1 \le \alpha_2 \le \alpha_3$ such that the function values $s(\alpha_1) \ge s(\alpha_2)$ and $s(\alpha_3) \ge s(\alpha_2)$. Then, the function $s(\alpha)$ can be approximated on the interval $[\alpha_1, \alpha_3]$ by quadratic interpolation as

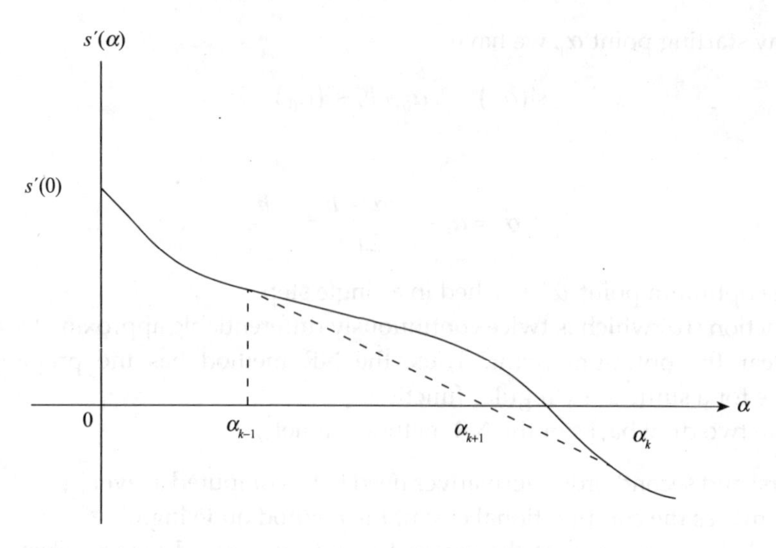

FIGURE 5.5 The Secant method

$$q(\alpha)=\sum_{i=1}^{3}s(\alpha_i)\frac{\prod_{j\neq i}(\alpha-\alpha_j)}{\prod_{j\neq i}(\alpha_i-\alpha_j)} \tag{5.65}$$

The minimum of $q(\alpha)$ is found at $\alpha_4\in[\alpha_1,\alpha_3]$ given as

$$\alpha_4=\frac{1}{2}\frac{u_{23}s(\alpha_1)+u_{31}s(\alpha_2)+u_{12}s(\alpha_3)}{v_{23}s(\alpha_1)+v_{31}s(\alpha_2)+v_{12}s(\alpha_3)} \tag{5.66}$$

where $u_{ij}=\alpha_i^2-\alpha_j^2$ and $v_{ij}=\alpha_i-\alpha_j$.

Next, the above procedure is repeated with the three new points $(\alpha_1',\alpha_2',\alpha_3')$ chosen as

$$\left.\begin{array}{lll}(\alpha_2,\alpha_4,\alpha_3) & \text{if } \alpha_2\le\alpha_4\le\alpha_3 & \text{and } s(\alpha_4)\le s(\alpha_2)\\ (\alpha_1,\alpha_2,\alpha_4) & \text{if } \alpha_2\le\alpha_4\le\alpha_3 & \text{and } s(\alpha_4)> s(\alpha_2)\\ (\alpha_1,\alpha_4,\alpha_2) & \text{if } \alpha_1\le\alpha_4\le\alpha_2 & \text{and } s(\alpha_4)\le s(\alpha_2)\\ (\alpha_4,\alpha_2,\alpha_3) & \text{if } \alpha_1\le\alpha_4\le\alpha_2 & \text{and } s(\alpha_4)> s(\alpha_2)\end{array}\right\} \tag{5.67}$$

If $s(\alpha)$ is continuous and unimodal with a unique minimum on the original interval, then the global convergence of the method can be proved.

5.5.4 Direct Search Methods

A direct search method is based on evaluating $s(\alpha)$ at a sequence of points $\{\alpha_1,\alpha_2,\ldots\}$ and comparing values, and in the process, the optimum point α^* is reached in finite number of steps.

Assume that the function $s(\alpha)$ is unimodal and its values are obtained at four distinct points $\alpha_1\le\alpha_2\le\alpha_3\le\alpha_4$. Further, assume that the unique minimum α^* is a point in the original interval $[\alpha_1, \alpha_4]$, whose length is $\Delta_1 = \alpha_4 - \alpha_1$, and the two intermediate points α_2 and α_3 are symmetrically placed in the interval.

After comparing the values of $s(\alpha)$ at the four points and utilizing the property of $s(\alpha)$ being unimodal, we can eliminate the subinterval $[\alpha_3, \alpha_4]$, as shown in Figure 5.6, leaving the reduced interval $[\alpha_1, \alpha_3]$, whose length is $\Delta_2 = \alpha_3 - \alpha_1$, for next stage of processing.

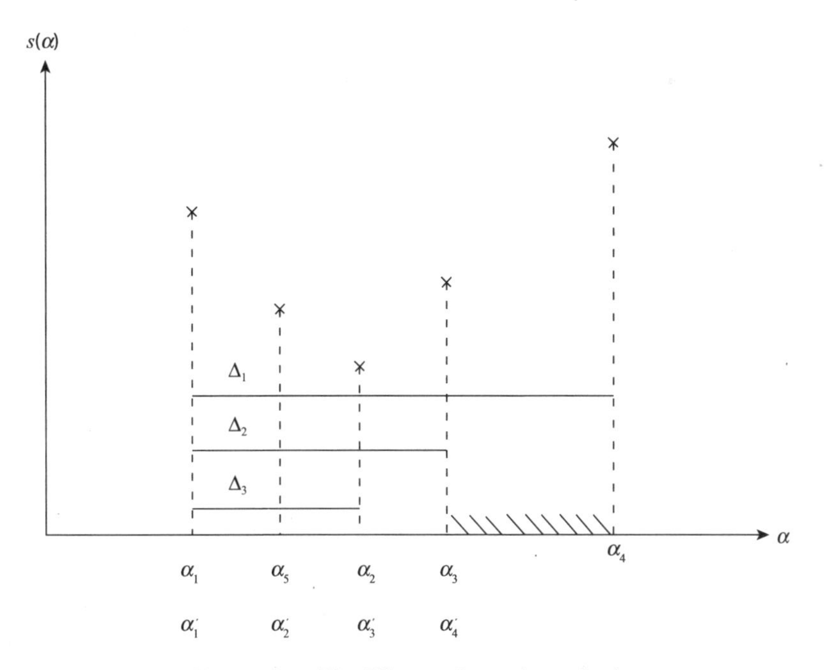

FIGURE 5.6 The Fibonacci search method

Note that for some other function, we may have to eliminate the subinterval $[\alpha_1, \alpha_2]$ leaving the reduced interval $[\alpha_2, \alpha_4]$, whose length is also $\Delta_2 = \alpha_4 - \alpha_2$. Thus, the method becomes independent of the function being processed by placing the two intermediate points α_2 and α_3 symmetrically in the original interval $[\alpha_1, \alpha_4]$.

Next, we choose the point α_5 symmetrically together with α_2 in the reduced interval $[\alpha_1, \alpha_3]$ as shown in Figure 5.6, and repeat the same procedure as above with the four new points $(\alpha_1', \alpha_2', \alpha_3', \alpha_4') = (\alpha_1, \alpha_5, \alpha_2, \alpha_3)$.

It follows that

$$\Delta_1 = \Delta_2 + \Delta_3 \tag{5.68}$$

and generalizing the above reasoning for all k,

$$\Delta_k = \Delta_{k+1} + \Delta_{k+2} \tag{5.69}$$

where k is the number of iteration.

Let Δ_{N-1} be the length of the interval after $(N+1)$ computations of the function $s(\alpha)$. Then defining a set of numbers $\{\phi_1, \phi_2, \ldots, \phi_{N-1}\}$ as follows

$$\begin{aligned} \Delta_1 &= \phi_{N-1}\Delta_{N-1} \\ \Delta_2 &= \phi_{N-2}\Delta_{N-1} \\ &\vdots \\ \Delta_{N-1} &= \phi_1\Delta_{N-1} \end{aligned} \quad \Rightarrow \quad \Delta_k = \phi_{N-k}\Delta_{N-1} \quad \text{for } k = 1, \ldots, N-1 \tag{5.70}$$

with $\phi_1 = 1$, we want to determine ϕ_n; $n = 2, \ldots, N-1$, such that the interval Δ_{N-1} is as small as possible.

Note that writing

$$\frac{\Delta_k}{\Delta_{N-1}} = \frac{\Delta_{k+1}}{\Delta_{N-1}} + \frac{\Delta_{k+2}}{\Delta_{N-1}}$$

we get the recurrence relation for ϕ_n,

$$\phi_n = \phi_{n-1} + \phi_{n-2} \quad \text{for } n = 3, 4, \ldots, N-1 \tag{5.71}$$

with $\phi_1 = 1$ and $\phi_2 = 2$ (to be proved below).

Now, observe that

$$\Delta_{N-1} = \frac{\Delta_1}{\phi_{N-1}} \tag{5.72}$$

and in order to get Δ_{N-1} as small as possible, we must have ϕ_{N-1} as large as possible. Since ϕ_{N-1} is generated from the recurrence relation (5.71), we must have ϕ_2 as large as possible. Note that from the relation $\Delta_{N-1} \geq \Delta_{N-2}/2$, we get $\phi_1 \geq \phi_2/2$. Thus, the largest value for ϕ_2 is $\phi_2 = 2$. Therefore, we generate the sequence of numbers $\{1, 2, 3, 5, 8, 13, \ldots\}$ known as the Fibonacci sequence, and accordingly, the reduced interval is obtained where the optimum point belongs. For instance, if we choose $N = 16$, then $\phi_{N-1} = 987 \simeq 10^3$, and $\Delta_{N-1} = 10^{-3}\Delta_1$, which provides a reduction of the original interval by a factor of 10^3. This procedure is called the Fibonacci search method.

Note that to start the Fibonacci search method, the value of N should be preassigned because the ratio of lengths

$$\frac{\Delta_1}{\Delta_2} = \frac{\phi_{N-1}}{\phi_{N-2}} \tag{5.73}$$

is to be used to get the first set of points $(\alpha_1, \alpha_2, \alpha_3, \alpha_4)$ on the original interval.

If the number N is not preassigned, then an alternative approach has to take the successive intervals in a given ratio of lengths

$$\frac{\Delta_1}{\Delta_2} = \frac{\Delta_2}{\Delta_3} = \cdots = \gamma \tag{5.74}$$

Since we have $\Delta_k = \Delta_{k+1} + \Delta_{k+2}$, and if we impose

$$\frac{\Delta_k}{\Delta_{k+1}} = \frac{\Delta_{k+1}}{\Delta_{k+2}} = \gamma$$

then,

$$\frac{\Delta_k}{\Delta_{k+1}} = 1 + \frac{\Delta_{k+2}}{\Delta_{k=1}}$$

$$\text{or,} \quad \gamma = 1 + \frac{1}{\gamma} \quad \Rightarrow \quad \gamma^2 - \gamma - 1 = 0$$

The positive roots of $\gamma^2 - \gamma - 1 = 0$ is the golden section number $\gamma = \frac{\sqrt{5}+1}{2} \simeq 1.618.$

The procedure described above is called the Golden section search method. The Golden section search method is not optimal like the Fibonacci search method. However, for sufficiently large N, both the methods perform similarly because

$$\lim_{N \to \infty} \frac{\phi_N}{\phi_{N-1}} = \gamma \tag{5.75}$$

5.6 Mathematical Programming: Unconstrained Optimization

Assuming that the function f is continuous and has continuous first derivatives, all unconstrained optimization techniques in $\mathbb{R}^n$ look for a stationary point $\mathbf{x}^*$, where the gradient $\nabla f(\mathbf{x}^*) = \mathbf{0}$. In an iterative procedure, one starts with an initial point $\mathbf{x}_0$ and generates a sequence of points $\mathbf{x}_1, \mathbf{x}_2, \ldots, \mathbf{x}$, which converges to a local optimum of f.

At each stage k, we compute

$$\mathbf{x}_{k+1} = \mathbf{x}_k + \alpha_k \mathbf{d}_k \tag{5.76}$$

where $\mathbf{d}_k$ is a descent direction such that $\left[\nabla f(\mathbf{x}_k)\right]^T \mathbf{d}_k < 0$. In particular, $\mathbf{d}_k$ can be the negative gradient computed at $\mathbf{x}_k$, i.e., $\mathbf{d}_k = -\nabla f(\mathbf{x}_k)$.

5.6.1 Steepest Descent Method

Using one-dimensional minimization, α is chosen to minimize the function

$$s(\alpha) = f(\mathbf{x}_k - \alpha \nabla f(\mathbf{x}_k)) \tag{5.77}$$

on the set of $\alpha \geq 0$.

Algorithm

(a) Initialization: Choose the starting point $\mathbf{x}_0$; Set $k = 0$.

(b) At step k: Compute $\mathbf{d}_k = -\nabla f(\mathbf{x}_k)$; Find α_k such that $f(\mathbf{x}_k + \alpha_k \mathbf{d}_k) = \min_{\alpha \geq 0} f(\mathbf{x}_k + \alpha \mathbf{d}_k)$; Set $\mathbf{x}_{k+1} = \mathbf{x}_k + \alpha_k \mathbf{d}_k$.

(c) Stopping condition: If satisfied, End; Otherwise, Set $k \leftarrow k+1$; Return to (b).

Two frequently used stopping conditions are:

$$\|\nabla f\|^2 < \varepsilon \quad \text{for a given } \varepsilon > 0 \tag{5.78}$$

$$|f(\mathbf{x}_{k+1}) - f(\mathbf{x}_k)| < \eta \quad \text{for a given } \eta > 0 \tag{5.79}$$

It may be a good precaution to set some maximum number p of iterations. If the stopping condition is not satisfied over p iterations, then the program is terminated for debugging.

In contrast to the exact steepest descent method as described above, when the step size α_k is chosen in some predefined manner such that $f(\mathbf{x}_k + \alpha_k \mathbf{d}_k) < f(\mathbf{x}_k)$, we get the approximate steepest descent method. It can be shown that in this case, the sequence $\mathbf{x}_k$ converges provided the following relations hold:

$$\begin{cases} \alpha_k \to 0 \quad \text{as } k \to \infty \\ \displaystyle\sum_{k=0}^{\infty} \alpha_k < +\infty \end{cases} \tag{5.80}$$

Alternatively, we can set the fixed step size $\alpha_s > 0$ sufficiently small to ensure that at each step $f(\mathbf{x}_k + \alpha_s \mathbf{d}_k) < f(\mathbf{x}_k)$. The global convergence of the steepest descent method (exact or approximate) is guaranteed by the following theorem.

Theorem 5.22 (Minoux)

If the function f is continuously differentiable with the property $f(\mathbf{x}) \to \infty$ for $\|\mathbf{x}\| \to \infty$, then with any starting point $\mathbf{x}_0$, the method of steepest descent converges to a stationary point of f.

Note that the stationary point will be a global or local minimum of f only if f is twice continuously differentiable, and the Hessian matrix is positive definite at the point.

The main disadvantage of the steepest descent method is that the convergence can be very slow for certain types of functions. Note that the one-dimensional minimization of $s(\alpha) = f(\mathbf{x}_k + \alpha \mathbf{d}_k)$ gives

$$\frac{ds}{d\alpha}(\alpha_k) = \mathbf{d}_k^T \nabla f(\mathbf{x}_k + \alpha_k \mathbf{d}_k) = \mathbf{d}_k^T \nabla f(\mathbf{x}_{k+1}) = 0$$

which yields to

$$\mathbf{d}_k^T \mathbf{d}_{k+1} = 0.$$

Thus, the successive displacement directions are orthogonal as shown on the contour plot of Figure 5.7. As a consequence, the number of steps necessary to reach the minimum point of an ill-conditioned function with 'narrow and elongated valley' can be very large.

5.6.2 Accelerated Steepest Descent Method

In the accelerated steepest descent method of order p, we carry out p stages of the steepest descent method, starting at $\mathbf{x}_k$, to reach a point $\mathbf{y}_k$. Then, the point $\mathbf{x}_{k+1}$ is obtained by one-dimensional minimization in the direction $\mathbf{d}_k = \mathbf{y}_k - \mathbf{x}_k$.

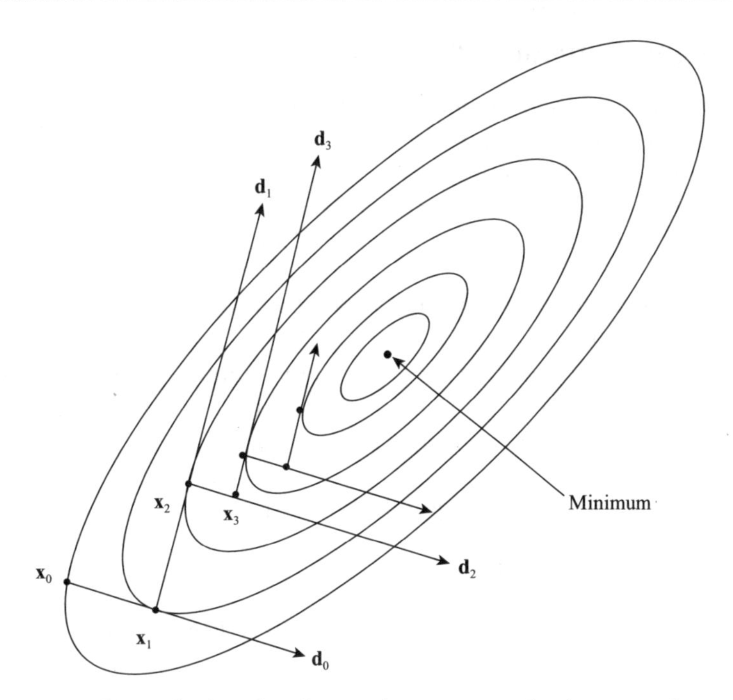

FIGURE 5.7 SD method with orthogonal consecutive displacement directions

The accelerated steepest descent method of order $p = 2$ is illustrated in Figure 5.8. For minimization in $\mathbb{R}^n$, if we set $p = n$, then the method shows convergence in at most n iterations. However, since every iteration requires $(p + 1)$ one-dimensional minimizations, the method is rather expensive in computational cost.

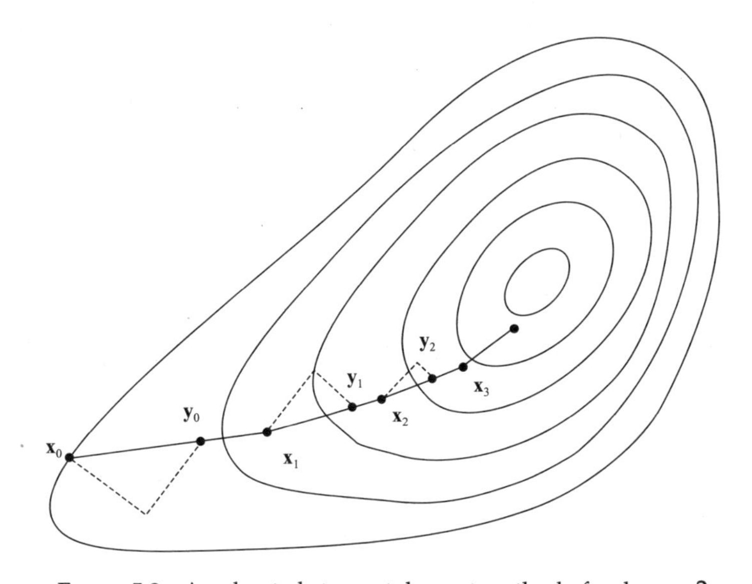

FIGURE 5.8 Accelerated steepest descent method of order $p = 2$

5.6.3 Conjugate Direction Methods

The conjugate direction methods, when applied to a quadratic function of n variables, lead to the optimum in at most n iterations.

Consider any quadratic function

$$q(\mathbf{x}) = \frac{1}{2}\mathbf{x}^T\mathbf{A}\mathbf{x} + \mathbf{b}^T\mathbf{x} + c \tag{5.81}$$

where $\mathbf{A}$ is a positive definite symmetric matrix of dimension $n \times n$, $\mathbf{b}$ is a vector of dimension $n \times 1$, and c is a scalar value.

In the conjugate direction method, we carry out minimizations of $q(\mathbf{x})$, starting at $\mathbf{x}_0$, along n linearly independent directions $\{\mathbf{d}_0, \mathbf{d}_1, \ldots, \mathbf{d}_{n-1}\}$ having the property of being mutually conjugate with respect to the matrix $\mathbf{A}$, that is,

$$\mathbf{d}_i^T\mathbf{A}\mathbf{d}_j = 0 \quad \text{for } i, j \in [0, n-1] \text{ and } i \neq j \tag{5.82}$$

Since α_k minimizes q in the direction $\mathbf{d}_k$:

$$\min_{\alpha \geq 0} q(\mathbf{x}_k + \alpha\mathbf{d}_k) = q(\mathbf{x}_k + \alpha_k\mathbf{d}_k) = q(\mathbf{x}_{k+1}),$$

we have for all $k \in [0, n-1]$ as follows,

$$\mathbf{d}_k^T\nabla q(\mathbf{x}_{k+1}) = \mathbf{d}_k^T(\mathbf{A}\mathbf{x}_{k+1} + \mathbf{b}) = 0$$

or,

$$\mathbf{d}_k^T\mathbf{A}(\mathbf{x}_k + \alpha_k\mathbf{d}_k) + \mathbf{d}_k^T\mathbf{b} = 0$$

which yields to

$$\alpha_k = -\frac{\mathbf{d}_k^T(\mathbf{A}\mathbf{x}_k + \mathbf{b})}{\mathbf{d}_k^T\mathbf{A}\mathbf{d}_k} \tag{5.83}$$

Note that the denominator of the above expression is a positive number.

Further, noticing that

$$\mathbf{x}_k = \mathbf{x}_0 + \sum_{j=0}^{k-1}\alpha_j\mathbf{d}_j \tag{5.84}$$

we can write

$$\mathbf{d}_k^T\mathbf{A}\mathbf{x}_k = \mathbf{d}_k^T\mathbf{A}\mathbf{x}_0 + \sum_{j=0}^{k-1}\alpha_j\mathbf{d}_k^T\mathbf{A}\mathbf{d}_j = \mathbf{d}_k^T\mathbf{A}\mathbf{x}_0$$

and hence α_k can be expressed as

$$\alpha_k = -\frac{\mathbf{d}_k^T(\mathbf{A}\mathbf{x}_0 + \mathbf{b})}{\mathbf{d}_k^T\mathbf{A}\mathbf{d}_k} \tag{5.85}$$

We can now state the following characteristic property of all conjugate direction methods.

Proposition 5.23 (Minoux)

For all $k \in [0, n]$, the point

$$\mathbf{x}_k = \mathbf{x}_0 + \sum_{j=0}^{k-1} \alpha_j \mathbf{d}_j$$

is the optimum of $q(\mathbf{x})$ in the k-dimensional subspace $\mathscr{V}_k$ generated by $\{\mathbf{d}_0, \mathbf{d}_1, \ldots, \mathbf{d}_{n-1}\}$ and passing $\mathbf{x}_0$.

In particular,

$$\mathbf{x}_n = \mathbf{x}_0 + \sum_{j=0}^{n-1} \alpha_j \mathbf{d}_j \tag{5.86}$$

is the optimum of $q(\mathbf{x})$ in $\mathbb{R}^n$, where the necessary condition for optimum

$$\nabla q(\mathbf{x}_n) = \mathbf{A}\mathbf{x}_n + \mathbf{b} = 0$$

is satisfied.

5.6.4 Conjugate Gradient Method

The conjugate gradient method belongs to the class of conjugate direction methods.

Consider the function to be minimized as quadratic of the form

$$q(\mathbf{x}) = \frac{1}{2}\mathbf{x}^T \mathbf{A}\mathbf{x} + \mathbf{b}^T \mathbf{x} + c.$$

At each iteration k, the direction $\mathbf{d}_k$ is obtained by linearly combining the negative gradient at $\mathbf{x}_k$, $-\nabla q(\mathbf{x}_k)$ and the previous directions $\mathbf{d}_0, \mathbf{d}_1, \ldots, \mathbf{d}_{k-1}$ in such a way that $\mathbf{d}_k$ is conjugate to all previous directions.

Denoting $\mathbf{g}_k = \nabla q(\mathbf{x}_k)$, the gradient of q at $\mathbf{x}_k$, we get the following algorithm.

Algorithm

(a) Initialization: Choose the starting point $\mathbf{x}_0$; $\mathbf{g}_0 = \nabla q(\mathbf{x}_0) = \mathbf{A}\mathbf{x}_0 + \mathbf{b}$;
Set $\mathbf{d}_0 = -\mathbf{g}_0$, $k = 0$.

(b) At step k : Define $\mathbf{x}_{k+1} = \mathbf{x}_k + \alpha_k \mathbf{d}_k$

with $\alpha_k = -\dfrac{\mathbf{g}_k^T \mathbf{d}_k}{\mathbf{d}_k^T \mathbf{A}\mathbf{d}_k}$;

Then, $\mathbf{d}_{k+1} = -\mathbf{g}_{k+1} + \beta_k \mathbf{d}_k$

with $\beta_k = \dfrac{\mathbf{g}_{k+1}^T \mathbf{A}\mathbf{d}_k}{\mathbf{d}_k^T \mathbf{A}\mathbf{d}_k}$.

(c) Stopping condition: If satisfied, End;
Otherwise, Set $k \leftarrow k+1$; Return to (b).

The algorithm converges in at most n iterations.

The following theorem given for the quadratic function paves the way for developing conjugate gradient methods for arbitrary functions.

Theorem 5.24

At kth iteration of the conjugate gradient method, denoting $\mathbf{g}_k = \nabla q(\mathbf{x}_k)$, the gradient of $q(\mathbf{x}) = \frac{1}{2}\mathbf{x}^T\mathbf{A}\mathbf{x} + \mathbf{b}^T\mathbf{x} + c$ evaluated at $\mathbf{x}_k$, we have

(i) $$\alpha_k = \frac{\mathbf{g}_k^T \mathbf{g}_k}{\mathbf{d}_k^T \mathbf{A} \mathbf{d}_k} \tag{5.87}$$

(ii) $$\beta_k = \frac{\mathbf{g}_{k+1}^T (\mathbf{g}_{k+1} - \mathbf{g}_k)}{\mathbf{g}_k^T \mathbf{g}_k} \tag{5.88}$$

$$= \frac{\mathbf{g}_{k+1}^T \mathbf{g}_{k+1}}{\mathbf{g}_k^T \mathbf{g}_k} \tag{5.89}$$

(iii) the directions $\mathbf{d}_0, \mathbf{d}_1, \ldots, \mathbf{d}_{k+1}$ generated by the method are mutually conjugate.

5.6.5 Method of Fletcher and Reeves

The method of Fletcher and Reeves is an extension of the conjugate gradient method for an arbitrary function $f(\mathbf{x})$.

Algorithm

(a) Initialization: Choose the starting point $\mathbf{x}_0$; Set $\mathbf{d}_0 = -\nabla f(\mathbf{x}_0)$, $k = 0$.

(b) At step k: Find α_k by one-dimensional minimization of $s(\alpha) = f(\mathbf{x}_k + \alpha \mathbf{d}_k)$;

Set $\mathbf{x}_{k=1} = \mathbf{x}_k + \alpha_k \mathbf{d}_k$ and $\mathbf{d}_{k+1} = -\nabla f(\mathbf{x}_{k+1}) + \beta_k \mathbf{d}_k$

with $\beta_k = \dfrac{\|\nabla f(\mathbf{x}_{k+1})\|^2}{\|\nabla f(\mathbf{x}_k)\|^2}$.

(c) Stopping condition: If satisfied, End.
Otherwise, Set $k \leftarrow k+1$; Return to (b).

A variant of the above method is the method of Polak and Ribière, where β_k is computed as

$$\beta_k = \frac{[\nabla f(\mathbf{x}_{k+1})]^T [\nabla f(\mathbf{x}_{k+1}) - \nabla f(\mathbf{x}_k)]}{\|\nabla f(\mathbf{x}_k)\|^2}.$$

The two methods are identical for the case of any quadratic function (refer to (5.88) and (5.89)), but when applied for an arbitrary function, the methods generally produce different sequences of direction vectors. It is important to note that to ensure global convergence of the methods, it may be necessary to restart the algorithms periodically after (say) n iterations. The global convergence of the restarted methods follows from the global convergence of the steepest descent method.

5.6.6 Newton's Method

We assume that the function f is twice continuously differentiable and that all first and second derivatives can be computed.

Further, assuming that at step k, the function f can be approximated by a quadratic function q in the neighbourhood of $\mathbf{x}_k$,

$$f(\mathbf{x}) \simeq q(\mathbf{x}) = f(\mathbf{x}_k) + [\nabla f(\mathbf{x}_k)]^T (\mathbf{x} - \mathbf{x}_k) + \frac{1}{2}(\mathbf{x} - \mathbf{x}_k)^T [\nabla^2 f(\mathbf{x}_k)](\mathbf{x} - \mathbf{x}_k) \tag{5.90}$$

we look for a minimum of $q(\mathbf{x})$.

When $\nabla^2 f(\mathbf{x}_k)$ is a positive definite matrix and $q(\mathbf{x})$ is a strictly convex function, a unique minimum $\mathbf{x}_{k+1}$ is given by $\nabla q(\mathbf{x}_{k+1}) = \mathbf{0}$. This leads to the equation

$$\nabla f(\mathbf{x}_k) = -[\nabla^2 f(\mathbf{x}_k)](\mathbf{x}_{k+1} - \mathbf{x}_k)$$

and the recurrence relation

$$\mathbf{x}_{k+1} = \mathbf{x}_k - [\nabla^2 f(\mathbf{x}_k)]^{-1} \nabla f(\mathbf{x}_k) \tag{5.91}$$

When applied to a strictly convex quadratic function, the method converges to the optimum, starting at any point, in a single step. However, the global convergence of the method is not guaranteed for any arbitrary function. In a modification of the Newton's method, we write the recurrence formula as

$$\mathbf{x}_{k+1} = \mathbf{x}_k - \alpha_k [\nabla^2 f(\mathbf{x}_k)]^{-1} \nabla f(\mathbf{x}_k) \tag{5.92}$$

where α_k is determined by one-dimensional optimization in the direction of $\mathbf{d}_k = -[\nabla^2 f(\mathbf{x}_k)]^{-1} \nabla f(\mathbf{x}_k)$.

Further, when the Hessian is not positive definite, we perturb the Hessian slightly so that a descent direction is obtained again for one-dimensional optimization, and we have the recurrence formula,

$$\mathbf{x}_{k+1} = \mathbf{x}_k - \alpha_k [\mu_k \mathbf{I} + \nabla^2 f(\mathbf{x}_k)]^{-1} \nabla f(\mathbf{x}_k), \quad \mu_k > 0 \tag{5.93}$$

5.6.7 Quasi-Newton Methods

The major drawback of the Newton's method is the necessity for computation of the second derivative at every point of iteration and the requirement that the Hessian matrix be positive definite. In the quasi-Newton methods, we replace the displacement direction $\mathbf{d}_k = -[\nabla^2 f(\mathbf{x}_k)]^{-1} \nabla f(\mathbf{x}_k)$ by the direction $\mathbf{d}_k = -\mathbf{H}_k \nabla f(\mathbf{x}_k)$, where $\mathbf{H}_k$ is a positive definite matrix iteratively computed by a correction formula, and then apply one-dimensional optimization. Thus, we get an iterative formula as follows,

$$\mathbf{x}_{k+1} = \mathbf{x}_k - \alpha_k \mathbf{H}_k \nabla f(\mathbf{x}_k) \tag{5.94}$$

These methods are also known as the Variable Metric methods.

There are many choices for computing the matrix $\mathbf{H}_k$. In general, we impose the condition

$$\mathbf{H}_k\left[\nabla f(\mathbf{x}_k)-\nabla f(\mathbf{x}_{k-1})\right]=\mathbf{x}_k-\mathbf{x}_{k-1} \tag{5.95}$$

We use the correction formula

$$\mathbf{H}_{k+1}=\mathbf{H}_k+\Delta_k \tag{5.96}$$

where $\Delta_k=\lambda_k\mathbf{u}_k\mathbf{u}_k^T$, $\text{rank}(\Delta_k)=1$, and find λ_k, $\mathbf{u}_k$ such that

$$\mathbf{H}_{k+1}\left[\nabla f(\mathbf{x}_{k+1})-\nabla f(\mathbf{x}_k)\right]=\mathbf{x}_{k+1}-\mathbf{x}_k.$$

Thus, we obtain the following correction formula

$$\mathbf{H}_{k+1}=\mathbf{H}_k+\frac{(\delta_k-\mathbf{H}_k\gamma_k)(\delta_k-\mathbf{H}_k\gamma_k)^T}{\gamma_k^T(\delta_k-\mathbf{H}_k\gamma_k)} \tag{5.97}$$

where $\delta_k=\mathbf{x}_{k+1}-\mathbf{x}_k$ and $\gamma_k=\nabla f(\mathbf{x}_{k+1})-\nabla f(\mathbf{x}_k)$.

Notice that if the matrix $\mathbf{H}_0$ is chosen to be symmetric, then the above correction formula preserves the symmetry of the matrices $\mathbf{H}_k$. Further, when applied to a quadratic function f, $\mathbf{H}_k$ converges to the inverse of the Hessian of f in at most n iterations.

However, the method can show numerical instability when the term $\gamma_k^T(\delta_k-\mathbf{H}_k\gamma_k)$ is very small or close to zero. We may set $\mathbf{H}_{k+1}=\mathbf{H}_k$ in such case, but then the convergence of $\mathbf{H}_k$ to the inverse of the Hessian is no longer guaranteed. Another problem is that $\mathbf{H}_k$ may not be positive definite even when $\mathbf{H}_0$ is positive definite and the Hessian of f is positive definite.

5.6.8 Method of Davidon–Fletcher–Powell

The Variable Metric methods that use a correction formula $\mathbf{H}_{k+1}=\mathbf{H}_k+\Delta_k$, where $\text{rank}(\Delta_k)=2$, do not have the drawbacks of the formula that uses a correction matrix Δ_k of rank 1.

The method of Davidon–Fletcher–Powell (DFP) uses the correction formula

$$\mathbf{H}_{k+1}=\mathbf{H}_k+\frac{\delta_k\delta_k^T}{\delta_k^T\gamma_k}-\frac{\mathbf{H}_k\gamma_k\gamma_k^T\mathbf{H}_k}{\gamma_k^T\mathbf{H}_k\gamma_k} \tag{5.98}$$

Algorithm

(a) Initialization: Choose the starting point $\mathbf{x}_0$, any positive definite $\mathbf{H}_0$; Set $k=0$.

(b) At step k: Find the direction $\mathbf{d}_k=-\mathbf{H}_k\nabla f(\mathbf{x}_k)$; Find $\mathbf{x}_{k+1}$ as the minimum of $f(\mathbf{x}_k+\alpha\mathbf{d}_k)$, Set $\delta_k=\mathbf{x}_{k+1}-\mathbf{x}_k$; Compute $\gamma_k=\nabla f(\mathbf{x}_{k+1})-\nabla f(\mathbf{x}_k)$,

$$\mathbf{H}_{k+1}=\mathbf{H}_k+\frac{\delta_k\delta_k^T}{\delta_k^T\gamma_k}-\frac{\mathbf{H}_k\gamma_k\gamma_k^T\mathbf{H}_k}{\gamma_k^T\mathbf{H}_k\gamma_k}.$$

(c) Stopping condition: If satisfied, End;
Otherwise, Set $k \leftarrow k+1$; Return to (b).

When applied to a quadratic function with positive definite Hessian **A**, the above algorithm generates the directions $\delta_0, \delta_1, \ldots, \delta_k$ which satisfy

$$\delta_i^T \mathbf{A} \delta_j = 0 \quad \text{for } 0 \le i < j \le k \tag{5.99}$$

and

$$\mathbf{H}_{k+1} \mathbf{A} \delta_i = \delta_i \quad \text{for } 0 \le i \le k \tag{5.100}$$

As a consequence of the above results, the algorithm generates the directions $\delta_0, \delta_1, \ldots, \delta_k$ which are mutually conjugate, and $\mathbf{H}_k$ converges to the inverse of the Hessian in at most n steps. Further, if the condition $\delta_k^T \gamma_k > 0$ holds and $\mathbf{H}_k$ is positive definite, then the matrix $\mathbf{H}_{k+1}$ given by the correction formula is positive definite. In this way, a direction of descent is always generated.

In order to ensure the global convergence of the method for an arbitrary function, the algorithm needs to be periodically restarted. The global convergence of the restarted method is followed from the global convergence of the steepest method.

A variant of the above method is the method of Broyden-Fletcher-Goldfarb-Shanno (BFGS) which uses the correction formula

$$\mathbf{H}_{k+1} = \mathbf{H}_k + \left[1 + \frac{\delta_k^T \mathbf{H}_k \delta_k}{\delta_k^T \gamma_k} \right] \frac{\delta_k \delta_k^T}{\delta_k^T \gamma_k} - \frac{\delta_k \gamma_k^T \mathbf{H}_k + \mathbf{H}_k \gamma_k \delta_k^T}{\delta_k^T \gamma_k} \tag{5.101}$$

The BFGS method has all the desirable properties of the DFP method. In fact, the convergence properties of the DFP method are rather sensitive to the accuracy of one-dimensional optimizations, whereas the BFGS method is found to be less sensitive to lack of accuracy in one-dimensional searches.

5.7 Mathematical Programming: Constrained Optimization – Primal Methods

Most of the constrained optimization techniques belong to one of the two broad families:

Primal methods which operate directly on the given or primal problem, generate a sequence of feasible solutions with monotone decrease (minimization problem) of the objective function. In general, these methods are difficult to implement, and the property of global convergence is difficult to achieve.

Dual methods which reduce the given problem to a sequence of unconstrained optimization problems, are robust in implementations, and their global convergence is easily obtained. However, these methods produce a feasible solution only at the end of the iterative process.

5.7.1 Method of Changing Variables

Various inequality and equality constraints can be handled by change of variables.

For instance, a constraint of the form

$$x \le c$$

can be transformed as

$$x = c - y^2,$$

where y is an unconstrained variable.

Similarly, a bounding constraint

$$a \le x \le b$$

can be transformed as

$$x = a + (b - a)\sin^2 y,$$

and an unconstrained optimization is carried out with respect to y.

In the similar fashion, at least in theory, when m equality constraints $g_i(\mathbf{x}) = 0$ ($i = 1, \ldots, m$) are given, it is possible to eliminate m dependent variables by the implicit functions

$$x_j = \psi_i\left(x_1, \ldots, x_{j-1}, x_{j+1}, \ldots, x_n\right) \tag{5.102}$$

and then, carry out unconstrained optimization of the objective function with respect to $(n - m)$ independent variables. However, this approach may not be applicable in practice due to the difficulty of finding such functions ψ_i.

5.7.2 Methods of Feasible Directions

The main difference between an unconstrained optimization technique and a constrained optimization technique is that in the constrained case, we start with a feasible point $\mathbf{x}_0$ belonging to the feasible set of solutions $\mathscr{S}$, and choose a feasible direction of displacement $\mathbf{d}$ such that a small step along $\mathbf{d}$ does not lead to a point outside $\mathscr{S}$ and the objective function decreases (minimization problem) strictly.

When the constraints are given in the form

$$\begin{aligned} g_i(\mathbf{x}) &\le 0, \quad i \in I_1 \\ g_i(\mathbf{x}) &= 0, \quad i \in I_2 \end{aligned}$$

starting at some non-feasible point, we minimize the scalar function

$$z = \sum_{i \in I_1'} g_i(\mathbf{x}) + \sum_{i \in I_2} \left[g_i(\mathbf{x})\right]^2 \tag{5.103}$$

where $I_1' \subseteq I_1$, is the subset of subscripts of the inequality constraints which are violated at the current point. In order to find a feasible point, the function z is reduced to zero by a suitable algorithm.

Consider the constrained optimization problem

$$
(\text{CP}) \quad \begin{cases} \text{Minimize } f(\mathbf{x}),\ \mathbf{x} \in \mathbb{R}^n \\ \text{subject to} \\ g_i(\mathbf{x}) \le 0,\ i \in I_1 \\ g_i(\mathbf{x}) = 0,\ i \in I_2 \end{cases}
$$

We start from a feasible point $\mathbf{x}_0 \in \mathcal{S} = \{\mathbf{x} \mid g_i(\mathbf{x}) \le 0,\ i \in I_1 \ \&\ g_i(\mathbf{x}) = 0,\ i \in I_2\}$ and choose a direction of displacement $\mathbf{d}$ that is feasible.

Let $I_0 = \{i \in I \mid g_i(\mathbf{x}_0) = 0\}$ be the collection of subscripts of the constraints which are saturated at $\mathbf{x}_0$. Then, the direction $\mathbf{d}$ must be such that a small step in that direction does not make any saturated constraint positive, that is

$$
\left[\nabla g_i(\mathbf{x}_0)\right]^T \mathbf{d} \le 0 \quad \text{for } i \in I_0 \tag{5.104}
$$

Moreover, a small displacement in the direction $\mathbf{d}$ must decrease the objective function f,

$$
\left[\nabla f(\mathbf{x}_0)\right]^T \mathbf{d} < 0 \tag{5.105}
$$

This leads to the equivalent problem

$$
(\text{EP}) \quad \begin{cases} \text{Minimize } \left[\nabla f(\mathbf{x}_0)\right]^T \mathbf{d} \\ \text{subject to} \\ g_i(\mathbf{x}_0) + \left[\nabla g_i(\mathbf{x}_0)\right]^T \mathbf{d} \le 0,\ i \in I \\ \mathbf{d}^T \mathbf{d} \le \alpha \end{cases}
$$

where all the constraints are approximated by their first order Taylor expansion to ensure that no constraint is violated by the displacement $\mathbf{d}$, and α is chosen sufficiently small to have a small step size.

Incidentally, we can notice that if for a saturated non-linear constraint $i \in I_0$, the displacement direction $\mathbf{d}$ satisfies $\left[\nabla g_i(\mathbf{x}_0)\right]^T \mathbf{d} = 0$, then it is possible to have the solution of (EP) leading out of the solution set $\mathcal{S}$ as illustrated in Figure 5.9. We need a special procedure to return into $\mathcal{S}$ after each such movement.

When several constraints are violated simultaneously, the return direction is given by

$$
\mathbf{r} = -\sum_{i \in I'} \lambda_i \nabla g_i(\mathbf{x}) \tag{5.106}
$$

where I' is the subset of subscripts of the constraints that are violated at the current point and λ_i are the positive weighting factors which are proportional to the amounts by which the constraints are violated.

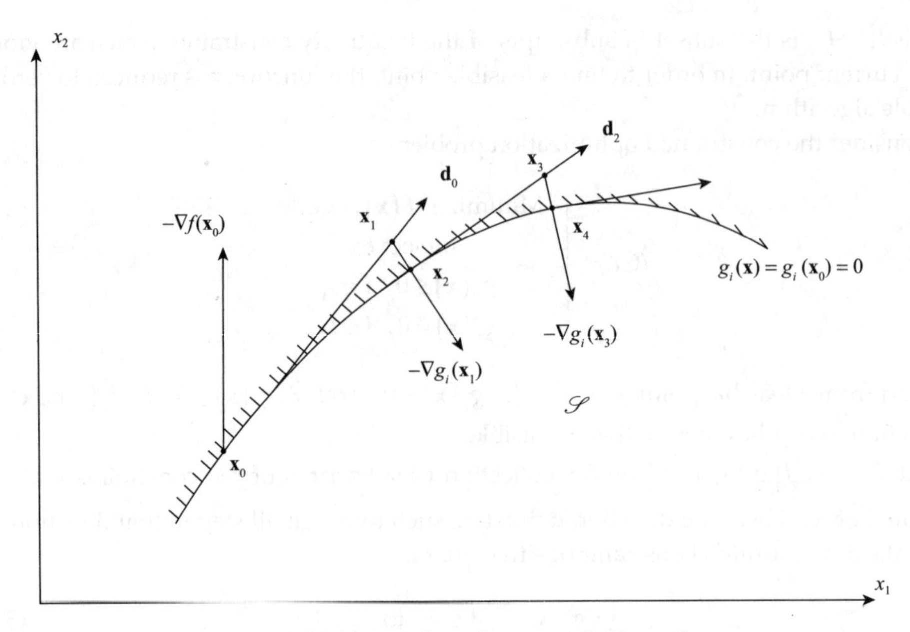

FIGURE 5.9 Moving along the constraint boundary

The modified problem as given below

$$\text{(MP)} \begin{cases} \text{Minimize } \xi \\ \text{subject to} \\ \left[\nabla f(\mathbf{x}_0)\right]^T \mathbf{d} \le \xi \\ g_i(\mathbf{x}_0) + \left[\nabla g_i(\mathbf{x}_0)\right]^T \mathbf{d} \le u_i \xi,\ u_i > 0,\, i \in I \end{cases}$$

requires that the displacement direction $\mathbf{d}$ lead away from the boundary of the feasible region $\mathscr{S}$. If the optimum solution of (MP) corresponds to a value $\xi^* < 0$, then $\left[\nabla f(\mathbf{x}_0)\right]^T \mathbf{d}^* < \mathbf{0}$, hence $\mathbf{d}^*$ is a direction of descent. Moreover, $g_i(\mathbf{x}_0) + \left[\nabla g_i(\mathbf{x}_0)\right]^T \mathbf{d}^* \le u_i \xi^* < 0$, which ensures that no constraint is violated by the displacement.

The parameters u_i are the normalization coefficients. The values are chosen depending on the curvature of the functions g_i at $\mathbf{x}_0$ (hence on the eigenvalues of the Hessians at $\mathbf{x}_0$). When this information is not available, set all $u_i = 1$.

5.7.3 The Gradient Projection Method

The idea is that starting from a feasible point $\mathbf{x}_0$, we choose a displacement direction $\mathbf{d}$ which decreases f along the boundaries of the active (saturated) constraints, thus making sure that the solution of (CP) belongs to the feasible region $\mathscr{S}$.

Let $I_0 = \{i_1, \ldots, i_l\}$ be the collection of the subscripts of the constraints which are saturated at $\mathbf{x}_0$. Assuming that the constraints surface normal vectors $\{\nabla g_i(\mathbf{x}_0),\ i \in I_0\}$ constitute a linearly independent set, we form the spanning matrix $\mathbf{G}$ given by

$$\mathbf{G} = \left[\nabla g_{i_1}(\mathbf{x}_0), \ldots, \nabla g_{i_l}(\mathbf{x}_0)\right] \tag{5.107}$$

In this case, we look for a displacement direction **d** which is tangential to the boundaries of the active constraints, that is

$$\mathbf{G}^T\mathbf{d} = \mathbf{0} \tag{5.108}$$

Thus, the second equivalent problem becomes

$$(\text{EP}')\quad \begin{cases} \text{Minimize } \left[\nabla f(\mathbf{x}_0)\right]^T \mathbf{d} \\ \text{subject to} \\ \mathbf{G}^T\mathbf{d} = \mathbf{0} \\ \|\mathbf{d}\| = 1 \end{cases}$$

In order to find the solution of (EP′), the orthogonal projection matrix **P** given by

$$\mathbf{P} = \mathbf{I} - \mathbf{G}\left(\mathbf{G}^T\mathbf{G}\right)^{-1}\mathbf{G}^T \tag{5.109}$$

is to be computed, and the displacement direction $\bar{\mathbf{d}}$ is then expressed as

$$\bar{\mathbf{d}} = -\mathbf{P}\left[\nabla f(\mathbf{x}_0)\right] \tag{5.110}$$

where $\nabla f(\mathbf{x}_0)$ is the gradient of the objective function evaluated at $\mathbf{x}_0$. The displacement direction **d** is computed as $\mathbf{d} = \bar{\mathbf{d}}/\|\bar{\mathbf{d}}\|$.

5.7.4 The Reduced Gradient Method

The reduced gradient method works in the similar principle of the gradient projection method, but with better efficiency. The methods are similar in the sense that in both the methods, we choose a feasible displacement direction along the boundary of the feasible region.

Consider the constrained optimization problem

$$(\text{CP}')\quad \begin{cases} \text{Minimize } f(\mathbf{x}),\ \mathbf{x} \in \mathbb{R}^n \\ \text{subject to} \\ \mathbf{g}(\mathbf{x}) = \left[g_1(\mathbf{x}), \ldots, g_m(\mathbf{x})\right]^T = \mathbf{0} \\ \mathbf{x} \geq 0 \end{cases}$$

We start from a feasible point $\mathbf{x}_0$ such that $\mathbf{g}(\mathbf{x}) = \mathbf{0}$ and $\mathbf{x}_0 \geq 0$.

Let $J = \{1, \ldots, n\}$ be the set of the subscripts of the variables. A subset of the variables with subscripts $J_B = \{j_1, \ldots, j_m\} \subset J$ constitute a basis if and only if the Jacobian matrix

$$\left[\frac{\partial \mathbf{g}^T}{\partial \mathbf{x}_B}(\mathbf{x}_0)\right] \text{ is regular (non-singular),}$$

where $\mathbf{x}_B = \left[x_{j_1}, \ldots, x_{j_m}\right]^T$.

Now, the vector $\mathbf{x}$ can be partitioned as

$$\mathbf{x}^T = \left[\mathbf{x}_B^T \quad \mathbf{x}_N^T\right]$$

where $\mathbf{x}_B$ comprises of the basic (dependent) variables and $\mathbf{x}_N$ comprises of the non-basic (independent) variables. Thus, the subset of the subscripts $J_N = J - J_B$ relate to the non-basic variables.

Then, in order to make an infinitesimal displacement $\mathbf{dx}^T = \left[\mathbf{dx}_B^T \quad \mathbf{dx}_N^T\right]$ compatible with the constraints $\mathbf{g}(\mathbf{x}) = \mathbf{0}$, we must have

$$\begin{aligned} \mathbf{dg} &= \left(\frac{\partial \mathbf{g}^T}{\partial \mathbf{x}_B}\right)^T \mathbf{dx}_B + \left(\frac{\partial \mathbf{g}^T}{\partial \mathbf{x}_N}\right)^T \mathbf{dx}_N = \mathbf{0} \\ \text{or,} \quad \mathbf{dx}_B &= -\left(\frac{\partial \mathbf{g}^T}{\partial \mathbf{x}_B}\right)^{-T} \left(\frac{\partial \mathbf{g}^T}{\partial \mathbf{x}_N}\right)^T \mathbf{dx}_N \end{aligned} \tag{5.111}$$

where the Jacobian matrices are evaluated at $\mathbf{x}_0$.

Therefore, the variation df of the objective function f for the displacement $\mathbf{dx}^T = \left[\mathbf{dx}_B^T \quad \mathbf{dx}_N^T\right]$ is given by

$$\begin{aligned} df &= \left[\left(\frac{\partial f}{\partial \mathbf{x}_N}\right)^T - \left(\frac{\partial f}{\partial \mathbf{x}_B}\right)^T \left(\frac{\partial \mathbf{g}^T}{\partial \mathbf{x}_B}\right)^{-T} \left(\frac{\partial \mathbf{g}^T}{\partial \mathbf{x}_N}\right)^T\right] \mathbf{dx}_N \\ &= \mathbf{u}_N^T \mathbf{dx}_N \end{aligned} \tag{5.112}$$

where the vector $\mathbf{u}_N$ is the reduced gradient, and all derivatives are computed at $\mathbf{x}_0$.

Now, we choose the displacement direction $\mathbf{d}_N$ relative to the independent variables as

$$\text{For all } j \in J_N \quad \begin{cases} d_j = 0 & \text{if } u_j > 0 \text{ and } x_j = 0 \\ d_j = -u_j & \text{otherwise} \end{cases} \tag{5.113}$$

Note that in the subspace of the non-basic variables, if a displacement along the negative reduced gradient makes a variable x_j negative, then we impose $\mathbf{d}_j = 0$ (no displacement) to be compatible with the non-negativity constraints of (CP').

The displacement direction $\mathbf{d}_B$ is given by

$$\mathbf{d}_B = -\left(\frac{\partial \mathbf{g}^T}{\partial \mathbf{x}_B}\right)^{-T} \left(\frac{\partial \mathbf{g}^T}{\partial \mathbf{x}_N}\right)^T \mathbf{d}_N \tag{5.114}$$

If $\mathbf{d}_N = \mathbf{0}$, then $\mathbf{x}_0$ is the optimum point for the solution of (CP').

If $\mathbf{d}_N \neq \mathbf{0}$, then we find $\hat{\theta} \geq 0$ which minimizes $\psi(\theta) = f(\mathbf{x}_0 + \theta\,\mathbf{d})$.

Due to the non-negativity constraints of the variables, we must have

$$x_j + \theta\, d_j \geq 0 \quad \text{for all } j,$$

which restricts θ as

$$\theta \le \theta_{\max} = \min_{d_j<0}\left(-\frac{x_j}{d_j}\right) \tag{5.115}$$

Finding $\hat{\theta}$ that minimizes $\psi(\theta)$ on $[0, \theta_{\max}]$, the point

$$\hat{\mathbf{x}} = \mathbf{x}_0 + \hat{\theta}\,\mathbf{d} \tag{5.116}$$

is obtained, which satisfies the non-negativity constraints ($\hat{\mathbf{x}} \ge \mathbf{0}$). However, $\hat{\mathbf{x}}$ may not satisfy the constraints $\mathbf{g}(\mathbf{x}) = \mathbf{0}$.

To obtain a feasible point $\mathbf{x}_1$, starting from $\hat{\mathbf{x}}^T = \begin{bmatrix}\hat{\mathbf{x}}_B^T & \hat{\mathbf{x}}_N^T\end{bmatrix}$, we solve the non-linear system of equations

$\mathbf{g}(\hat{\mathbf{x}}) = \mathbf{0}$ by the Newton's method,

where $\hat{\mathbf{x}}_N = \mathbf{x}_{N,0} + \hat{\theta}\,\mathbf{d}_N$ remains fixed, and $\hat{\mathbf{x}}_B = \mathbf{x}_{B,0} + \hat{\theta}\,\mathbf{d}_B = \overline{\mathbf{x}}_{B,0}$ is the starting point for the iterations.

This leads to the recurrence relation

$$\overline{\mathbf{x}}_{B,\,k+1} = \overline{\mathbf{x}}_{B,\,k} - \left[\frac{\partial \mathbf{g}^T}{\partial \mathbf{x}_B}(\overline{\mathbf{x}}_k)\right]^{-1} \mathbf{g}(\overline{\mathbf{x}}_k) \tag{5.117}$$

where $\overline{\mathbf{x}}_k^T = \begin{bmatrix}\overline{\mathbf{x}}_{B,\,k}^T & \hat{\mathbf{x}}_N^T\end{bmatrix}$, and k is the iteration number. In most cases, the inverse of the Jacobian matrix is to be computed only once at $\overline{\mathbf{x}}_0^T = \begin{bmatrix}\overline{\mathbf{x}}_{B,0}^T & \hat{\mathbf{x}}_N^T\end{bmatrix}$ for the iteration to converge.

At this stage, two situations can arise:

(i) The Newton's method converges to $\tilde{\mathbf{x}}_B \ge \mathbf{0}$. Then, we set $\mathbf{x}_1^T = \begin{bmatrix}\tilde{\mathbf{x}}_B^T & \hat{\mathbf{x}}_N^T\end{bmatrix}$, and the process is repeated with the same basis.

(ii) If one of the basic variables x_r vanishes or becomes negative after a number of steps, then x_r is replaced, and substituted by a non-basic variable x_s, and the method is restarted from $\mathbf{x}_0$.

5.8 Mathematical Programming: Constrained Optimization – Dual Methods

The common principle of the dual methods is to reduce the original constrained optimization (primal) problem into a sequence of unconstrained optimization problems.

5.8.1 Penalty Function Methods

The methods of penalty function belong to the family of dual methods. Here the constrained optimization problem is converted into a series of unconstrained problems with varied degree of penalty for violating the constraints. The series of unconstrained problems is then solved iteratively.

Consider the constrained optimization problem

$$(\mathrm{P}) \quad \begin{cases} \text{Minimize } f(\mathbf{x}), \ \mathbf{x} \in \mathbb{R}^n \\ \text{subject to} \\ g_i(\mathbf{x}) \le 0 \ (i = 1, \ldots, m) \end{cases}$$

Let the function $h : \mathbb{R} \to \mathbb{R}$ be defined by

$$\begin{aligned} h(y) &= 0 \quad \text{if } y \le 0 \ (y \in \mathbb{R}) \\ h(y) &= \infty \quad \text{if } y > 0 \end{aligned} \tag{5.118}$$

We form the penalty function $H(\mathbf{x})$ as

$$\mathrm{H}(\mathbf{x}) = \sum_{i=1}^{m} h(g_i(\mathbf{x})) \quad \text{for all } \mathbf{x} \tag{5.119}$$

and consider the unconstrained problem

$$(\mathrm{UP}) \quad \text{Minimize } \varphi(\mathbf{x}) = f(\mathbf{x}) + H(\mathbf{x}), \quad \mathbf{x} \in \mathbb{R}^n$$

where the penalty function replaces the constraint by a steep hill ensuring that the solution is always a feasible point.

In fact, if the solution set $\mathscr{S} = \{\mathbf{x} \mid g_i(\mathbf{x}) \le 0 \ (i = 1, \ldots, m)\}$ is non-empty, then the solution of the constrained problem (P) and the solution of the unconstrained problem (UP) will be the same.

In practice, we use a continuous penalty function with continuous derivatives to avoid numerical ill-conditioning.

Let us define the function h as

$$\begin{aligned} h(y) &= 0 \quad \text{for } y \le 0 \\ h(y) &= y^{2M} \text{ for } y > 0, \ M \text{ integer} \end{aligned} \tag{5.120}$$

and form the penalty function H as

$$H(\mathbf{x}) = \sum_{i=1}^{m} \sigma_i h(g_i(\mathbf{x})) = \sum_{i=1}^{m} \sigma_i \left(g_i^+(\mathbf{x})\right)^{2M} \tag{5.121}$$

where $g_i^+(\mathbf{x}) = \max\{0, g_i(\mathbf{x})\}$ and $\sigma_i > 0$ are the weighting factors.

Then, we solve the unconstrained penalized problem

$$(\mathrm{PP}) \quad \text{Minimize } \varphi(\mathbf{x}, r) = f(\mathbf{x}) + rH(\mathbf{x}), \quad \mathbf{x} \in \mathbb{R}^n$$

where $r > 0$ is the penalty coefficient.

The unconstrained problem (PP) is solved for the point $\overline{\mathbf{x}}(r)$ iteratively by setting increasing values of r, the gradual increase of r helps to avoid numerical difficulty when the current point is far away from the desired solution.

5.8.2 Barrier Function Methods

The penalty function methods provide various intermediate solutions $\mathbf{x}_1, \ldots, \mathbf{x}_k$ for the penalty coefficients $r_1, \ldots, r_k$ which do not belong to the solution set $\mathscr{S} = \{\mathbf{x} \mid g_i(\mathbf{x}) \le 0\ (i = 1, \ldots, m)\}$.

Suppose that $\mathscr{S}$ is closed, that the interior of $\mathscr{S}$, $\text{int}(\mathscr{S})$ is not empty, and that each point of the boundary of $\mathscr{S}$ is the limit of a sequence of points belonging to $\text{int}(\mathscr{S})$.

We form the barrier function $B(\mathbf{x})$ as

$$B(\mathbf{x}) = -\sum_{i=1}^{m} \frac{\sigma_i}{g_i(\mathbf{x})} \tag{5.122}$$

where $\sigma_i > 0$ are the weighting factors chosen in such a way that the initial values of $\sigma_i / g_i(\mathbf{x})$ are of the same order of magnitude, and consider the unconstrained optimization problem

$$\text{(BP)} \quad \text{Minimize } \psi(\mathbf{x},t) = f(\mathbf{x}) + tB(\mathbf{x}), \quad \mathbf{x} \in \mathbb{R}^n$$

where $t > 0$ is the barrier coefficient. In this way, the constraints are replaced by a steep-sided ridge as explained below.

The unconstrained problem (BP) is solved for the point $\bar{\mathbf{x}}(t)$ iteratively by setting decreasing values of t. Notice that starting from $\mathbf{x}_0 \in \text{int}(\mathscr{S})$ and minimizing $\psi(\mathbf{x}, t)$, it is impossible to cross the boundary of $\mathscr{S}$, because in the vicinity of the boundary of $\mathscr{S}$, $\psi(\mathbf{x},t) \to \infty$. Thus, the successive points are obtained from the inside of the solution set giving all intermediate solutions as feasible points.

The barrier function methods are very convenient to deal with optimization problems with strongly non-linear constraints.

5.8.3 Classical Lagrangian Methods

These methods are developed based on the duality of the Lagrange function. We present the relevant theory in the sequel.

Consider the optimization problem

$$\text{(P)} \quad \begin{cases} \text{Minimize } f(\mathbf{x}), \ \mathbf{x} \in \mathbb{R}^n \\ \qquad \text{subject to} \\ g_i(\mathbf{x}) \le 0 \ \ (i = 1, \ldots, m) \end{cases}$$

where the functions f and g_i are continuous and differentiable.

We form the Lagrange function given by

$$F(\mathbf{x}, \boldsymbol{\lambda}) = f(\mathbf{x}) + \sum_{i=1}^{m} \lambda_i g_i(\mathbf{x}), \quad \boldsymbol{\lambda} = [\lambda_1, \ldots, \lambda_m]^T$$

If we can find a saddle point $(\bar{\mathbf{x}}, \bar{\boldsymbol{\lambda}})$ of the Lagrange function with $\bar{\boldsymbol{\lambda}} \ge \mathbf{0}$ satisfying

$$(\text{SP}) \begin{cases} F(\bar{\mathbf{x}}, \bar{\lambda}) = \underset{\mathbf{x}}{\text{Min}} \{F(\mathbf{x}, \bar{\lambda})\} \\ g_i(\bar{\mathbf{x}}) \leq 0 \ (i = 1, \ldots, m) \\ \bar{\lambda}_i g_i(\bar{\mathbf{x}}) = 0 \end{cases}$$

then the problem (P) can be solved.

Let us define the dual function $w(\boldsymbol{\lambda})$ for $\boldsymbol{\lambda} \geq \mathbf{0}$ by

$$w(\boldsymbol{\lambda}) = \underset{\mathbf{x}}{\text{Min}} \{F(\mathbf{x}, \boldsymbol{\lambda})\} \tag{5.123}$$

Assume that at every point $\boldsymbol{\lambda} \geq \mathbf{0}$ in the neighbourhood of $\bar{\boldsymbol{\lambda}}$, $w(\boldsymbol{\lambda})$ has a finite value. Then, there exists $\bar{\mathbf{x}}$ such that $w(\boldsymbol{\lambda}) = F(\bar{\mathbf{x}}, \boldsymbol{\lambda})$. However, $\bar{\mathbf{x}}$ will be unique if, for instance, $F(\mathbf{x}, \boldsymbol{\lambda})$ is strictly convex in the neighbourhood of $\bar{\mathbf{x}}$.

In order to find a saddle point, when it exists, we solve the min–max problem given by

$$(\text{D}) \quad \underset{\boldsymbol{\lambda} \geq \mathbf{0}}{\text{Max}}\, w(\boldsymbol{\lambda}) = \underset{\boldsymbol{\lambda} \geq \mathbf{0}}{\text{Max}} \left\{ \underset{\mathbf{x}}{\text{Min}}\, F(\mathbf{x}, \boldsymbol{\lambda}) \right\}$$

Note that the dual function w and the dual problem (D) are defined even if there is no saddle point of $F(\mathbf{x}, \boldsymbol{\lambda})$.

Proposition 5.25 (Minoux)

For all $\lambda_i \geq 0 \ (i = 1, \ldots, m)$, the value of the dual function $w(\boldsymbol{\lambda})$ is a lower bound of the optimum value $f(\mathbf{x}^*)$ of (P), that is

$$w(\boldsymbol{\lambda}) \leq w(\boldsymbol{\lambda}^*) \leq f(\mathbf{x}^*) \tag{5.124}$$

where $w(\boldsymbol{\lambda}^*)$ is the optimal value of (D).

Proposition 5.26

The dual function w is a concave function of $\boldsymbol{\lambda}$.

The above property of the dual function requires no assumption about convexity of the functions f and g_i or convexity of the Lagrange function F in **x**.

Theorem 5.27 **(Duality theorem)**

If the problem (P) has a saddle point $(\mathbf{x}^*, \boldsymbol{\lambda}^*)$, then we have

$$w(\boldsymbol{\lambda}^*) = f(\mathbf{x}^*) \tag{5.125}$$

that is, the optimal value of the primal problem (P) equals to the optimal value of the dual problem (D).

Conversely, if there is a solution $\mathbf{x}^*$ of (P) and $\boldsymbol{\lambda}^* \geq \mathbf{0}$ such that

$$w(\boldsymbol{\lambda}^*) = f(\mathbf{x}^*)$$

then (P) has a saddle point at $(\mathbf{x}^*, \boldsymbol{\lambda}^*)$.

If the problem (P) has no saddle point or the minimum point $\mathbf{x}$ of $F(\mathbf{x}, \bar{\boldsymbol{\lambda}})$ is not unique, then the dual function w is not differentiable at the optimum $\bar{\boldsymbol{\lambda}}$. However, we can use a subgradient technique to solve the problem (D) because w is a concave function in general. In this case, the solution of (D) will be a good approximation of the solution of (P).

Definition

For a convex (concave) differentiable function h, the gradient $\nabla h(\bar{\mathbf{y}})$ satisfies at every point $\bar{\mathbf{y}}$ the fundamental inequality given by

$$h(\mathbf{y}) \geq h(\bar{\mathbf{y}}) + [\nabla h(\bar{\mathbf{y}})]^T (\mathbf{y} - \bar{\mathbf{y}}) \quad (\leq \text{ for concave}) \tag{5.126}$$

For a convex (concave), not everywhere differentiable function, a subgradient at the point $\bar{\mathbf{y}}$ is any vector $\boldsymbol{\gamma} = [\gamma_1, \ldots, \gamma_p]^T \in \mathbb{R}^p$ which satisfies

$$h(\mathbf{y}) \geq h(\bar{\mathbf{y}}) + \boldsymbol{\gamma}^T (\mathbf{y} - \bar{\mathbf{y}}) \quad (\leq \text{ for concave}) \tag{5.127}$$

The concept of subgradient is useful in developing a subgradient technique for optimization of convex (concave) functions which are not everywhere differentiable.

Incidentally, a subgradient $\mathbf{g}(\bar{\mathbf{x}})$ of $w(\bar{\boldsymbol{\lambda}})$ is readily available when we compute the dual function $w(\bar{\boldsymbol{\lambda}}) = f(\bar{\mathbf{x}}) + \boldsymbol{\lambda}^T \mathbf{g}(\bar{\mathbf{x}})$ by minimization.

The method of Uzawa uses a classical gradient technique (subgradient technique in the non-differentiable case) to solve the dual problem.

Algorithm

(a) Initialization: Choose a starting point $\boldsymbol{\lambda}_0 \geq \mathbf{0}$; Set $k = 0$.

(b) At step k: Compute

$$\begin{aligned} w(\boldsymbol{\lambda}_k) &= \underset{\mathbf{x}}{\text{Min}}\{f(\mathbf{x}) + \boldsymbol{\lambda}_k^T \mathbf{g}(\mathbf{x})\} \\ &= f(\mathbf{x}_k) + \boldsymbol{\lambda}_k^T \mathbf{g}(\mathbf{x}_k) \end{aligned}$$

where $\mathbf{g}(\mathbf{x}_k)$ is a subgradient of w at the point $\boldsymbol{\lambda}_k$ (the gradient in the differentiable case);

Compute

$$w(\boldsymbol{\lambda}_k + \rho_k \mathbf{g}(\mathbf{x}_k)) = \underset{\rho \geq 0}{\text{Max}}\{w(\boldsymbol{\lambda}_k + \rho \mathbf{g}(\mathbf{x}_k))\};$$

Define $\boldsymbol{\lambda}_{k+1}$ by

$$\boldsymbol{\lambda}_{k+1} = \underset{\boldsymbol{\lambda} \geq \mathbf{0}}{\text{Proj}}\{\boldsymbol{\lambda}_k + \rho_k \mathbf{g}(\mathbf{x}_k)\}$$

where the projection operator sets $\lambda_i = 0$ when it becomes negative.

(c) Stopping condition: If satisfied, End;
Otherwise, Set $k \leftarrow k+1$, Return to (b).

The method of Arrow-Hurwicz uses a differential gradient technique for finding a saddle point of the Lagrange function $F(\mathbf{x}, \boldsymbol{\lambda}) = f(\mathbf{x}) + \boldsymbol{\lambda}^T \mathbf{g}(\mathbf{x}),\ \boldsymbol{\lambda} \geq \mathbf{0}$ corresponding to the primal problem

$$(\text{P}') \begin{cases} \text{Minimize } f(\mathbf{x}) \\ \quad \text{subject to} \\ \quad \mathbf{g}(\mathbf{x}) \leq \mathbf{0} \\ \quad \mathbf{x} \geq \mathbf{0} \end{cases}$$

If there exists a saddle point $(\bar{\mathbf{x}}, \bar{\boldsymbol{\lambda}})$ for which

$$F(\bar{\mathbf{x}}, \boldsymbol{\lambda}) \leq F(\bar{\mathbf{x}}, \bar{\boldsymbol{\lambda}}) \leq F(\mathbf{x}, \bar{\boldsymbol{\lambda}}),\ \bar{\mathbf{x}}, \bar{\boldsymbol{\lambda}} \geq \mathbf{0}$$

then we have

$$f(\mathbf{x}) \leq f(\mathbf{x}),\ \mathbf{x} \in \mathcal{S}\{\mathbf{x} | \mathbf{g}(\mathbf{x}) \leq \mathbf{0}, \mathbf{x} \geq \mathbf{0}\}$$

In other words, the saddle point problem and the constrained extremum problem are formally equivalent. The following theorem states how to find a saddle point of the Lagrange function, if it exists.

Theorem 5.28 (Arrow-Hurwicz)

Let the Lagrange function $F(\mathbf{x}, \boldsymbol{\lambda})$ be strictly convex in $\mathbf{x}$, concave in $\boldsymbol{\lambda}$, and differentiable in the neighbourhood of $(\bar{\mathbf{x}}, \bar{\boldsymbol{\lambda}})$.

If F has a global saddle point at $(\bar{\mathbf{x}}, \bar{\boldsymbol{\lambda}})$, then starting from a non-negative point $(\mathbf{x}_0, \boldsymbol{\lambda}_0)$, the solution $(\mathbf{x}_k, \boldsymbol{\lambda}_k)$ of the system of recurrence equations

$$x_{j,k+1} = \begin{cases} 0 \ \text{ if } x_{j,k} = 0,\ \dfrac{\partial \mathrm{F}}{\partial x_j}(\mathbf{x}_k, \boldsymbol{\lambda}_k) > 0 \\ x_{j,k} - \alpha_k \dfrac{\partial \mathrm{F}}{\partial x_j}(\mathbf{x}_k, \boldsymbol{\lambda}_k), \ \text{otherwise} \end{cases}$$

$$\lambda_{i,k+1} = \begin{cases} 0 \ \text{ if } \lambda_{i,k} = 0,\ g_i(\mathbf{x}_k) < 0 \\ \lambda_{i,k} + \alpha_k g_i(\mathbf{x}_k), \ \text{otherwise} \end{cases}$$

where the step size $\alpha_k > 0$ is appropriately chosen, converges to $(\bar{\mathbf{x}}, \bar{\boldsymbol{\lambda}})$ as $k \to \infty$.

5.8.4 Generalized Lagrangian Methods

In order to solve the constrained optimization problem (P), we have used the iterative penalty function method where at each iteration a solution of the unconstrained optimization.

$$\underset{\mathbf{x}}{\text{Min}}\ \varphi(\mathbf{x},r)=\underset{\mathbf{x}}{\text{Min}}\left\{f(\mathbf{x})+r\sum_{i=1}^{m}\sigma_i\left[\max\{0,g_i(\mathbf{x})\}\right]^2\right\},\quad r>0 \tag{5.128}$$

is computed. As r tends to ∞, the solution converges to a feasible point. However, for a large value of r, the function $\varphi(\mathbf{x},r)$ becomes ill-conditioned and the unconstrained optimization becomes difficult.

On the other hand, when we use the classical Lagrangian method to solve the problem (P), we form the Lagrange function $F(\mathbf{x},\boldsymbol{\lambda})=f(\mathbf{x})+\sum_{i=1}^{m}\lambda_i g_i(\mathbf{x})$, and at each iteration compute the solution of $\underset{\mathbf{x}}{\text{Min}}\ F(\mathbf{x},\boldsymbol{\lambda})$, $\boldsymbol{\lambda}\geq\mathbf{0}$. The unconstrained optimization here may not have a unique minimum, unless $F(\mathbf{x},\boldsymbol{\lambda})$ is strictly convex in $\mathbf{x}$. In this case, the iterative method can get only an approximate optimum of the problem (P) when the functions f and g_i are non-convex.

The method of Hestenes–Powell combines the above two concepts in the case of equality constraints $\mathbf{g}(\mathbf{x})=\mathbf{0}$ and the generalized Lagrangian is formed as

$$F(\mathbf{x},\boldsymbol{\lambda},r)=f(\mathbf{x})+\sum_{i=1}^{m}\lambda_i g_i(\mathbf{x})+r\sum_{i=1}^{m}\sigma_i\left[g_i(\mathbf{x})\right]^2,\quad r>0 \tag{5.129}$$

The unconstrained optimization $\underset{\mathbf{x}}{\text{Min}}\ F(\mathbf{x},\boldsymbol{\lambda},r)$ for $\boldsymbol{\lambda}$ unrestricted in sign depends on parameters λ_i and r. This problem can generate a suitable sequence of solutions which converges to the optimum $\mathbf{x}^*$ with a moderate value of r when $\boldsymbol{\lambda}$ converges to the optimum of the dual problem $\boldsymbol{\lambda}^*$.

A natural extension of the above method has been proposed in the case of inequality constraints $\mathbf{g}(\mathbf{x})\leq\mathbf{0}$ and the generalized Lagrangian is formed as

$$F(\mathbf{x},\boldsymbol{\lambda},r)=f(\mathbf{x})+\sum_{i=1}^{m}\lambda_i\left[g_i(\mathbf{x})+s_i\right]+r\sum_{i=1}^{m}\sigma_i\left[g_i(\mathbf{x})+s_i\right]^2,\quad r>0 \tag{5.130}$$

where $s_i\geq 0$ are the slack variables.

Minimizing F with respect to s_i, we get

$$\frac{\partial F}{\partial s_i}=\lambda_i+2r\sigma_i\left[g_i(\mathbf{x})+s_i\right]=0$$

$$\Rightarrow\quad s_i=\begin{cases}-g_i(\mathbf{x})-\dfrac{\lambda_i}{2r\sigma_i} & \text{if } g_i(\mathbf{x})\leq-\dfrac{\lambda_i}{2r\sigma_i}\\[2ex] 0 & \text{if } g_i(\mathbf{x})\geq-\dfrac{\lambda_i}{2r\sigma_i}\end{cases} \tag{5.131}$$

Accordingly, the generalized Lagrangian becomes

$$F(\mathbf{x},\boldsymbol{\lambda},r)=f(\mathbf{x})+\sum_{i=1}^{m}\begin{cases}-\dfrac{\lambda_i^2}{4r\sigma_i} & \text{if } g_i(\mathbf{x})\leq-\dfrac{\lambda_i}{2r\sigma_i}\\[2ex] \lambda_i g_i(\mathbf{x})+r\sigma_i\left[g_i(\mathbf{x})\right]^2 & \text{if } g_i(\mathbf{x})\geq-\dfrac{\lambda_i}{2r\sigma_i}\end{cases} \tag{5.132}$$

which is minimized in $\mathbf{x}$ for $\boldsymbol{\lambda}$ unrestricted in sign and $r > 0$.

The method of Nakayama–Sayama–Sawaragi uses the function

$$F(\mathbf{x}, \lambda, t) = f(\mathbf{x}) + \sum_{i=1}^{m} \begin{cases} t\left[g_i(\mathbf{x})\right]^2 + \lambda_i g_i(\mathbf{x}) & \text{if } g_i(\mathbf{x}) \geq 0 \\ \dfrac{\lambda_i^2 g_i(\mathbf{x})}{\lambda_i - t g_i(\mathbf{x})} & \text{if } g_i(\mathbf{x}) < 0 \end{cases} \tag{5.133}$$

which is minimized in $\mathbf{x}$ for $\lambda \geq \mathbf{0}$ and $t > 0$.

The generalized Lagrangian methods are found to be most efficient and fail-safe methods for solving non-convex programming with strongly non-linear objective functions and constraints.

5.9 Signal Processing Applications VI: Adaptive Filters

An adaptive filter is a self-optimizing system which updates its internal design by a recursive algorithm with the availability of new data samples. In the process, an adaptive filter achieves the usual operations of a filter, namely, signal prediction, noise elimination, signal decomposition, signal classification in time-varying environment where the signal or noise statistics may be changing with time.

A tapped-delay-line (transversal) adaptive filter employed to model an unknown dynamic system is shown in Figure 5.10.

An adaptive linear combiner, as shown in Figure 5.11, is the main part of an adaptive filter.

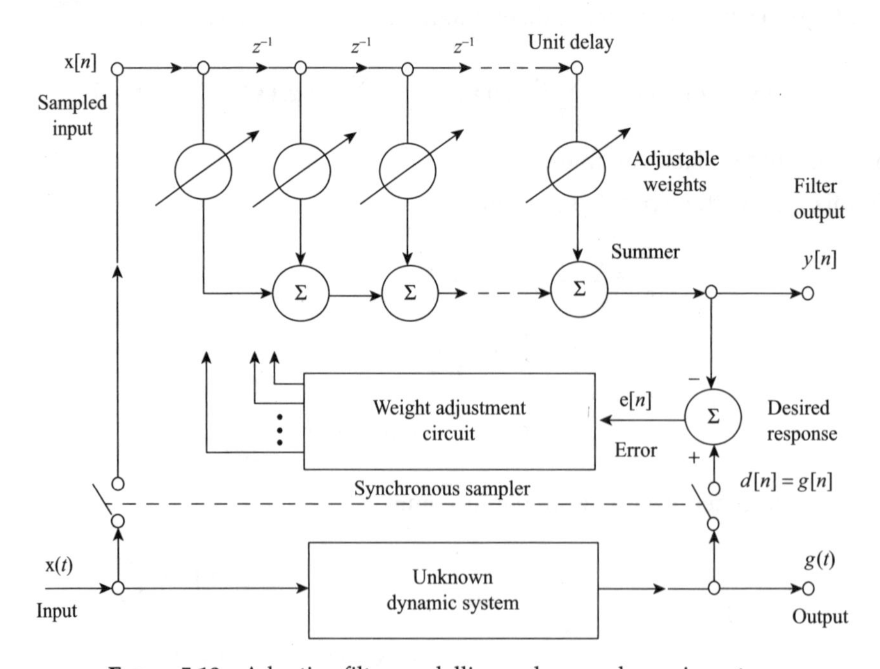

FIGURE 5.10 Adaptive filter modelling unknown dynamic system

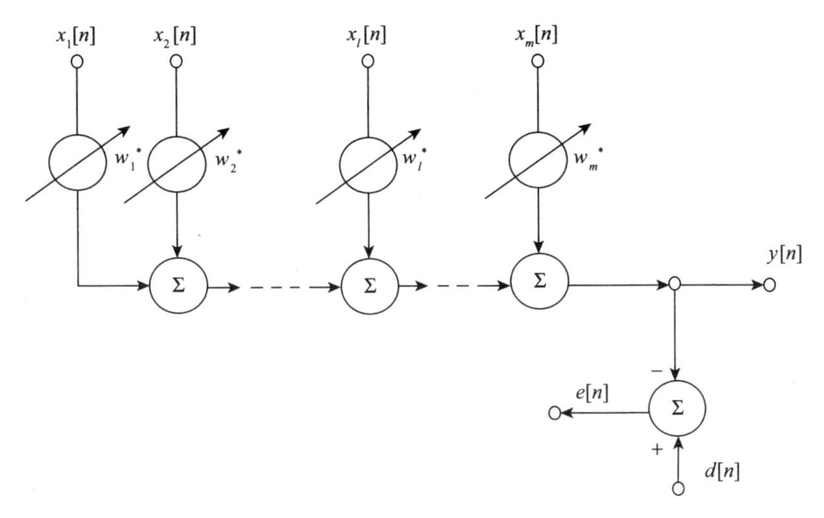

FIGURE 5.11 Adaptive linear combiner

The filter output $y[n]$ is given by

$$\begin{aligned} y[n] &= \sum_{l=1}^{m} w_l^*[n]x_l[n] \\ &= \mathbf{w}^H[n]\mathbf{x}[n] \end{aligned} \tag{5.134}$$

where $\mathbf{w}[n]=[w_1[n],\ldots,w_m[n]]^T$ and $\mathbf{x}[n]=[x_1[n],\ldots,x_m[n]]^T$.

The error signal $e[n]$ is

$$\begin{aligned} e[n] &= d[n]-y[n] \\ &= d[n]-\mathbf{w}^H[n]\mathbf{x}[n] \end{aligned} \tag{5.135}$$

The square magnitude of this error is

$$|e[n]|^2 = |d[n]|^2 - \mathbf{w}^H[n]\mathbf{x}[n]d^*[n] - d[n]\mathbf{x}^H[n]\mathbf{w}[n] + \mathbf{w}^H[n]\mathbf{x}[n]\mathbf{x}^H[n]\mathbf{w}[n] \tag{5.136}$$

and the mean-square error-magnitude (MSE) or the expected value of $|e[n]|^2$ is given by

$$\begin{aligned} E\left\{|e[n]|^2\right\} &= E\left\{|d[n]|^2\right\} - \mathbf{w}^H[n]E\{\mathbf{x}[n]d^*[n]\} - E\{d[n]\mathbf{x}^H[n]\}\mathbf{w}[n] \\ &\quad + \mathbf{w}^H[n]E\{\mathbf{x}[n]\mathbf{x}^H[n]\}\mathbf{w}[n] \end{aligned}$$

or

$$\overline{|e[n]|^2} = \overline{|d[n]|^2} - \mathbf{w}^H[n]\theta_{xd} - \theta_{xd}^H\mathbf{w}[n] + \mathbf{w}^H[n]\Phi_{xx}\mathbf{w}[n] \tag{5.137}$$

where $\theta_{xd} \triangleq E\{\mathbf{x}[n]d^*[n]\}$ is the cross-correlation vector between the input signals and the desired response, and $\Phi_{xx} \triangleq E\{\mathbf{x}[n]\mathbf{x}^H[n]\}$ is the correlation matrix of the input signals.

The gradient of the MSE in the m-dimensional space of the weight variables is expressed as

$$\nabla\left(\overline{|e[n]|^2}\right) = -2\boldsymbol{\theta}_{xd} + 2\boldsymbol{\Phi}_{xx}\mathbf{w}[n] \tag{5.138}$$

where it is assumed that the correlation matrix $\boldsymbol{\Phi}_{xx}$ to be Hermitian and positive definite.

The optimal weight vector that yields the least mean-square (LMS) error-magnitude is obtained by setting the gradient to zero,

$$\mathbf{w}_{\text{opt}} = \boldsymbol{\Phi}_{xx}^{-1}\boldsymbol{\theta}_{xd} \tag{5.139}$$

The above equation is the Wiener–Hopf equation in matrix form. The least mean-square error-magnitude is given by

$$\overline{|e|^2}_{\text{min}} = \overline{|d|^2} - \mathbf{w}_{\text{opt}}^H\boldsymbol{\theta}_{xd} \tag{5.140}$$

where $\overline{|d|^2}$ is the variance of the desired response.

5.9.1 The LMS Adaptation Algorithm

In order to find the optimal weight vector by an iterative approach, we start with an initial weight vector $\mathbf{w}[0]$, and the method of steepest descent is used to modify the weight vector by the relation,

$$\begin{aligned}\mathbf{w}[n+1] &= \mathbf{w}[n] - \alpha_s\nabla\left(\overline{|e[n]|^2}\right)\\ &= \mathbf{w}[n] - 2\alpha_s\boldsymbol{\Phi}_{xx}\mathbf{w}[n] + 2\alpha_s\boldsymbol{\theta}_{xd}\end{aligned} \tag{5.141}$$

where $\alpha_s > 0$ is the fixed step size, sufficiently small.

The feedback flow graph of the method of steepest descent is shown in Figure 5.12.

For the analysis of adaptive transients, we expand the Hermitian and positive definite matrix $\boldsymbol{\Phi}_{xx}$ in normal form,

$$\boldsymbol{\Phi}_{xx} = \mathbf{Q}\boldsymbol{\Lambda}\mathbf{Q}^H \tag{5.142}$$

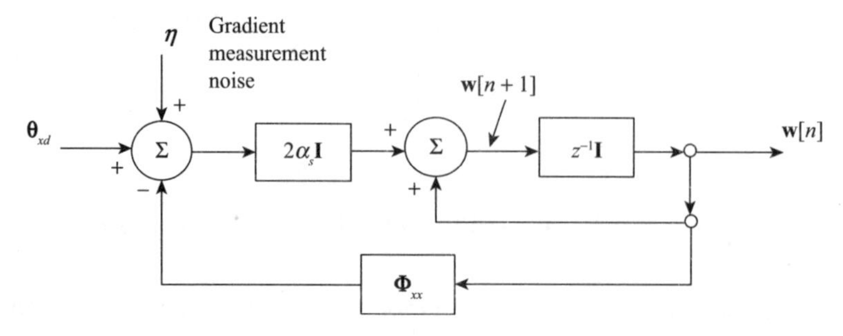

FIGURE 5.12 Feedback model of the method of steepest descent

where $\mathbf{\Lambda}$ is the diagonal matrix of real positive eigenvalues and the modal matrix $\mathbf{Q}$ is the square matrix of orthonormal eigenvectors.

Substituting for $\mathbf{\Phi}_{xx}$ we get the following iterative formula for the transformed weight vector $\mathbf{w}'[n] = \mathbf{Q}^H \mathbf{w}[n]$ in diagonalized form,

$$\mathbf{w}'[n+1] = \mathbf{w}'[n] - 2\alpha_s \mathbf{\Lambda} \mathbf{w}'[n] + 2\alpha_s \mathbf{Q}^H \mathbf{\theta}_{xd} \tag{5.143}$$

Note that all cross-couplings in the feedback paths are now eliminated, and the one-dimensional feedback model for the pth normal weight variable w' can be drawn in the absence of measurement noise as shown in Figure 5.13.

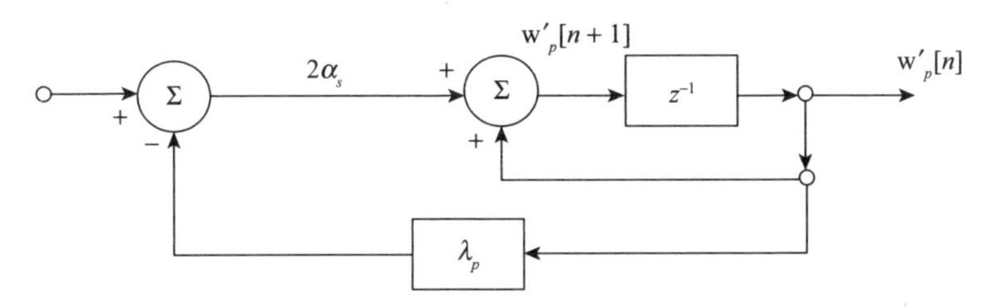

FIGURE 5.13 One-dimensional feedback model

The transfer function of this feedback system is

$$T(z) = \frac{\alpha_s z^{-1}}{1 - \left(1 - 2\alpha_s \lambda_p\right) z^{-1}} \tag{5.144}$$

and the stability of the feedback system is assured when

$$\left|1 - 2\alpha_s \lambda_p\right| < 1$$

Since $\lambda_p > 0$ for all p, the condition of stability can be met for $\alpha_s > 0$ when

$$\left|\alpha_s \lambda_{\max}\right| < 1$$

where $\lambda_{\max}$ is the maximum eigenvalue of $\mathbf{\Phi}_{xx}$.

Thus, the stability condition in the absence of measurement noise becomes

$$0 < \alpha_s < \frac{1}{\lambda_{\max}} \tag{5.145}$$

Note that the rate of adaptation depends on the value of α_s.

We now describe the LMS adaptation algorithm which can be used to find a solution of the Wiener–Hopf equation.

In the implementation of the method of steepest descent for updating the weight vector, we need to obtain an estimate of the gradient of the MSE. We approximate the gradient estimate $\hat{\nabla}\left(\overline{\left|e[n]\right|^2}\right)$ by the gradient of a single time sample of the squared error, that is

$$\hat{\nabla}\left(\overline{|e[n]|^2}\right) = \nabla\left(|e[n]|^2\right) = -2\mathbf{x}[n]e^*[n] \tag{5.146}$$

This estimate of the gradient vector is unbiased because we see that

$$E\left\{\hat{\nabla}\left(\overline{|e[n]|^2}\right)\right\} = -2E\left\{\mathbf{x}[n]\left(d^*[n]-\mathbf{x}^H[n]\mathbf{w}[n]\right)\right\} = -2\left(\boldsymbol{\theta}_{xd} - \boldsymbol{\Phi}_{xx}\mathbf{w}[n]\right)$$

which is the true gradient of the MSE.

Thus, the LMS adaptation formula becomes

$$\mathbf{w}[n+1] = \mathbf{w}[n] + 2\alpha_s\mathbf{x}[n]e^*[n] \tag{5.147}$$

The LMS adaptation process in terms of one weight variable $w_l[n]$ is now shown in Figure 5.14.

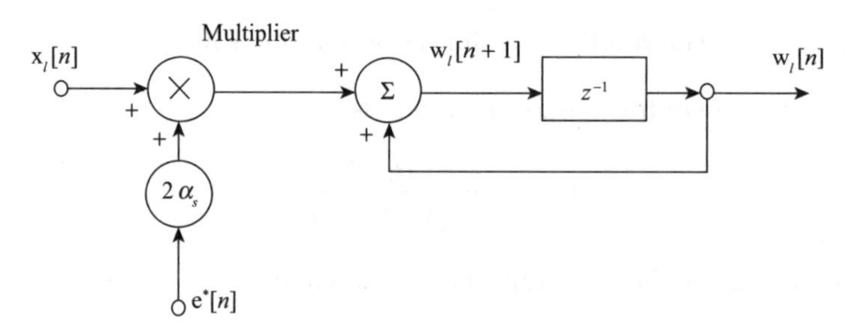

FIGURE 5.14 LMS algorithm in block diagram form

To study the convergence of the statistical mean of the weight vector, we assume that the input vectors $\mathbf{x}[n]$ and $\mathbf{x}[n+1]$ are uncorrelated. Thus, the weight vector $\mathbf{w}[n]$ which depends on $\{\mathbf{x}[n-1], \mathbf{x}[n-2], \ldots, \mathbf{x}[0]\}$ is independent of $\mathbf{x}[n]$.

The expected value of the weight vector $\mathbf{w}[n+1]$ is given by

$$E\{\mathbf{w}[n+1]\} = E\{\mathbf{w}[n]\} + 2\alpha_s E\left\{\mathbf{x}[n]\left(d^*[n]-\mathbf{x}^H[n]\mathbf{w}[n]\right)\right\} = (\mathbf{I}-2\alpha_s\boldsymbol{\Phi}_{xx})E\{\mathbf{w}[n]\} + 2\alpha_s\boldsymbol{\theta}_{xd}$$

Therefore, starting with an initial weight vector $\mathbf{w}[0]$ and going through $(n+1)$ iterations, we get

$$E\{\mathbf{w}[n+1]\} = (\mathbf{I}-2\alpha_s\boldsymbol{\Phi}_{xx})^{n+1}\mathbf{w}[0] + 2\alpha_s\sum_{i=0}^{n}(\mathbf{I}-2\alpha_s\boldsymbol{\Phi}_{xx})^i\boldsymbol{\theta}_{xd} \tag{5.148}$$

Substituting $\boldsymbol{\Phi}_{xx} = \mathbf{Q}\boldsymbol{\Lambda}\mathbf{Q}^H$ and simplifying, the above equation can be written as

$$E\{\mathbf{w}[n+1]\} = \mathbf{Q}(\mathbf{I}-2\alpha_s\Lambda)^{n+1}\mathbf{Q}^H\mathbf{w}[0] + 2\alpha_s\mathbf{Q}\sum_{i=0}^{n}(\mathbf{I}-2\alpha_s\Lambda)^i\mathbf{Q}^H\theta_{xd} \tag{5.149}$$

Note that for the diagonal matrix in (5.149)

$$\lim_{n\to\infty}(\mathbf{I}-2\alpha_s\Lambda)^{n+1} \to \mathbf{0}$$

as long as the diagonal elements are of magnitude less than unity,

$$|1-2\alpha_s\lambda_{\max}| < 1$$

which implies

$$0 < \alpha_s < \frac{1}{\lambda_{\max}}$$

This is the same condition as the condition of stability.

Further, notice that for the summation term in (5.149)

$$\lim_{n\to\infty}\sum_{i=0}^{n}(\mathbf{I}-2\alpha_s\Lambda)^i = \frac{1}{2\alpha_s}\Lambda^{-1} \tag{5.150}$$

which can be shown easily for the pth element of the diagonal matrix as a geometric series,

$$\sum_{i=0}^{\infty}\left(1-2\alpha_s\lambda_p\right)^i = \frac{1}{1-\left(1-2\alpha_s\lambda_p\right)} = \frac{1}{2\alpha_s\lambda_p}$$

Thus in the limit, (5.149) becomes

$$\begin{aligned}\lim_{n\to\infty}E\{\mathbf{w}[n+1]\} &= \mathbf{Q}\Lambda^{-1}\mathbf{Q}^H\,\theta_{xd}\\ &= \Phi_{xx}^{-1}\,\theta_{xd}\end{aligned}$$

Therefore, the expected value of the weight vector converges to the Wiener solution as the number of iterations tends to ∞.

The condition of convergence can be related to the total input signal power by observing that the maximum eigenvalue $\lambda_{\max}$ of $\mathbf{\Phi}_{xx}$ satisfies

$$\lambda_{\max} \le \text{trace}(\Phi_{xx}) = \sum_{l=1}^{m}\lambda_l \tag{5.151}$$

where λ_l are the eigenvalues of Φ_{xx}. Moreover, we have

$$\text{trace}(\Phi_{xx}) = E\{\mathbf{x}^H[n]\mathbf{x}[n]\} = \sum_{l=1}^{m}E\{|x_l|^2\}$$

Thus, the convergence can be assured when the step size $\alpha_s > 0$ is chosen such that

$$\alpha_s < \frac{1}{\sum_{l=1}^{m}E\{|x_l|^2\}} \tag{5.152}$$

where $\sum_{l=1}^{m}E\{|x_l|^2\} = P$ is the total input signal power.

The LMS Algorithm

(a) Initialization: Set $\hat{\mathbf{w}}[0]=\mathbf{0}$; Choose $0<\alpha_s<1/P,\quad P=\text{Total input signal power}$.

(b) At step $n=1,2,\ldots$ Compute

$$e[n-1]=d[n-1]-\hat{\mathbf{w}}^H[n-1]\mathbf{x}[n-1],$$

$$\hat{\mathbf{w}}[n]=\hat{\mathbf{w}}[n-1]+2\alpha_s\mathbf{x}[n-1]e^*[n-1],$$

$$y[n]=\hat{\mathbf{w}}^H[n]\mathbf{x}[n].$$

(c) Stopping condition: If end of data sequence, Stop; Otherwise, Go to (b).

5.9.2 Simulation Study: Example 1 (Adaptive Predictor)

We now include a simple example to demonstrate how the LMS algorithm works to implement an FIR adaptive filter. Suppose that the filter to be designed is an adaptive predictor, that is, by processing an input sequence $x[n]$, the filter is to produce an output sequence $y[n]=x[n+k]$ in real-time, where k is a specified time-lead. Accordingly, we set the desired response $d[n]=x[n+k]$ or effectively, $d[n-k]=x[n]$ which is available at present time [Widrow].

To see how the LMS algorithm can be used in the given situation, examine the recurrence relation for the weight vector

$$\mathbf{w}[n+1]=\mathbf{w}[n]+2\alpha_s\mathbf{x}[n]e^*[n].$$

Now, note that with the availability of $d[n-k]$, we can compute the error signal $e[n-k]$, and accordingly, the iterative formula needs to be modified as

$$\begin{aligned}\mathbf{w}[n+1-k]&=\mathbf{w}[n-k]+2\alpha_s\mathbf{x}[n-k]e^*[n-k]\\&=\mathbf{w}[n-k]+2\alpha_s\mathbf{x}[n-k]\left(d^*[n-k]-\mathbf{x}^H[n-k]\mathbf{w}[n-k]\right)\end{aligned} \tag{5.153}$$

An adaptive system based on the above formula is shown in Figure 5.15, where the weight and input vectors are defined as

$$\mathbf{w}^T[n]=[w_1[n],\ldots,\mathrm{w}_m[n]] \text{ and}$$
$$\mathbf{x}^T[n]=[x[n],\ldots,x[n-m+1]], \text{ respectively.}$$

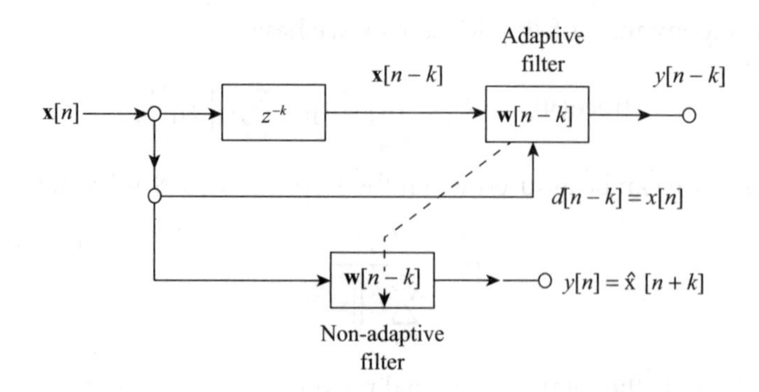

FIGURE 5.15 Signal flow-graph for adaptive predictor

The output of the adaptive system is $y[n-k]$ which is an estimate of the present input $x[n]$. The predictive output $y[n]$ which is an estimate of $x[n+k]$, is generated by a non-adaptive filter whose design is identical with the same weights as that of the adaptive filter. We are assuming that the weight vector is slowly varying, and $\mathbf{w}[n-k]$ is as good as $\mathbf{w}[n]$ to generate $y[n]$.

Suppose now that the input sequence is noise-corrupted, that is, $x[n] = g[n] + r[n]$ where $g[n]$ is the signal and $r[n]$ is the random noise. We assume that the noise is zero-mean white and uncorrelated with the signal:

$$\begin{aligned} &E\{r[n]\} = 0, \quad E\{r^*[n]r[n+k]\} = 0 \ \text{ for } k \neq 0, \text{ and} \\ &E\left[g^*[n]r[n+k]\right] = E\left[r^*[n]g[n+k]\right] = 0 \ \text{ for all } k \end{aligned} \tag{5.154}$$

The correlation matrix Φ_{xx} and the vector θ_{xd} are then computed as

$$\begin{aligned} \Phi_{xx} &= E\left\{(\mathbf{g}[n] + \mathbf{r}[n])(\mathbf{g}[n] + \mathbf{r}[n])^H\right\} \\ &= \Phi_{gg} + \sigma_r^2 \mathbf{I} \\ \theta_{xd} &= E\left\{(\mathbf{g}[n] + \mathbf{r}[n])(g[n+k] + r[n+k])^*\right\} \\ &= E\{\mathbf{g}[n] g^*[n+k]\} \end{aligned}$$

where Φ_{gg} is the signal correlation matrix and σ_r^2 is the variance of the noise process.

Therefore, the adaptive system produces a weight vector whose mean converges to

$$\mathbf{w}_{\text{opt}} = \left(\Phi_{gg} + \sigma_r^2 \mathbf{I}\right)^{-1} E\{\mathbf{g}[n] g^*[n+k]\} \tag{5.155}$$

Note that although we have fed the desired response $d[n-k] = x[n]$, the optimum weight is the same as if the desired response given were $d[n-k] = g[n]$. This feature of the adaptive predictor can be used effectively for noise elimination.

We consider the signal $g(t) = \mathrm{e}^{-0.2t}\cos 4t + 0.5\mathrm{e}^{-0.1t}\cos(2t + \pi/4) + 0.5\mathrm{e}^{-0.3t}\cos(t + \pi/6)$ for simulation study. A total of 256 points sampled with the interval $T = 0.05$ and mixed with zero-mean white noise is available for processing. A 5-weight (m = 5) adaptive predictor is used to generate the signal one unit ahead of time (k = 1). The LMS algorithm with the step size $\alpha_s = 0.05$ is used for weight adjustment. The noisy input $x[n]$ and the underlying signal $g[n]$ are plotted in Figure 5.16. The predictor output $y[n] = \hat{g}[n+1]$ and the underlying signal $g[n]$ are plotted in Figure 5.17. The input signal-to-noise ratio (SNR) is set to be 3.7 dB, whereas the output SNR is found to be 9.6 dB. Thus, the adaptive predictor provides an enhancement of the signal by 5.9 dB.

5.9.3 The RLS Adaptation Algorithm

We now develop the recursive least-squares (RLS) algorithm to find the optimal weight for the adaptive filter. Let the cost function to be minimized be given by

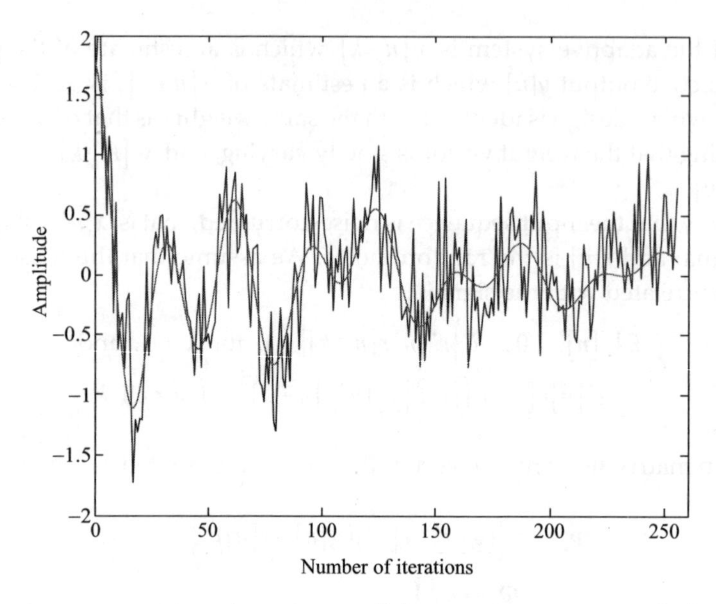

Figure 5.16 Plots of noisy input $x[n]$ and underlying signal $g[n]$

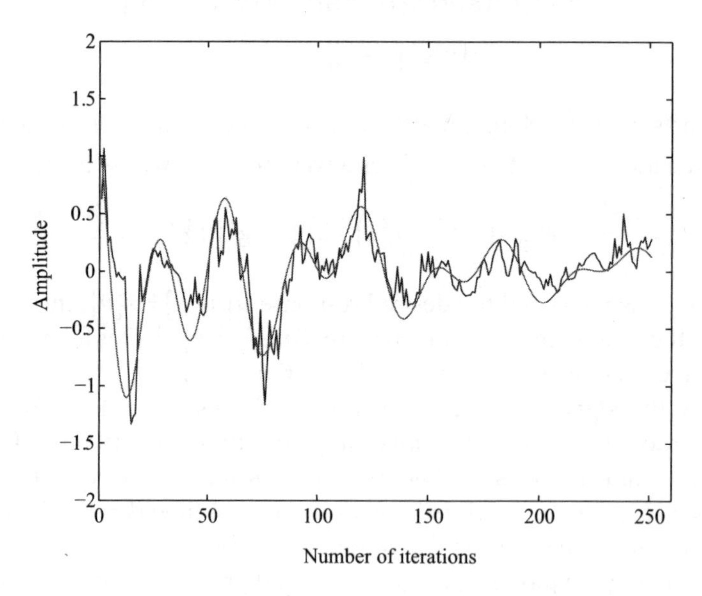

Figure 5.17 Plots of predictor output $y[n]$ and underlying signal $g[n]$

$$c[n]=\sum_{i=1}^{n}\varphi^{n-i}\left|e[n]\right|^{2},\quad 0<\varphi<1 \tag{5.156}$$

where φ^{n-i} is the forgetting factor introduced to give lesser weightage to error values of distant past. The exponentially weighted least-squares estimate, as it is called, has the feature which fits well in non-stationary environments.

The estimated weight vector $\hat{\mathbf{w}}[n]$ is now expressed as

$$\hat{\mathbf{w}}[n] = \mathbf{\Phi}^{-1}[n]\,\mathbf{\theta}[n] \tag{5.157}$$

where the correlation matrix $\mathbf{\Phi}[n]$ is defined by

$$\begin{aligned} \mathbf{\Phi}[n] &= \sum_{i=1}^{n} \varphi^{n-i}\mathbf{x}[i]\mathbf{x}^H[i] \\ &= \varphi\left(\sum_{i=1}^{n-1} \varphi^{n-1-i}\mathbf{x}[i]\mathbf{x}^H[i]\right) + \mathbf{x}[n]\mathbf{x}^H[n] \\ &= \varphi\,\mathbf{\Phi}[n-1] + \mathbf{x}[n]\mathbf{x}^H[n] \end{aligned} \tag{5.158}$$

and the cross-correlation vector $\mathbf{\theta}[n]$ is similarly given by

$$\begin{aligned} \mathbf{\theta}[n] &= \sum_{i=1}^{n} \varphi^{n-i}\mathbf{x}[i]d^*[i] \\ &= \varphi\,\mathbf{\theta}[n-1] + \mathbf{x}[n]d^*[n] \end{aligned} \tag{5.159}$$

To compute the weight vector $\hat{\mathbf{w}}[n]$, the inverse of the correlation matrix $\mathbf{\Phi}[n]$ is to be found. The inverse of $\mathbf{\Phi}[n]$ can be calculated by using the matrix inversion lemma:

If $\mathbf{A} = \mathbf{B}^{-1} + \mathbf{C}\mathbf{D}^{-1}\mathbf{C}^H$, $\mathbf{A}, \mathbf{B}, \mathbf{D}$ square positive definite

Then, $\mathbf{A}^{-1} = \mathbf{B} + \mathbf{BC}(\mathbf{D} + \mathbf{C}^H\mathbf{BC})^{-1}\,\mathbf{C}^H\mathbf{B}$
Now, by identifying as given below

$$\mathbf{A} = \mathbf{\Phi}[n],\quad \mathbf{B}^{-1} = \varphi\,\mathbf{\Phi}[n-1],\quad \mathbf{C} = x[n],\quad \mathrm{D}^{-1} = 1,$$

we have the following recursive equation for the inverse of the correlation matrix:

$$\mathbf{\Phi}^{-1}[n] = \varphi^{-1}\mathbf{\Phi}^{-1}[n-1] - \frac{\varphi^{-2}\mathbf{\Phi}^{-1}[n-1]\mathbf{x}[n]\mathbf{x}^H[n]\mathbf{\Phi}^{-1}[n-1]}{1+\varphi^{-1}\mathbf{x}^H[n]\mathbf{\Phi}^{-1}[n-1]\mathbf{x}[n]} \tag{5.160}$$

Let us denote the inverse matrix as
$\mathbf{P}[n] = \mathbf{\Phi}^{-1}[n]$, and the gain vector as

$$\mathbf{\gamma}[n] = \frac{\varphi^{-1}\mathbf{P}[n-1]\mathbf{x}[n]}{1+\varphi^{-1}\mathbf{x}^H[n]\mathbf{P}[n-1]\mathbf{x}[n]} \tag{5.161}$$

Then, the recursive formula can be rewritten as

$$\mathbf{P}[n] = \varphi^{-1}\mathbf{P}[n-1] - \varphi^{-1}\mathbf{\gamma}[n]\mathbf{x}^H[n]\mathbf{P}[n-1] \tag{5.162}$$

Now rearranging (5.161), we get

$$\begin{aligned} \mathbf{\gamma}[n] &= \varphi^{-1}\mathbf{P}[n-1]\mathbf{x}[n] - \varphi^{-1}\mathbf{\gamma}[n]\mathbf{x}^H[n]\mathbf{P}[n-1]\mathbf{x}[n] \\ &= \left(\varphi^{-1}\mathbf{P}[n-1] - \varphi^{-1}\mathbf{\gamma}[n]\mathbf{x}^H[n]\mathbf{P}[n-1]\right)\mathbf{x}[n] \\ &= \mathbf{P}[n]\mathbf{x}[n] \end{aligned} \tag{5.163}$$

Next, we develop a recursive formula for updating $\hat{\mathbf{w}}[n]$. The weight vector estimate at time n is given by

$$\begin{aligned}\hat{\mathbf{w}}[n] &= \mathbf{P}[n]\,\theta[n] \\ &= \varphi\,\mathbf{P}[n]\,\theta[n-1] + \mathbf{P}[n]\mathbf{x}[n]d^*[n]\end{aligned} \tag{5.164}$$

where (5.159) is substituted.

Note that the first term of the above equation can be expanded as

$$\begin{aligned}\varphi\mathbf{P}[n]\,\theta[n-1] &= \mathbf{P}[n-1]\,\theta[n-1] - \gamma[n]\mathbf{x}^H[n]\mathbf{P}[n-1]\,\theta[n-1] \\ &= \hat{\mathbf{w}}[n-1] - \gamma[n]\mathbf{x}^H[n]\hat{\mathbf{w}}[n-1]\end{aligned}$$

where we have used (5.162) and the expression for the weight vector at time (n–1).

Therefore, the formula for the weight vector update reduces to

$$\begin{aligned}\hat{\mathbf{w}}[n] &= \hat{\mathbf{w}}[n-1] - \gamma[n]\mathbf{x}^H[n]\hat{\mathbf{w}}[n-1] + \mathbf{P}[n]\mathbf{x}[n]d^*[n] \\ &= \hat{\mathbf{w}}[n-1] + \gamma[n]\big(d^*[n] - \mathbf{x}^H[n]\hat{\mathbf{w}}[n-1]\big)\end{aligned}$$

where (5.163) is substituted.

We define the innovation $\alpha[n]$ by the relation

$$\alpha[n] = d[n] - \hat{\mathbf{w}}^H[n-1]\mathbf{x}[n] \tag{5.165}$$

and express the above update formula as

$$\hat{\mathbf{w}}[n] = \hat{\mathbf{w}}[n-1] + \gamma[n]\alpha^*[n] \tag{5.166}$$

Note that the innovation $\alpha[n]$ is available before the weight vector update, whereas the error $e[n] = d[n] - \mathbf{w}^H[n]\,\mathbf{x}[n]$ is available only after the weight update. It is to be pointed out here that the cost function to be minimized for the least-squares estimate is based on the error $e[n]$. But for implementing the RLS algorithm, we need to use the innovation $\alpha[n]$.

The RLS algorithm in signal-flow representation is shown in Figure 5.18.

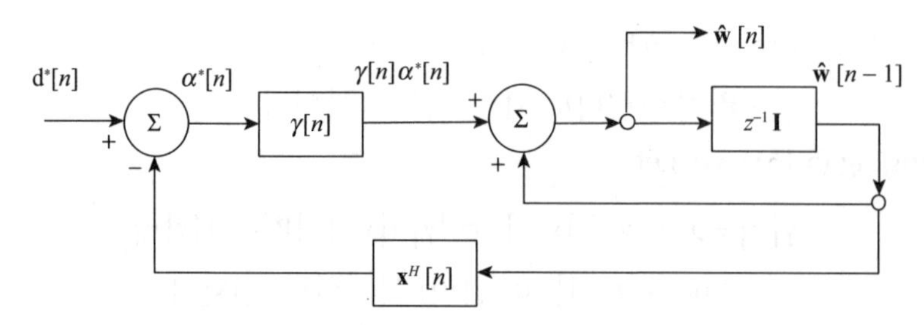

FIGURE 5.18 Signal-flow representation of RLS algorithm

In order to ensure that the correlation matrix $\mathbf{\Phi}[n]$ is non-singular at each step, we modify the expression of $\mathbf{\Phi}[n]$ as

$$\mathbf{\Phi}[n] = \sum_{i=1}^{n} \varphi^{n-i}\mathbf{x}[i]\mathbf{x}^H[i] + \delta\varphi^n\mathbf{I} \tag{5.167}$$

where δ is a small positive constant. This modification affects the starting value, leaving the recursions in the RLS algorithm intact. Thus, the matrix $\mathbf{P}[n] = \mathbf{\Phi}^{-1}[n]$ remains Hermitian and positive definite throughout the iterations.

The RLS Algorithm

(a) Initialization: Set $\hat{\mathbf{w}}[0] = \mathbf{0}$, $\mathbf{\Phi}[0] = \delta\mathbf{I}$, $\mathbf{P}[0] = \delta^{-1}\mathbf{I}$, δ small positive constant;
Choose $0 < \varphi < 1$.

(b) At step $n = 1,\ 2,\ \ldots$ Compute

$\xi[n] = \mathbf{P}[n-1]\mathbf{x}[n]$,

$\rho[n] = \varphi + \mathbf{x}^H[n]\xi[n]$,

$\gamma[n] = \dfrac{\xi[n]}{\rho[n]}$,

$\alpha[n] = d[n] - \hat{\mathbf{w}}^H[n-1]\mathbf{x}[n]$,

$\hat{\mathbf{w}}[n] = \hat{\mathbf{w}}[n-1] + \gamma[n]\alpha^*[n]$,

$y[n] = \hat{\mathbf{w}}^H[n]\mathbf{x}[n]$.

(c) Stopping condition: If end of data sequence, Stop;
Otherwise, Go to (b).

5.9.4 Simulation Study: Example 2 (Non-linear System Modelling)

In this example, we investigate the comparative performance of the LMS and RLS algorithm in adaptation to the input–output relationship of a non-linear dynamic system. The linear adaptive filter will exhibit the same transfer characteristics of a non-linear dynamic system. This simulation example will show how non-linear system characteristics can be produced by time-varying linear system.

Let the transfer characteristics of the non-linear system be given by

$$y[n] = 1.4x[n] + 0.5x[n-2] + 0.8x^2[n] + x^2[n-1], \quad n = 1, \ldots, 200 \tag{5.168}$$

where $y[n]$ is the output sequence and $x[n]$ is the random input sequence uniformly distributed over [0, 1]. The input sequence is plotted in Figure 5.19. We choose a 2-weight adaptive filter with the same input $x[n]$, and the output $y[n]$ generated by the non-linear system is used as the desired response for the adaptation process.

For the LMS algorithm, the weight updates are implemented using the fixed step size $\alpha_s = 0.35$. The output of the adaptive filter is generated with the updated weight vector and the input vector. The output of the adaptive filter is plotted in Figure 5.20, where the

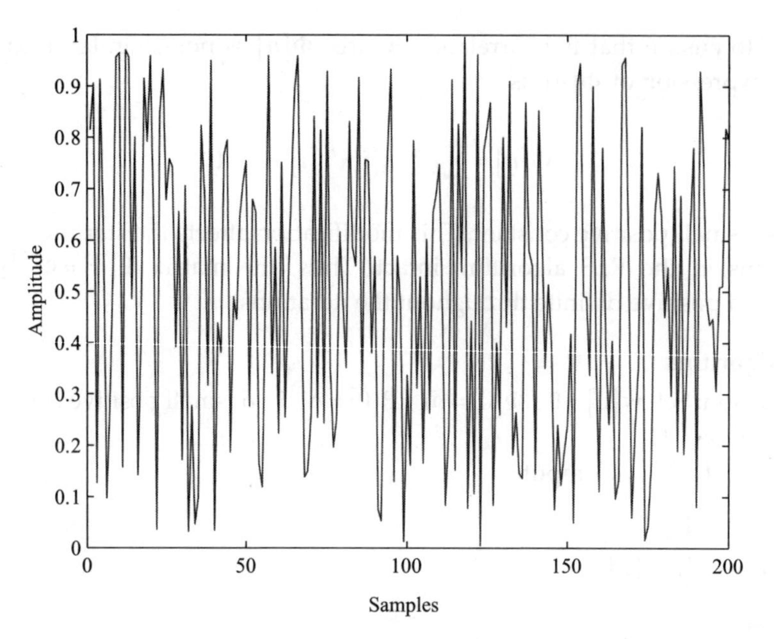

FIGURE 5.19 Input sequence generated by random number

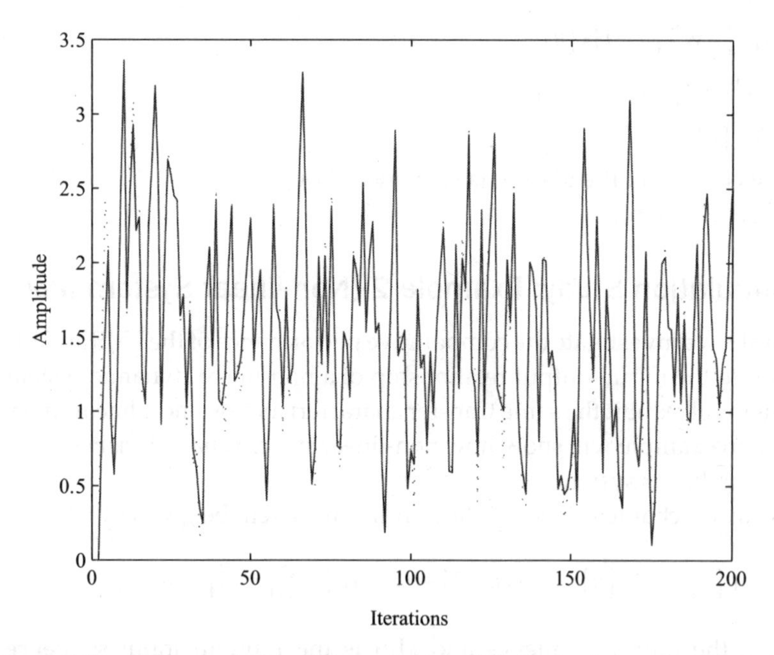

FIGURE 5.20 LMS adaptation; Filter output (solid), Desired response (dotted)

desired response is also plotted for comparison. The error variance σ_e^2 between the actual and desired responses is computed to be $\sigma_e^2 = 1.79\times10^{-2}$. The weight variations in the LMS adaptation are plotted in Figure 5.21.

For the RLS algorithm, we choose the forgetting coefficient $\varphi = 0.05$ and initialize the inverse correlation matrix $\mathbf{P}[0] = \delta\mathbf{I}$ with $\delta = 0.05$. The output of the adaptive filter is

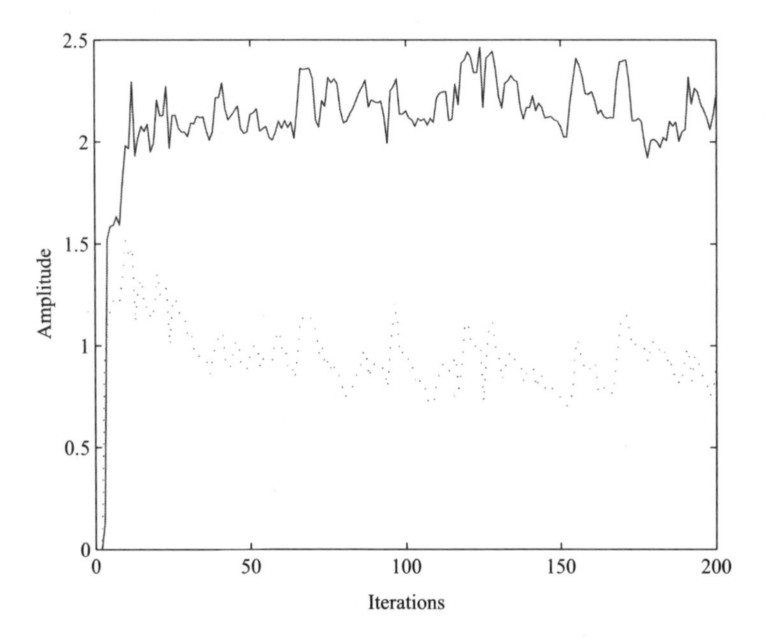

FIGURE 5.21 LMS adaptation; Weight variations, w_1 (solid), w_2 (dotted)

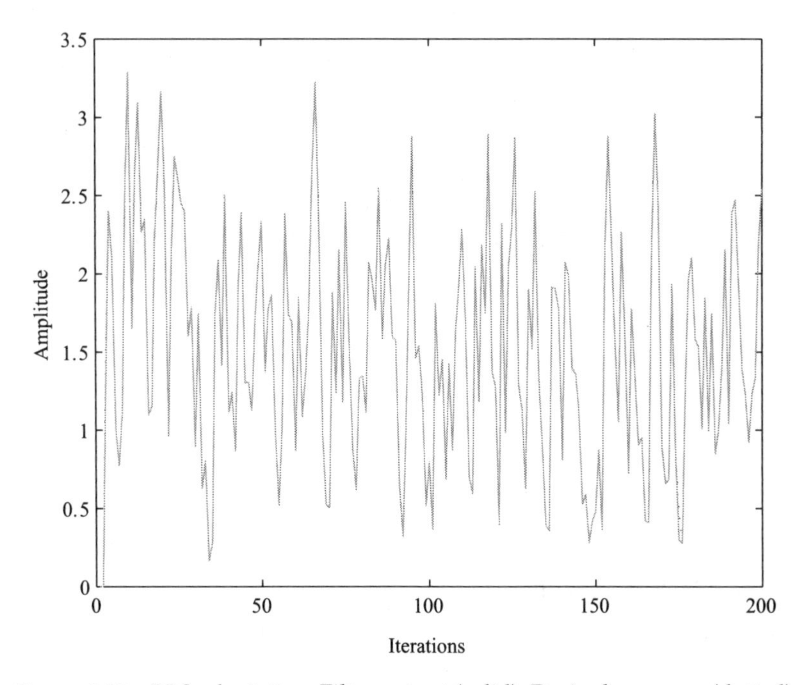

FIGURE 5.22 RLS adaptation; Filter output (solid), Desired response (dotted)

generated through iterations and plotted in Figure 5.22. The desired response is also plotted in the same figure for comparison. The error variance is computed to be $\sigma_e^2 = 2.63 \times 10^{-4}$. The weight variations in the RLS adaptation are plotted in Figure 5.23.

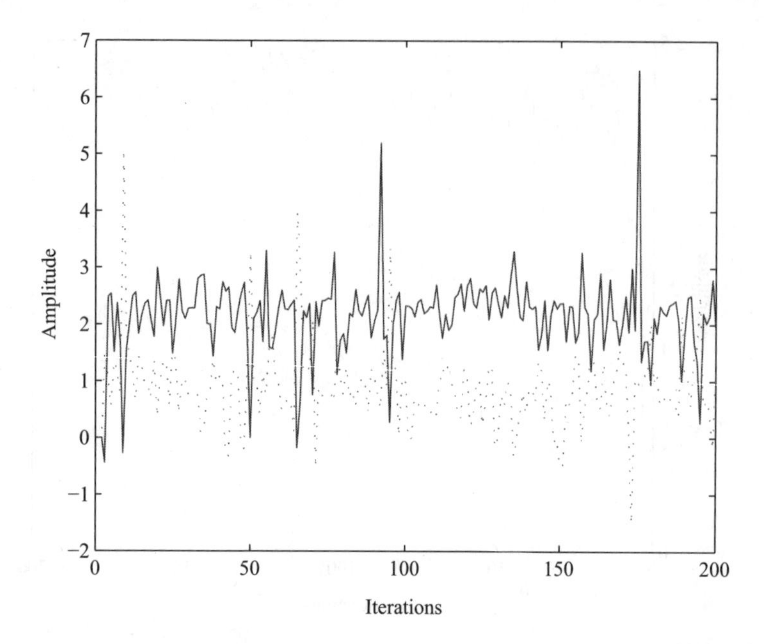

FIGURE 5.23 RLS adaptation; Weight variations, w_1 (solid), w_2 (dotted)

It is seen in the above results that the RLS algorithm has better performance in adaptation as compared to that of the LMS algorithm. It is to be noted that the step size α_s for the LMS algorithm and the forgetting coefficient φ for the RLS algorithm are chosen in the simulation study for having optimal performances of the respective algorithms. The weight variables for the RLS adaptation show rapid variations in their values which enable the filter to have fast adaptability. In contrast, the weight variables for the LMS adaptation show slow variations in their values, and as a result, the adaptive filter can track the desired response with less accuracy.

Problems

Some sample problems are given below. Additional problems can be obtained from standard texts on optimization, and mathematical programming.

P5.1 Consider the matrix equation $\mathbf{Ax} = \mathbf{b}$, $\mathbf{A} \in \mathbb{C}^{m \times n}$, $m < n$ and $\mathbf{A}$ full rank. Find the minimum-norm solution of $\mathbf{x}$.

P5.2 The prediction coefficients a_k of a p-order linear prediction model:

$\hat{x}[n] = -\sum_{k=1}^{p} a_k x[n-k]$ is estimated by minimizing the prediction error power,

$$\rho_1 = \frac{1}{N} \sum_{n=-\infty}^{\infty} \left| x[n] + \sum_{k=1}^{p} a_k x[n-k] \right|^2$$

for the Autocorrelation method, and by minimizing the error power,

$$\rho_2 = \frac{1}{N-p}\sum_{n=p}^{N-1}\left|x[n]+\sum_{k=1}^{p}a_k x[n-k]\right|^2$$

for the Covariance method.

Find the matrix equation to be solved for the prediction coefficients, and the minimum error power in each case.

Note that the data sequence $\{x[n];\ n=0,1,\dots,N-1\}$ is available for processing, and the samples which are not available are set equal to zero in the expression for ρ_1.

P5.3 Consider the data sequence $\{y[n];\ n=0,1,\dots,N-1\}$ satisfying the set of linear prediction equations given by

$y[n]=-\sum_{k=1}^{q}a_k y[n-k]+e[n],\ n=q,q+1,\dots,N-1$, where $e[n]$ is the prediction error.

Let the error criterion be to minimize the real function

$E_1=\sum_{n=n_1}^{n_2}|e[n]|^2$, which is the squared error-magnitude summed over a fixed frame.

The values of n_1 and n_2 are chosen so that there is no running off the data.

Show that the minimization of E_1 gives rise to the normal equation

$$\sum_{n=n_1}^{n_2}\sum_{k=0}^{q}a_k y^*[n-l]y[n-k]=0,\ l=0,1,\dots,q$$

or, $\mathbf{Pa}=\mathbf{0}$ where $[\mathbf{P}]_{lk}=\sum_{n=n_1}^{n_2}y^*[n-l]y[n-k],\ l,k=0,1,\dots,q$, which can be solved for $\mathbf{a}$.

Now, let the error criterion be to minimize the real function

$E_2=\sum_{n=n_1+l}^{n_2+l}|e[n]|^2,\ l=0,1,\dots,q$, which is the squared error-magnitude summed over a moving frame of fixed length. Show that the minimization of E_2 leads to the normal equation

$$\sum_{n=n_1}^{n_2}\sum_{k=0}^{q}a_k y^*[n]y[n+(l-k)]=0,\ l=0,1,\dots,q$$

or, $\mathbf{Qa}=\mathbf{0}$ where $[\mathbf{Q}]_{lk}=\sum_{n=n_1}^{n_2}y^*[n]y[n+(l-k)],\ l,k=0,1,\dots,q$, which can be solved for $\mathbf{a}$.

Note that the matrix Q, unlike P, has the Toeplitz structure which brings considerable computational advantage.

P5.4 Derive an iterative method to compute the vector $\mathbf{x}$ which minimizes

$$L(\alpha,\mathbf{x})=\|\mathbf{Ax}-\mathbf{b}\|^2+\alpha\|\mathbf{Cx}\|^2.$$

Starting with the initial guess $\mathbf{x}_0 = \mathbf{0}$, the successive approximations may be obtained by adding a correction factor which is proportional to the residue of that stage.
Comment on the rate of convergence of the iterative method as $\alpha \geq 0$ is varied.

P5.5 Let $\mathbf{x}$ be the true solution of minimizing $\|\mathbf{b} - \mathbf{A}\mathbf{x}\|$, and the computed solution $\hat{\mathbf{x}}$ exactly minimizes $\|\mathbf{b} - (\mathbf{A} + \mathbf{E}_i)\hat{\mathbf{x}}\|$; $i = 1,2$. Derive the expressions of $\mathbf{E}_i$, and find a procedure for software testing of stability of least-squares solutions.

P5.6 Find the maximum value of $f = \sin a \sin b \sin c$, given that $a + b + c = \pi$.

P5.7 Find the maximum value of $f(x_1, x_2) = x_1^2 + x_2^2$, $x_1 \geq 0, x_2 \geq 0$,
subject to the inequality constraints $(x_1 - 4)^2 + x_2^2 \leq 1$ and $(x_1 - 1)^2 + x_2^2 \leq 4$.

Bibliography

The list is indicative, not exhaustive. Similar references can be searched in the library.

Aoki, M. 1971. *Introduction to Optimization Techniques: Fundamentals and Applications of Non-linear Programming*. New York: The Macmillan Company.

Arrow, K. J., L. Hurwicz, and H. Uzawa. 1972. *Studies in Linear and Non-Linear Programming*. Stanford: Stanford University Press.

Beveridge, G. S. G., and R. S. Schechter. 1970. *Optimization: Theory and Practice*. New York: McGraw-Hill.

Fiacco, A. V., and G. P. McCormick. 1968. *Non-linear Programming: Sequential Unconstrained Minimization Techniques*. New York: John Wiley.

Fletcher, R. 1975. 'An Ideal Penalty Function for Constrained Optimization'. *IMA J Appl Math* 15 (3): 319–42.

———. 1987. *Practical Methods of Optimization*. Chichester: John Wiley.

Fletcher, R., and C. M. Reeves. 1964. 'Function Minimization by Conjugate Gradients.' *Computer J* 7 (2): 149–54.

Gill, P. E., and W. Murray. 1972. 'Quasi-Newton Methods for Unconstrained Optimization.' *IMA J Appl Math* 9 (1): 91–108.

Haykin, S. 1991. *Adaptive Filter Theory*. 2nd ed., Englewood Cliffs, NJ: Prentice Hall.

Hestenes, M. R. 1969. 'Multiplier and Gradient Methods'. *J Opt Theory Appl* 4 (5): 303–20.

Honig, M. L., and D. G. Messerschmitt. 1984. *Adaptive Filters: Structures, Algorithms, and Applications*, Boston: Kluwer Academic.

Jacoby, S. L. S., J. S. Kowalik, and J. T. Pizzo. 1972. *Iterative Methods for Non-linear Optimization Problems*. Englewood Cliffs, NJ: Prentice-Hall.

Kay, S. M. 1988. *Modern Spectral Estimation: Theory and Application*. Englewood Cliffs, NJ: Prentice Hall.

Kuhn, H. W., and A. W. Tucker, 1951. 'Non-linear Programming.' In *Proc. of the Second Berkeley Sym. on Mathematical Statistics and Probability*, edited by J. Neyman, 481–492. Berkeley, CA: University of California Press.

Luenberger, D. G. 1973. *Introduction to Linear and Non-linear Programming*. Boston, MA: Addison-Wesley.

Minoux, M. 1986. *Mathematical Programming: Theory and Algorithms*. Chichester: John Wiley.

Nakayama, H., H. Sayama, and Y. Sawaragi. 1975. 'A Generalized Lagrangian Function and Multiplier Method'. *J Opt Theory Appl* 17 (3/4): 211–27.

Panik, M. J. 1976. *Classical Optimization: Foundations and Extensions*. Amsterdam: North-Holland.

Powell, M. J. D. 1969. 'A Method for Non-linear Constraints in Minimization Problems.' In *Optimization*, edited by R. Fletcher, 283–298. London: Academic Press.

——— 1978. 'A Fast Algorithm for Non-linearly Constrained Optimization Calculations.' In *Numerical Analysis*, edited by G. A. Watson, 144–157. Berlin: Springer.

Ralston, A., and P. Rabinowitz. 1978. *A First Course in Numerical Analysis*. 2nd ed., Singapore: McGraw-Hill.

Rao, S. S. 1996. *Engineering Optimization: Theory and Practice*. New York: John Wiley

Ravindran, A., K. M. Ragsdell, and G. V. Reklaitis. 2006. *Engineering Optimization: Methods and Applications*. 2nd ed., Hoboken, NJ: John Wiley.

Rockafellar, R. T. 1974. 'Augmented Lagrange Multiplier Functions and Duality in Nonconvex Programming'. *SIAM J Control* 12 (2): 268–85.

Rosen, J. B. 1960. 'The Gradient Projection Method for Non-linear Programming - Part I: Linear Constraints.' *J Soc Indust Appl Math* 8 (1):181–217.

———. 1961. 'The Gradient Projection Method for Non-linear Programming - Part II: Non-linear Constraints.' *J Soc Indust Appl Math* 9 (4): 514–32.

Sage, A. P., and C. C. White, III. 1977. *Optimum Systems Control*. 2nd ed., Englewood Cliffs, NJ: Prentice-Hall.

Sayed, A. H., and T. Kailath. 1998. 'Recursive Least-squares Adaptive Filters'. In *The Digital Signal Processing Handbook*, edited by V. Madisetti and D. B. Williams. Boca Raton, FL: CRC Press.

Sircar, P., and S. Mukhopadhyay. 1995. 'Accumulated Moment Method for Estimating Parameters of the Complex Exponential Signal Models in Noise'. *Signal Process* 45 (2): 231–43.

Smeers, Y. 1977. 'Generalized Reduced Gradient Method as an Extension of Feasible Direction Methods'. *J. Opt Theory Appl* 22 (2): 209–26.

Snyman, J. A. 2005. *Practical Mathematical Optimization: An Introduction to Basic Optimization Theory and Classical and New Gradient-Based Algorithms*. New York: Springer Science.

Stewart, G. W. 1973. *Introduction to Matrix Computation*. New York: Academic Press.

Walsh, G. R. 1975. *Methods of Optimization*. London: John Wiley.

Widrow, B. 1971. 'Adaptive Filters'. In *Aspects of Network and System Theory*, edited by R. E. Kalman and N. DeClaris. New York: Holt, Rinehart and Winston.

Wismer, D. A., and R. Chattergy. 1978. *Introduction to Non-linear Optimization: A Problem Solving Approach*. New York: Elsevier North-Holland.

Solutions of Problems

P2.5 Zeros of T_n: $x_i = \cos\left[\dfrac{(2i+1)\pi}{2n}\right], \quad i = 0, 1, \cdots, n-1$

Extrema of T_n: $x_i = \cos\left[\dfrac{(n-i)\pi}{n}\right], \quad i = 0, 1, \cdots, n$

P2.7 $a_1 = -\dfrac{1}{5},\ b_1 = \dfrac{9}{10};\ a_2 = \dfrac{1}{20},\ b_2 = -\dfrac{3}{5},\ c_2 = \dfrac{3}{2}$

P3.4 $\mathbf{A}^+ = \dfrac{1}{84}\begin{bmatrix} 1 & 2 & 3 \\ 2 & 4 & 6 \\ 1 & 2 & 3 \end{bmatrix}$

P4.1 Eigenvalues are $2\mathrm{e}, 1, 2\mathrm{e}^{-1}$

P4.2 $\mathbf{A} = \mathbf{U}\Sigma\mathbf{U}^H$

P5.1 $\mathbf{x} = \mathbf{A}^H\left(\mathbf{A}\mathbf{A}^H\right)^{-1}\mathbf{b}$

P5.2 Autocorrelation method:

$$\begin{bmatrix} r_{xx}[0] & r_{xx}[-1] & \cdots & r_{xx}[-(p-1)] \\ r_{xx}[1] & r_{xx}[0] & \cdots & r_{xx}[-(p-2)] \\ \vdots & \vdots & & \vdots \\ r_{xx}[p-1] & r_{xx}[p-2] & \cdots & r_{xx}[0] \end{bmatrix}\begin{bmatrix} a_1 \\ a_2 \\ \vdots \\ a_p \end{bmatrix} = -\begin{bmatrix} r_{xx}[1] \\ r_{xx}[2] \\ \vdots \\ r_{xx}[p] \end{bmatrix}$$

where $r_{xx}[k] = \begin{cases} \dfrac{1}{N}\displaystyle\sum_{n=0}^{N-1-k} x^*[n]x[n+k] & \text{for } k = 0, 1, \cdots, p \\ r_{xx}^*[-k] & \text{for } k = -(p-1), -(p-2), \cdots, -1 \end{cases}$

$$\rho_{1,\min} = r_{xx}[0] + \sum_{k=1}^{p} a_k r_{xx}[-k]$$

Covariance method:

$$\begin{bmatrix} c_{xx}[1,1] & c_{xx}[1,2] & \cdots & c_{xx}[1,p] \\ c_{xx}[2,1] & c_{xx}[2,2] & \cdots & c_{xx}[2,p] \\ \vdots & \vdots & & \vdots \\ c_{xx}[p,1] & c_{xx}[p,2] & \cdots & c_{xx}[p,p] \end{bmatrix}\begin{bmatrix} a_1 \\ a_2 \\ \vdots \\ a_p \end{bmatrix} = -\begin{bmatrix} c_{xx}[1,0] \\ c_{xx}[2,0] \\ \vdots \\ c_{xx}[p,0] \end{bmatrix}$$

where $c_{xx}[l,k] = \frac{1}{N-p}\sum_{n=p}^{N-1} x^*[n-l]x[n-k]$

$$\rho_{2,\min} = c_{xx}[0,0] + \sum_{k=1}^{p} a_k c_{xx}[0,k]$$

P5.5 $\mathbf{E}_1 = -\frac{\hat{\mathbf{r}}\hat{\mathbf{r}}^H \mathbf{A}}{\|\hat{\mathbf{r}}\|^2}, \quad \mathbf{E}_2 = \frac{(\hat{\mathbf{r}} - \mathbf{r})\hat{\mathbf{x}}^H}{\|\hat{\mathbf{x}}\|^2}$ with $\mathbf{r} = \mathbf{b} - \mathbf{A}\mathbf{x}, \quad \hat{\mathbf{r}} = \mathbf{b} - \mathbf{A}\hat{\mathbf{x}}$

P5.6 $f_{\max} = \frac{3\sqrt{3}}{8}$

P5.7 $f_{\max} = 9$

Index